Brain Ischemia

Brain Ischemia

Eugene Gusev and Veronica I. Skvortsova

Russian State Medical University
Moscow, Russia

Kluwer Academic / Plenum Publishers
New York, Boston, Dordrecht, London, Moscow

BS

Library of Congress Cataloging-in-Publication Data

Gusev, E. I.
  Brain ischemia / Eugene I. Gusev and Veronica I. Skvortsova.
      p.; cm.
  Includes bibliographical references and index.
  ISBN 0-306-47694-0
  1. Cerebral ischemia–Pathophysiology. I. Skvortsova, Veronica I. II. Title
  [DNLM: 1. Brain Ischemia–physiopathology. WL 355 G982b 2003]
  RC388.5.G87 2003
  616.8'1–dc21

                                                                          2002043274

ISBN: 0-306-47694-0

©2003 Kluwer Academic/Plenum Publishers, New York
233 Spring Street, New York, New York 10013

http://www.wkap.com

10 9 8 7 6 5 4 3 2 1

A C.I.P. record for this book is available from the Library of Congress

Printed in Hong Kong

6/23/04

Authors

Eugene I. Gusev — MD, Professor of Neurology, Academician of the Russian Academy of Medical Sciences, Head of the Department of Neurology and Neurosurgery of the Russian State Medical University, Chairman of the All-Russian Society of Neurologists, President of the National Stroke Association, corresponding member of German Neurological Society.

Veronica I. Skvortsova — MD, Professor of Neurology, Head of the Department of Fundamental and Clinical Neurology of the Russian State Medical University, General Secretary of the All-Russian Society of Neurologists, Vice-president of the National Stroke Association, member of Board of European Stroke Society and the Stroke Scientific Panel of the European Federation of Neurological Societies.

Abbreviations

α-1-A – α-1-antitrypsin
AB – antibodies
ACTH – adrenocorticotropic hormone
ADP – adenosine diphosphate
AEP – acoustic brain stem evoked potentials
AIF – apoptosis-inducing factor
AMP – adenosine monophosphate
AMPA – alpha-amino-3-(3-hydroxy-5-methyl-4-isoxazol)-propionic acid
ANT – adenine nucleotide translocator
anti-ICAM – anti-intercellular adhesion molecule antibodies
Apaf-1 – activating factor of apoptotic protease-1
ATP – adenosine triphosphate
BBB – blood–brain barrier
BDNF – brain derived neurotrophic factor
BFGF – basic fibroblastic growth factor
BI – Barthel Index
BP – blood pressure
CAD – caspase-activated deoxyribonuclease
CBF – cerebral blood flow
cGMP – cyclic guanosine monophosphate
CIBD – chronic ischemic brain disease
CMP – central motor potential
CNS – central nervous system
CoQ – coenzyme Q
COX-2 – cyclooxygenase-2
CRH – corticotropin releasing hormone

CRP– C-reactive protein
CSA – compressed spectrum analysis
CSF – cerebrospinal fluid
CSFs – colony-stimulating factors
CT – computer tomography
DAG – diacylglycerol
DED – death effector domain
5′ D-I – type I iodothyronine 5′-deiodinase
DNA – deoxyribonucleic acid
ECG – electrocardiography
EEG – electroencephalography
ELISA – enzyme-linked immunosorbent assay
EMG – electromyography
ENMG – electroneuromyography
EPO – erythropoietin
Erks – extracellular signal-regulated kinases
FADD – Fas-associated death domain
FSH – follicle stimulating hormone
fT_3 – free triiodothyronine
fT_4 – free thyroxine
GABA – gamma-aminobutyric acid
GAPs – growth-associated phosphoproteins
GDNF – glial-derived neurotrophic factor
GFAP – glial fibrillar acid protein
GOBA – gamma-oxybutyric acid
GTP – guanosine triphosphate
HIF-1 – hypoxia-inducible factor-1
HO – heme oxygenase-1
HPLC – high performance liquid chromatography
HSF – transcription factor for HSP
HSP – heat-shock protein
ICAM – intercellular adhesion molecule
ICE – IL-1β converting enzyme
IFN – interferon
IGF – insulin-dependent growth factor
IHA – inter-hemispheric asymmetry
IL – interleukin
IL-1ra – IL-1 receptor antagonist
iNOS – inducible NO-synthase
IP_3 – inositol-1,4,5-triphosphate
ISU – Intensive Stroke Unit
JNK – c-Jun N-terminal protein kinases

K – kainic acid
L-AP$_4$ – L-2-amino-4-phosphobutyric acid
LDH – lactate dehydrogenase
LE – leukocytic elastase
LH – luteinizing hormone
LHRF – luteinizing hormone releasing factor
MAP2 – microtubule-associated protein 2
MAPKs – mitogen-activated protein kinases
MBP – myelin basic protein
MCAO – middle cerebral artery occlusion
mCCT – motor central conduction time
MEF2 – myocyte enhancer factor-2
MHC – major histocompatibility complex
mRNA – messenger RNA
MRI – magnetic resonance imaging
MRS – magnetic resonance spectroscopy
NAD/NADH – nicotinamide-dinucleotide
NADP – nicotinamide-dinucleotide phosphate
NANA – N-acetylneuraminic acid
NGF – nerve growth factor
NGFI-A – nerve growth factor inducer A
NMDA – N-methyl-D-aspartate
NO – nitric oxide
NOS – nitric oxide synthase
NT – neurotrophin
OP-1 – osteogenic protein-1
OS – the Original Scale
OSS – the Orgogozo Stroke Scale
PARP – poly(ADP-ribose) polymerase
PET – positron emission tomography
PGI$_2$ – prostaglandin I$_2$
PMNL – polymorphonuclear leukocytes
RNA – ribonucleic acid
sCCT – sensory central conduction time
SOD – superoxide dismutase
SSEP – somatosensory evoked potentials
SSS – the Scandinavian Stroke Scale
T$_3$ – triiodothyronine
T$_4$ – thyroxine
TBARS – thiobarbituric acid reactive substances
TES – transcranial electrical stimulation
TGF-β_1 – transforming growth factor β_1

TNF – tumor necrosis factor
TNF-R – TNF-receptor
tPA – tissue plasminogen activator
TRADD – TNF-R-associated death domain
TRH – thyrotropin releasing hormone
TTH – thyrotropic hormone
TxA_2 – thromboxane A_2
VCAM – vascular cell adhesion molecule
VEGF – vascular endothelial growth factor
VIP – vasoactive intestinal polypeptide

Contents

Introduction

Ischemia is localized tissue anemia due to obstruction of the inflow of arterial blood, thus **brain ischemia** is the condition where insufficient blood is delivered to the brain. Many physiological processes occurring in the brain critically depend on the state of its energy metabolism. The state of brain energy metabolism in turn depends on the delivery of oxygen and glucose to the brain via the bloodstream.

Although it comprises only 2% of the total body weight, the human brain consumes 20–25% of the oxygen and up to 70% of the free glucose taken in by the body. The brain respires more intensively than any other organ of the body. The intensity of oxygen consumption by cortical brain tissue much exceeds the demands of other tissues (5.43 mmol O_2/g per h versus 3.06 and 4.02 mmol for heart at rest and intensively working, respectively, 2.4 mmol for kidneys, and 1.8 mmol for liver). Oxidative phosphorylation in mitochondria generates 95% of the adenosine triphosphate (ATP) that is formed in the brain. Thus, it is clear why insufficiency of oxygen delivery to brain cells adversely affects brain function.

Glucose is the main energy-providing substrate in the brain. The basic pathway of its metabolism in neural tissue is aerobic glycolysis. Approximately 85–90% of glucose consumed by the brain is entirely oxidized to carbon dioxide and water. The intrinsic brain glucose supply is small compared to the rate of its consumption. The supply of glucose in the brain can be exhausted in 3–6 min even under conditions of its utilization only for oxidative processes. Thus, brain functioning is highly dependent on constant delivery of glucose. If the blood glucose concentration required for maintenance of energy-requiring processes critically decreases, brain tissue begins to use the free fraction of glycogen. However, due to its small supply complete oxidation of brain glycogen takes only 5–7 min.

Cardiac arrest, severe cardiac rhythm disorders, and severe systemic hypotension cause *global brain ischemia*. Its prevention and treatment take important places among the urgent problems of modern cardiology and cardiosurgery. However, the occurrence of global ischemia is lower than that of *focal (or local) brain ischemia* that develops in cases of acute disorder of brain circulation in the area of one artery (more rarely of several arteries) and clinically manifests as syndromes of ischemic stroke or transient ischemic attack.

In recent years there has been an increase in the reported morbidity due to vascular disorders and stroke. Yearly six million people suffer from stroke worldwide, more than 500,000 in Russia, i.e. each 1.5 min stroke strikes a new victim in Russia.

According to international epidemiological studies (*World Development Report*), 4.7 million people die from stroke each year. In the majority of countries stroke ranks 2nd or 3rd in total mortality; it is in 2nd place in Russia, yielding 1st place to cardiovascular pathology. In 1999 average early (30-day) stroke mortality in Russia was about 35%, and each second patient died within a year of the onset of stroke.

Stroke is a basic reason for disability. Approximately 55% of patients surviving for three years after stroke onset are to some extent dissatisfied with their life quality [1]. Only 20% of surviving patients can return to their previous work. Yearly loss of sufficiently high life quality due to post-stroke disability amounts to 20.3 and 22.9 million person-years for males and females, respectively, worldwide [1]. It should also be noted that stroke imposes additional duties on the patient's family members and becomes a severe social and economic burden for society. According to WHO estimates, the average direct and indirect cost per stroke patient is about USD 55,000–73,000 yearly. Thus, the total expense connected with stroke in Russia corresponds to USD 16.5–22 billion per year. Thus, stroke is a crucial medical and social problem.

Most strokes are of ischemic nature. According to international multi-center studies, the ratio between ischemic and hemorrhagic strokes is (5.0–5.5):1, i.e. 80–85 and 10–15% on average, respectively.

Decades-long studies of pathogenic mechanisms involved in acute cerebrovascular insufficiency have revealed four groups of basic pathogenic factors [2–13]:

- morphologic changes in vessels supplying the brain (stenosis and occlusions, vessel anomalies, changes in vessel shape and configuration);
- disturbances of global and cerebral hemodynamics leading to decrease in cerebral blood flow to a critical level with cerebrovascular insufficiency;
- changes in the physical and chemical properties of blood, in particular its coagulability, blood cell aggregation, changes in viscosity and other rheological properties, changes in protein fractions, electrolytes and under-oxidized metabolite content;
- individual and age-dependent peculiarities of brain metabolism and the reactivity of the common neuro–immune–endocrine system, the variability of which correlates with different reactions of brain tissue to local decrease in blood flow.

Population studies have shown that about 50% of acute ischemic cerebrovascular disorders appear to be thrombotic or embolic complications of atherosclerosis that affect large and middle-sized vessels (atherothrombotic stroke and arterio-arterial embolism), about 25% occur due to pathology of small arteries (lacunar stroke), about 20% due to cardiac embolism, and approximately 5% due to rare or unusual causes (hemodynamic stroke due to myocardial infarction or any other depression of systemic hemodynamics; vascular disorders of primarily inflammatory origin; fibro-muscular dysplasia; arterial dissection; hematological disorders; infections; cancer and radiation; Moya Moya syndrome, etc.). But whatever the reason or the pathogenic variant is the influence of the vascular pathological factor on brain tissue results in acute focal ischemia.

Brain ischemia is a process quintessential for the "metabolic stage" of ischemic stroke pathogenesis. The underlying mechanisms of this stage are the universal brain tissue response to damage of any character or "environmental" changes including disorders of cerebral blood supply.

Understanding of the mechanisms of changes occurring in cerebral ischemia has been gradually developing in recent decades. Up to the 1960s the concept of immediacy and irreversibility of ischemic damage dominated in angioneurology. The first evidence that cerebral ischemia is a dynamic process was revealed in the late 70s and early 80s. Researchers made the first attempts to determine the mechanisms of sequential functional and morphological changes in ischemized neural tissue. This allowed the formulation of the concept of cerebral ischemia as a dynamic process and the treatment of ischemic damage as potentially reversible; the term "cerebral ischemia" is not equal to the term "brain infarction", the latter reflecting the formation of an irreversible morphologic lesion.

The studies conducted in the last two decades open a new stage in understanding of the mechanisms of changes occurring in brain during cerebral ischemia. They delineated how reversible changes of blood flow and cell and molecular changes are transformed into a permanent morphological lesion, i.e. brain infarction. In spite of universal features, the process of cerebral ischemia is always individual and the peculiarities of its course are determined by the background (pre-stroke) state on brain metabolism including the energetic demands as well as the state and reactivity of the general neuro–immune–endocrine system.

It is hard to overestimate the significance of scientific discoveries connected with the problem of brain ischemia. On the base of stroke studies that have been conducted, the delayed character of irreversible post-onset brain damage has been proved. Stroke became strictly treated as an emergency condition that requires fast and pathogenically substantiated medical aid. The dominant concept of stroke treatment has been formulated—the "therapeutic window" concept. As a result, specialized wards and intensive stroke units have been designed. Opinions regarding approaches to stroke therapy have changed. New effective methods, such as thrombolysis and neuroprotection (synonyms: cytoprotection, brain metabolic defense), have been designed and put into practice.

This monograph highlights modern concepts of brain ischemia and the molecular mechanisms of its development leading to brain tissue changes. The book also describes strategies of neuroprotective therapy.

REFERENCES

1. Asplund, K., 1998, *European White Book on Stroke*.
2. Badalyan, L. O., 1975, *Neurological Syndromes in Heart Diseases*. Meditsina, Moscow, 416 (in Russian).
3. Baron, J. C., 1991, *Cerebrovasc Dis*. **1**: 22–31.
4. Bogolepov, N. K., 1962, *J Neurol Psychiatr*. **8**: 1137–1142 (in Russian).
5. Bogolepov, N. K., 1971, *Cerebral Crises and Stroke*. Meditsina, Moscow, 392 (in Russian).
6. Bogousslavsky, J., 1991, *Cerebrovasc Dis*. **1**: 61–68.
7. Fisher, M., Bogousslavsky, J , 1996, *Current Review of Cerebrovascular Disease*. CM, Philadelphia, 237.
8. Gusev, E. I., Burd, G. S., Bogolepov, N. N., 1979, *Cerebrovascular Diseases*. Meditsina, Moscow, 142 (in Russian).
9. Hennerici, M., Klemm, C., Rautenberg, W., 1988, *Neurology*. **38**: 669–673.

10. Schmidt, E. V., Lunev, D. K., Vereshchagin, N. V., 1976, *Vascular Diseases of the Brain and the Spinal Cord.* Meditsina, Moscow, 250 (in Russian).
11. Vereshchagin, N. V., 1980, *Pathology of Vertebral and Basilar System and Cerebrovascular Diseases*. Meditsina, Moscow, 180 (in Russian).
12. Voiculescu, V., 1989, *Neuropsychiatr Roum Med.* **27** (3): 175–186.
13. Warlow, C. P., Dennis, M. S., van Gijn, J., *et al.*, 1996, *Stroke. A Practical Guide to Management*. Blackwell Science Ltd, Oxford, 6.

PART I

MECHANISMS OF ISCHEMIC BRAIN DAMAGE

Chapter 1

Hemodynamic Events Associated with Acute Focal Brain Ischemia and Reperfusion. Ischemic Penumbra

Many studies have shown that the extent of injury caused by acute focal brain ischemia are determined by the severity and the duration of decreased cerebral blood flow (CBF).

In cases of short-term (not more than 1–1.5 min) and mild CBF decrease in patients with well developed collateral vascular net and unchanged vessel reactivity, hypoxia–ischemia can be absent due to compensatory increase in oxygen extraction from blood. However, when reperfusion comes several minutes later the tissue compensatory process is exhausted and an ischemic event develops with the induction of a universal algorithm of brain tissue response.

Experiments have shown the sequence of metabolic responses of rat brain tissue to a progressive decrease in blood flow (Fig. 1.1) [27, 28, 34, 35]. When blood flow is decreased to 70–80% (less than 50–55 ml/100 g brain tissue per min, i.e. *the first critical level*) brain first responds to it by the inhibition of protein synthesis. According to Hossmann [28], protein synthesis can be inhibited due to disaggregation of ribosomes, which is seen during the first stages of ischemia.

Further decrease in blood flow to 50% of its normal value (to 35 ml/100 g brain tissue per min, i.e. *the second critical level*) activates anaerobic glycolysis and increases lactate concentration, and then lactic acidosis and cytotoxic edema develop (Fig. 1.2).

CBF, ml/100 g
brain tissue per minute

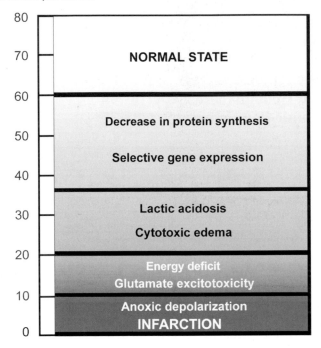

Figure 1.1. Brain tissue responses to cerebral blood flow (CBF) decrease.

When brain blood flow decreases to about 30% of its normal value (20 ml/100 g brain tissue per min, i.e. *the third critical level*) enhancing ischemia leads to depletion of ATP synthesis, energetic deficit, and, hence, to the dysfunction of active ion transport channels, cell membrane instability, and excessive efflux of excitatory neurotransmitter amino acids.

When CBF decreases to 20% of its normal value (10–15 ml/100 g brain tissue per min), neurons begin to lose ion gradients and the anoxic depolarization of membranes ensues [5, 20, 21]. Until now the latter has been believed to be the main criterion of irreversible cell damage [26, 27].

Mies *et al.* [45] experimentally showed that the mentioned stages are inter-dependent in their duration. The longer is a period of moderate CBF decrease and consequent "soft" ischemia, the shorter is the period required for severe ischemia to occur on further reduction in blood flow.

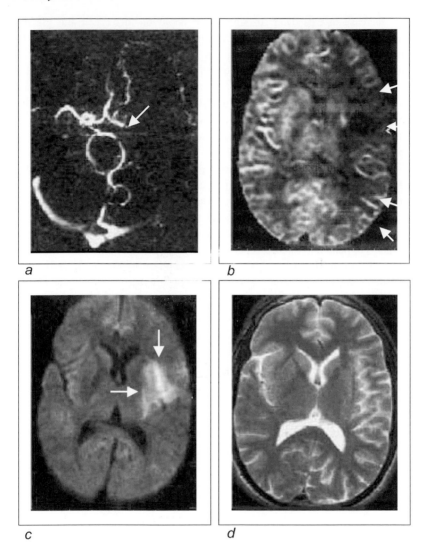

Figure 1.2. Early MRI diagnosis of ischemic stroke (within the first 4 h after the event): a) MR-angiography: middle cerebral artery occlusion (arrow); b) perfusion regimen: CBF decrease in left hemisphere (arrows); c) diffusion regimen: cytotoxic edema in the ischemized area (arrows); d) T_2-weighted image: the absence of morphological lesion in the brain.

Decrease in CBF leads to significant depletion in oxygen and glucose delivery to brain tissue. Oxygen and glucose metabolism was shown to be mostly changed in the central part of an ischemized territory, and to be minimally changed in a bordering area [50]. A brain area where blood flow is decreased to the utmost extent (<10 ml/100 g brain tissue per min) rapidly becomes irreversibly damaged within 6–8 min. This region is called the "ischemic core" [13]. Within several hours this central infarction becomes encircled by ischemized yet living tissue with blood flow more than 20 ml/100 g brain tissue per min called the "ischemic penumbra" [4] (Fig. 1.3).

Energy metabolism is preserved in the penumbral area to some extent. Only functional but not morphological changes are present there. This is the zone of "critical" perfusion, where the function of neurons is inhibited because their metabolic demand is not provided, but the cells remain alive and their homeostasis is preserved [60]. As the reserve for local perfusion is exhausted, penumbral neurons become very sensitive to any further decrease in perfusion, for instance, as caused by secondary hypovolemia [60] (after dehydration), by inadequate anti-hypotensive therapy or when the patient

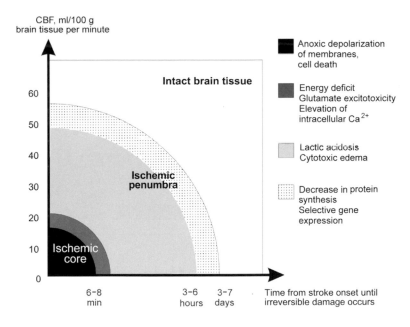

Figure 1.3. Formation of brain infarction during decreased CBF.

quickly stands up from bed. Next, the infarction zone gradually expands onto the penumbra. The penumbra can be rescued if we restore adequate perfusion or use neuroprotective drugs. In fact, the penumbra is the main therapeutic target during the first hours and days after stroke onset.

If the arterial occlusion is temporary or a thick collateral vascular network is quickly activated, blood flow in the ischemized area is completely or partially restored [54, 55]. However, restoration of blood flow later than 2 min after the occlusion occurs [60], i.e. within an already triggered ischemic process, does not mean that blood flow is completely normalized. Gradual disturbances of brain perfusion develop. The first stage of post-ischemic hyperemia (hyperperfusion) is replaced by a stage of post-ischemic hypoperfusion (Fig. 1.4).

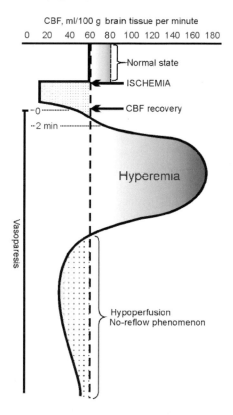

Figure 1.4. Hemodynamic events associated with brain ischemia and reperfusion.

Post-ischemic hyperemia [53] or "luxuriant perfusion" is connected not only with excessive delivery of blood via collateral vessels or with previously blocked vessel recanalization, but also with the release of pro-inflammatory and vasoactive metabolites from ischemized tissue, with blood viscosity decrease [19, 25, 26, 56], and with changes in neurogenic vasodilatory mechanisms [40, 60]. Excessive blood flow occurring under these conditions is out of correspondence with metabolic requirements of the brain tissue, and the extracted oxygen fraction decreases.

The stage of hyperemia is followed by a decrease to blood flow lower than the pre-ischemic level. Such post-ischemic hypoperfusion is a result of delayed metabolic changes in ischemized tissue caused by microglia activation, accompanied by synthesis of pro-inflammatory cytokines in excessive amount as well as the secondary activation of astroglia and synthesis of pro-inflammatory acute phase proteins. These factors lead to severe disturbances of blood circulation and obstruction of small vessels [19, 25, 26, 29–31, 56]. Thus, restoration of arterial blood flow leads to incomplete reperfusion of previously ischemized tissue. This phenomenon has been called "no-reflow" [1]. Its basic mechanisms are: 1) elevation of blood viscosity and intravascular coagulation; 2) microvascular occlusion due to compression of brain capillaries by adjacent edematous astrocytes; 3) elevation of intracranial pressure; 4) endothelial swelling and formation of endothelial micropili, and 5) post-ischemic hypotension [25, 26, 32]. Iadecola [29] showed that cerebral circulation in the post-ischemic period is paralyzed, brain vessel reactivity to hypercapnia (excessive carbon dioxide in the blood) is prominently reduced, and cerebrovascular regulation is altered. Whereas in the intact brain blood flow tightly depends upon metabolic demand, functional activation of the brain in a post-ischemic period never leads to adequate increase in blood flow.

Partial blood flow recovery by compensatory switching of collateral vessels causes a special dynamic state of ischemized tissue in which patchwork-like reperfusion occurs. This phenomenon has been called the "temporal–spatial dynamics of microcirculation" [48]. The "patches" of relative and/or absolute hyperemia and hypoperfusion alternate in the penumbral and the infarction areas respectively at this stage [60].

The no-reflow phenomenon is especially prominent under conditions of prolonged ischemia as well as in ischemia caused by the obstruction of veins, which leads to stagnation of blood in small vessels [25, 26]. Incomplete reperfusion limits the survival capabilities of ischemized tissue. According to experimental findings in rats [35, 51], therapeutic reperfusion conducted within 48 h after transient vessel occlusion may decrease the size of the lesion and improve recovery of brain functions. However, the problem of therapeutic reperfusion is rather complicated and, being started belatedly, is associated with risk of additional reperfusion damage to brain tissue. Such damage can be of oxidant origin, when oxygen begins to cause lipid peroxidation [44], and of osmotic origin, when excessive amounts of water and substances possessing osmotic properties begin to escalate edema [33] (see also Chapter 4).

The area of the ischemic penumbra should be regarded not only in terms of topography. This area undergoes a dynamic process in which energy disturbances spread downstream from the ischemic core to the periphery [16, 17]. How long the penumbra exists depends upon the affected individual in each separate case. The duration of its existence delineates the time limits (i.e. the therapeutic window) when therapeutic intervention may be of utmost efficacy. Experimental findings in animals (rats, mice) have shown that acute focal ischemia leaves a far narrower therapeutic window than global ischemia [18, 21, 37–39, 49].

Global brain ischemia occurs in cardiac arrest and in severe systemic hypotension. It leads to only selective neuronal changes in the brain areas most vulnerable and susceptible to ischemia (e.g., pyramidal neurons in the CA_1 hippocampal area or middle-sized neurons in the dorsolateral area of the striatum). The manifestation of neuropathological changes is delayed. They occur after 4–8 h in the striatum, and even after 2–3 days after the onset of global ischemia in the CA_1 hippocampal area. Thus, it is possible to rescue brain tissue under conditions of global ischemia if therapy is started even 24 h after the onset of ischemia [17].

The major part of a brain infarction is completely formed 3–6 h after the first clinical manifestations of stroke appear [10, 18, 22, 46, 49]. On the other hand, the time for the development of a morphological lesion strongly depends upon what brain area was damaged and what is the individual susceptibility of exposed neurons to ischemia [12]. Experimental findings obtained from the most sensitive immunohistochemical techniques [32, 42, 43] using metabolic dyes (e.g., triphenyl tetrazolium chloride) and data acquired from modern magnetic resonance imaging (MRI) and positron emission tomography (PET) visualization techniques [8, 23, 47] suggest that morphological completion of infarction takes a relatively long time span, about 48–72 h from the stroke onset. However, the time is sometimes extended [3] due to a continuing influence of brain edema and other delayed consequences of ischemia (see Chapters 6–10). According to Chiamulera *et al.* [11], edema causes the area of ischemic damage to expand within the 1st week from the stroke onset. Starting with the 2nd week of illness resorption of liquid begins, scavenger cells ablate dead neurons and glia, and the infarction zone decreases in size due to progressive atrophy.

MRI and PET techniques used to monitor events in an ischemized area confirmed that there exists a significant individual variability of limits given by the therapeutic window. Using PET to monitor experimental ischemic stroke in primates [6, 57, 59, 61], it was shown that the reversible changes in the penumbral area may last for many hours or even days after arterial occlusion. Based on PET results, Marchal *et al.* [41] found ischemized but possibly viable tissue within the necrotic zone of brain 18–24 h after stroke onset. This tissue showed very high rate of brain metabolism via oxygen and elevated fraction of extracted oxygen under conditions of blood flow reduction to 22 ml/100 g brain tissue per min. According to Jones *et al.* [36] and Furlan [15], tissue demonstrating such physiological parameters may avoid infarction and should be treated as being potentially at risk. In some patients [7, 14] the therapeutic window appeared to be significantly wider than previously established limits of 3–6 h. At the same time, Heiss *et al.* [24] showed that in certain cases critical hypoperfusion below the viability threshold in retrospective analysis, revealed by PET CBF studies within 3 h after stroke onset, accounted for the largest proportion (mean, 70%) of the final brain infarction volume, estimated by MRI 2 to 3 weeks after a hemispheric stroke. Penumbral tissue (18%) and initially sufficiently perfused tissue (12%) were responsible for considerably smaller portions of the final infarction.

Apparently, the individual variability of the therapeutic window is connected with adequate compensatory abilities of collateral circulation as well as with the background (pre-stroke) condition of brain metabolism and with the reactivity of the neuro–immune–endocrine system. The duration of the ischemic penumbra in separate cases provides hope for possible invention of effective neuroprotective drugs that would alter brain tissue susceptibility to ischemia, helping cells to survive.

During recent years the possibility of applying clinical criteria to define the condition of the ischemic penumbra and the limits of the therapeutic window has been studied. Some researchers suggest a connection between clinical progression of stroke within 12–24 h from its onset and worsening of the penumbra's condition resulting in transformation of reversible changes into an irreversible morphological lesion, especially in cases when no early formation of brain edema and hemorrhagic transformation of the lesion were observed. There is evidence for such a connection when regarding the negative impact of hyperglycemia, systemic acidosis, and delayed consequences of ischemia on the penumbral area. Along with this, clinical and functional criteria never give us a chance to reliably assess whether ischemic changes causing neurological symptoms are reversible. It is dangerous to apply interpretation of clinical symptoms to assess the limits of the therapeutic window because clinical manifestations always precede transformation of reversible tissue changes to irreversible [17]. Besides, it is known that even small damage in a functionally important area can manifest with severe neurological deficit. On the other hand, even in cases of major infarctions one can find partial or complete reversibility of functional deficit due to improvement of dysfunctional state caused by the spreading of trans-synaptic inhibition of neuronal activity (diaschisis) onto brain areas far from the ischemic lesion [2, 50, 52]. Other factors minimizing the dysfunctional state are the processes of regeneration and reparation, improvement of brain plasticity, and formation of new polysynaptic contacts [9, 17, 58].

Thus, studies of the 1980–90s proved the existence of penumbra within the first hours after stroke onset. The penumbra is the ischemized but still viable tissue that appears to be the main target for therapeutic intervention. The concept of "therapeutic window" has also been formed, and this markedly altered approaches to treatment of stroke patients in the acute period that require emergent pathogenically substantiated therapy especially desirable within the first 3 h after stroke onset. The absence of strict time limitations for the therapeutic window was also proven, and individual peculiarities of ischemic process in the penumbral area were disclosed. These two findings became the utmost stimuli to the design of novel strategies of reperfusion and neuroprotection.

REFERENCES

1. Ames, A., Wright., R., Kowada., M., *et al.*, 1968, *Amer J Pathol*. **52**: 437–453.
2. Andrews, R. J., 1991, *Stroke*. **22**: 943–949.
3. Asplund, K., 1992, *Cerebrovasc Dis*. **2**: 317–319.
4. Astrup, J., Siesjo, B. K., Symon, L., 1981, *Stroke*. **12**: 723–725.
5. Astrup, J., Symon, L., Branston, N. M., *et al.*, 1977, *Stroke*. **8**: 51–57.
6. Baird, A. E., Benfield, A., Schlaug, G., *et al.*, 1997, *Ann Neurol*. **41**: 581–589.
7. Baron, J. C., van Hummer, R., del Zoppo, G. J., 1995, *Stroke*. **26**: 2219–2221.
8. Baron, J. C., Frackowiak, R. S. J., Herholi, K., *et al.*, 1989, *J Cerebr Blood Flow Metab*. **9**: 723–742.

9. Bogolepov, N. N., Burd, G. S., 1981, *Abst VII All-Union Symp. of Neurologists and Psychiatrists.* Moscow. **2**: 32-35 (in Russian).

10. Butcher, S. P., Bullock, R., Graham, D. I., *et al.*, 1990, *Stroke.* **21**: 1727–1733.

11. Chiamulera, C., Terron, A., Reggiani, A., *et al.*, 1993, *Brain Res.* **606** (2): 251–258.

12. Dereski, M.O., Chopp, M., Knight, R. A., *et al.*, 1993, *Acta Neuropathol.* **85**: 327–333.

13. Fisher, M., Takano, K., 1995, In *Ballierie's Clinical Neurology, Cerebrovascular Disease* (Hachinski, V., ed.) London, pp. 279–296.

14. Fisher, M., Garcia, J. H., 1996, *Neurology.* **47**: 884–888.

15. Furlan, A. J., 1995, *Cleve Clin J Med.* **62** (1): 6–8.

16. Ginsberg, M. D., 1990, *Revieves.* **2**: 68–93.

17. Ginsberg, M. D., 1994, *New Strategies to Prevent Neural Damage from Ischemic Stroke.* New York, pp. 1–34.

18. Ginsberg, M. D., Pulsinelli, W. A., 1994, *Ann Neurol.* **36**: 553–554.

19. Hallenbeck, J. M., 1996, *Mechanisms of Secondary Brain Damage in Cerebral Ischemia and Trauma.* New York, pp. 27–31.

20. Hansen, A. J., 1985, *Physiol Rev.* **65**: 101–148.

21. Heiss, W.-D., Rosner G., 1983, *Ann Neurol.* **14**: 294–301.

22. Heiss, W.-D., Graf, R., 1994, *Curr Opin Neurobiol.* **1**: 11–19.

23. Heiss, W.-D., Huber, M., Fink, G. R., *et al.*, 1992, *J Cerebr Blood Flow Metab.* **12**: 193–203.

24. Heiss, W.-D., Thiel, A., Grond, M., Graf, R., 2000, *Stroke.* **31** (4): 984-986.

25. Hossmann, K. A,.1993, *Progr Brain Res.* **96**: 161–177.

26. Hossmann, K. A., 1993, *Resuscitation.* **26**: 225–235.

27. Hossmann, K. A., 1994, *Ann Neurol.* **36**: 557–565.

28. Hossmann, K. A., 1994, *Brain Pathol.* **4**: 23–36.

29. Iadecola, C., 1998, Cerebral circulatory dysregulation in ischemia. In *Cerebrovascular Diseases* (Ginsberg, M. D., Bogousslavsky, J., eds.) Blackwell, MA, Cambridge, pp. 319–332.

30. Iadecola, C., 1999, Mechanisms of cerebral ischemic damage. In *Cerebral Ischemia* (Walz, W., ed.) Humana Press, New Jersey, Totowa, pp. 3–33.

31. Iadecola, C., Zhang, F., Nogawa, S., Ross, M. E., 1998, Post-ischemic gene expression and cerebral ischemic damage: role of inducible nitric oxide synthase. In *Frontiers in Cerebrovascular Disease: Mechanisms, Diagnosis and Treatment* (Robertson, J. T., Nowak, T. S., eds.) Futura Publications, New York, pp. 299–313.

32. Isayama, K., Pilts, L. H., Nishimura, M. C., 1991, *Stroke.* **22**: 1394–1398.

33. Ito, U., Ohno, K., Nakamura, R., et al., 1979, *Stroke.* **10**: 542–547.

34. Jacewicz, M., Kiessling, M., Pulsinelli, W. A., 1986, *J Cerebr Blood Flow Metab.* **6**: 263–272.

35. Jones, T. H., Morawetz, R. B., Crowell, R. M., 1981, *J Neurosurg.* **54**: 773–782.

36. Jones, S. C., Perez-Trepichio, A. D., Xue, M., *et al.*, 1994, *Acta Neurochir Suppl.* **60** (-HD-): 207–210.

37. Kirino, T., 1982, *Brain Res.* **239**: 57–69.

38. Kirino, T., Sano, K., 1984, *Acta Neuropathol.* **62**: 209–218.

39. Kirino, T., Tamura, A., Sano, K., 1984, *Acta Neuropathol.* **64**: 139–147.

40. Macfarlane, R., Tasdemiroglu, E., Moskowitz, M. A., *et al.*, 1991, *J Cerebr Blood Flow Metab.* **II**: 261–271.

41. Marchal, G., Beaudoin, V., Rioux, P., *et al.*, 1996, *Stroke.* **27**: 599–606.

42. Matsumoto, K., Yamada, K., Kohmura, E., *et al.*, 1994, *Neurol Res.* **16** (6): 460–464.

43. Matsumoto, K., Yamamoto, K., Hamburger, H. A., *et al.*, 1987, *Mayo Clinic Proc.* **62**: 460–472.
44. McCord, J. M., 1985, *N Engl J Med.* **312**: 159–163.
45. Mies, G., Ishimaru, S., Xie, Y., *et al.*, 1991, *J Cerebr Blood Flow Metab.* II: 753–761.
46. Ozyurt, E., 1988, *J Cerebr Blood Flow Metab.* **8**: 138–143.
47. Pappata, S., Fiorelli, M., Ronnel, T., *et al.*, 1993, *J Cerebr Blood Flow Metab.* **13**: 416–424.
48. Pulsinelli, W. A., 1992, *Proc Natl Acad Sci USA.* **89**: 10499-10503.
49. Pulsinelli, W. A., Brierley, J., Plum, F., 1982, *Ann Neurol*, pp. 491–498.
50. Scheinberg, P., 1991, *Neurology.* **41**: 1867–1873.
51. Siesjo, B. K., 1992, *J Neurosurg.* **77**: 169–184.
52. Skvortsova, V. I., 1993, *Clinical and Neurophysiological Monitoring and Metabolic Therapy in Acute Period of Cerebral Ischemic Stroke.* Doctoral dissertation. Moscow, 379 (in Russian).
53. Tanahashi, N., 1988, Cerebral microvascular reserve for hyperemia. In *Cerebral Hyperemia and Ischemia* (Tomita, M., Sawada, T., Naritomi, H., Heiss, W.-D., eds.) Excerpta Medica, ICS 764, Amsterdam, pp. 173–182.
54. Tomita, M., 1988, Significance of cerebral blood volume. In *Cerebral Hyperemia and Ischemia* (Tomita, M., Sawada, T., Naritomi, H., Heiss, W.-D., eds.) Excerpta Medica, ICS 764, Amsterdam, pp. 3–31.
55. Tomita, M., 1993, Microcirculatory stasis in the brain. In *Microcirculatory Stasis in the Brain* (Tomita, M., Mchedlishvili, G., Rosenblum, W. I., Heiss, W.-D., Fukuuchi, Y., eds.) Excerpta Medica, ICS 1031, Amsterdam, pp. 1–7.
56. Tomita, M., Fukuuchi, Y., 1996, *Acta Neurochir.* **66**: 32–39.
57. Touzani, A., Young, A. R., Derlon, J. M., *et al.*, 1995, *Stroke.* **26**: 12–21.
58. Twichell, T. E., 1951, *Brain.* **74**: 443–480.
59. Warach, S., Gaa, J., Siewert, B., *et al.*, 1995, *Ann Neurol.* **37**: 231–241.
60. Warlow, C. P., Dennis, M. S., van Gijn, J., *et al.*, 1996, *Stroke. A Practical Guide to Management.* Blackwell Science Ltd, Oxford, 6.
61. Welch, K. M. A., Windham, J., Knight, R. A., *et al.*, 1995, *Stroke.* **26**: 1983–1989.

Chapter 2

Cellular Reactions in Response to Acute Focal Brain Ischemia

Acute focal brain ischemia triggers pathological biochemical reactions that take place in all neural tissue cell pools and cause neuronal dysfunction, astrocytosis, microglial activation, and accompanying changes in neutrophils, macrophages, and endothelial cells.

The size of the infarct and the subsequent cell response to ischemia that determines the condition of the penumbra depend on the severity and duration of ischemia. In mild or short-term focal ischemia, only selectively vulnerable cells are affected [14]. As the ischemic period becomes longer, the selectivity of cell damage decreases and tissue dependent changes occur [32] mediating the formation of a brain infarction by the mechanisms of necrosis and apoptosis.

In the penumbral area, glial cells are damaged faster and more extensively, whereas cortical neurons are damaged less severely [44]. However, the changes in all cell pools are interdependent.

Glial cells significantly dominate in number over neurons and occupy all the space between vessels and neurons. The majority of central neurons are encircled by neuroglia so tightly that sometimes it is hard to differentiate neuronal and glial fractions [21]. Close morphological interaction is one of the basic principles for both physiological and pathological relations between glia and neurons [39, 55]. These interactions have been studied for more than a century. As a result the concept of dynamic signal neuron–glia interactions in the central nervous system (CNS) has appeared. Earlier these contacts were regarded as a passive data transition between two pools [45]. The possibility of data transition from neurons via glia to other neurons may envisage many variants of such intercellular "cross-talk" [39]. Glia should not be treated as only the trophic support system. It actively participates in specific functions of neural tissue: 1) contributes to electrogenesis, physiologically inhibiting neuronal hyperactivity; 2) regulates adequate energy flow during the activation of neurons via glucose consumption [5, 37, 38] and production of lactate, which is an adequate energy substrate [8, 49, 53] (see Section 4.1.2). Due to selectively increased permeability for potassium ions, astroglia regulates the activity of enzymes required for the support of neuronal metabolism and removal of mediators and other agents that are released during neuronal activity [6, 21]. The active uptake of glutamic acid from synaptic clefts by astrocytes is very important because excessive concentration of this amino acid may cause irreversible damage to neurons (see Section 4.1). Glia is involved in the synthesis of cytokines, immune mediators, and other signal molecules (cyclic guanosine monophosphate, cGMP; nitric oxide, NO) that are then transmitted to neurons and in the synthesis of glial growth factors (GDNF) that play a role in trophic support and neuronal repair. During recent years, it has been found that astrocytes react to increase of the synaptic concentrations of neurotransmitters as their processes are tightly interconnected with neuronal dendrites [2, 38, 47] and respond to changes in electrical activity of neurons with their reciprocal changes in intracellular calcium concentrations. Thus, a "wave" of calcium migration is triggered between astrocytes leading to calcium content oscillation in certain brain areas, which in turn can modulate the functional state of many neurons [21]. In acute focal ischemia the mechanisms of delayed neuronal death are directly associated with an aggressive influence of ischemia-excited neuroglia on still viable cells in the peri-infarction area [23, 46] (see Section 8.1).

Despite the tightest interaction between neurons and glial cells, all metabolic processes taking place in these two pools are strictly compartmentalized. The essence of metabolic compartments is to spatially separate biosynthetic processes from those metabolic pathways being watchfully controlled by energetic demands [3, 21]. Under these conditions the composition of internal connections which use similar signal molecules (neurotransmitters, cytokines, growth factors, and hormones) and influence similar targets (receptors) promote the interaction between nervous, immune, and endocrine systems in order to design a common assembled response following pathological stimuli such as brain ischemia [39, 40].

According to findings acquired from experimental models of stroke after the occlusion of the middle cerebral artery in rats, the sequence of responses produced by different cell pools onto acute focal ischemia has been elucidated. Undoubtedly, there are differences in the development of brain changes in rodents and primates [52], but the observed common features allow the experimental results to be extrapolated to a certain extent for understanding of cellular mechanisms of stroke in humans (see Scheme 2.1).

Neurons exhibit rather variegated changes under the conditions of ischemia. They include chromatolysis of variable extent, hyperchromatism, homogenization of cytosol, the appearance of "shadow cells" and, finally neuronophagia and focal cell loss [30]. Cell damage usually starts with peripheral chromatolysis [10], which appears to be the marker of reactive changes of neurons and reflects the dysfunction of protein metabolism [13]. The first signs of neuronal chromatolysis in ischemia appear several minutes after vessel occlusion. Evident neuronal shrinkage is observed 30 min after induced ischemia. After an hour, one can observe in neurons a heterochromatin fraction, enlargement of endoplasmic reticulum, and vacuolization and swelling of the internal mitochondrial matrix [22, 24, 30]. These potentially reversible changes last for approximately 6 h. Ten to twelve hours later in the ischemic core one can observe features of irreversible cell damage such as destruction of cytosolic and nuclear membranes and deposition of calcium-rich salts in the internal mitochondrial membrane [22, 24], whereas the reversible changes in the penumbra do last longer. The first shadow cells (dead neurons) are found in brain tissue on the second or third day after the stroke onset [7, 48].

Destruction of intracellular membranes and damage to membranes that encircle cells lead to a situation when differentiation between damaged neuron and glial cells appears impossible. That is the clear sign of beginning neuronophagia, whose terminal stages comprise the "submerging" of glial cell into neuron [30].

Along with the mentioned features of necrotic death of neurons in the ischemized brain area, one can observe cells located alongside the internal border of the ischemic core dying by the mechanism of programmed cell death, i.e. apoptosis [16, 17, 35, 36, 47]. The first of these will be observed already 2 h after the onset of stroke. The maximal number of apoptotic cells is found 24–48 h after the stroke onset, and later their number tends to decrease [19]. However, in the penumbral area apoptotic neurons are found for 4 weeks. Morphological features of apoptosis and the differential criteria of apoptosis and necrosis are described in Chapter 10.

The most ischemia-sensitive synapses are the axo-dendritic axo-somatic ones [9, 56]. Ultrastructural changes in these synapses are manifested by swelling of synaptic terminals 5 min after the onset of ischemia [9, 50, 51], corresponding to the development of anoxic depolarization of the terminals and the dysfunction of ionic synaptic homeostasis. The swelling of the postsynaptic area starts a bit later (5–6 min after the onset of ischemia), and its extent depends on the location of the synapse. The most vivid swelling occurs in small dendrites, while high cytoskeletal neurofilament content in pili hampers their prominent edematous changes [1].

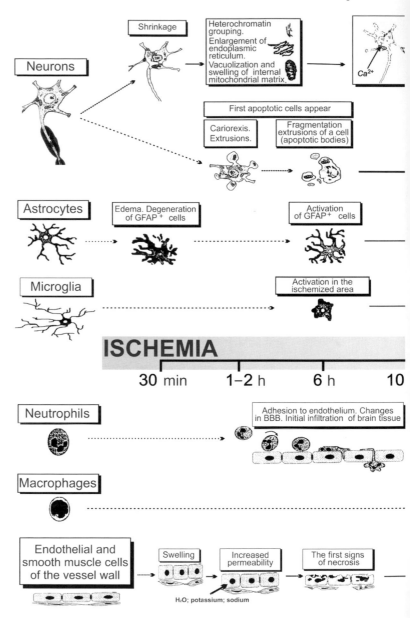

Scheme 2.1. The sequence of cellular reactions in acute focal ischemia.

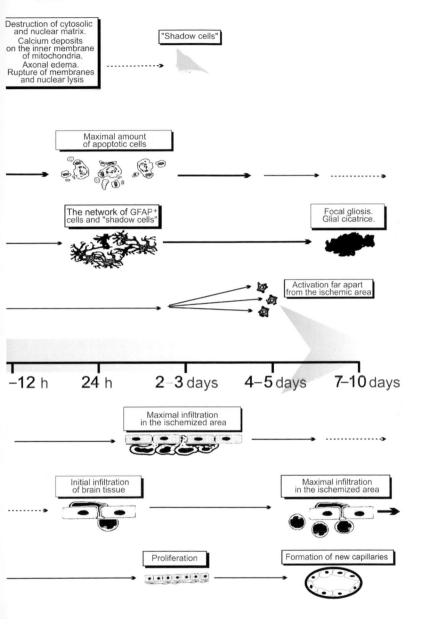

The destruction of earlier intensively functioning synapses is accompanied by activation of the passive flat synapses. The compensatory reorganization of the latter includes hypertrophy and splitting. Due to this, the overall area of the synaptic zone significantly enlarges [11].

The response of glial cells to ischemia can be classified as follows: 1) productive hypertrophy and hyperplasia; 2) appearance of edematous forms, and 3) destruction of glial cells, which starts with the fragmentation of their processes [10]. Swelling and process fragmentation of astrocytes is seen from the first minutes of ischemia, preceding neuronal changes [24]. These glial changes are accompanied by the decrease of such astrocytic marker as glial fibrillar acid protein (GFAP) [18, 24]. However, 4–6 h after the onset of stroke the activation of astrocytes encircling the ischemized area, which begin to synthesize GFAP more rapidly, can be detected [24]. Later (after 24 h), a network of GFAP-positive astrocytes forms around the infarction area. The reaction of astrocytes becomes increasingly aggressive and leads to the formation of a glial cicatrix at the end of the 1st or the beginning of the 2nd week after stroke onset [20]. When regeneration and reparative processes develop in brain the ultrastructure of glial cells considerably normalizes. However, signs of their functional activation persist, as confirmed by increase in the number of mitochondria, elements of granular and agranular reticulum, and polysomes [10].

Findings by Graeber and Streit [29] and by Wood [55] suggest that microglia is the only compartment having the property of immunocompetent cells able to defend brain from injuring factors. Microglial cells are widely distributed throughout human brain tissue and comprise 9–12% of the total glial population in gray matter and around 7.5–9% in white matter. In contrast to astrocytes, microglial cells originate from stem cells. Under normal conditions they are branched with many processes [27], remain in the state of functional rest, and can be temporarily activated to increase the production of various toxic pro-inflammatory mediators [4], but then rapidly return to the previous state of rest. Under pathological conditions (e.g., in ischemia), microglial cells pull their processes in and assume amoeboid shape, which is accompanied by their functional activation up to readiness for phagocytosis [27]. Under these conditions, the cells do not return to the state of rest but continue to synthesize a wide range of substances toxic for brain tissue, chronically supporting inflammatory response that leads to delayed neuronal loss, disturbances of microcirculation, and changes in the blood–brain barrier (BBB) [40, 42, 54] (see Section 8.2). By the end of the 1st day from the onset of stroke, activated microglia becomes widely distributed throughout the ischemized area, especially in the penumbral region [28, 43]. Later the area of microglial distribution continues to expand, and by the fifth day activated microglial cells are observed in brain areas far from the ischemic focus. The latter is probably caused via processes of antero- and retroneuronal trans-synaptic degeneration [12, 15, 43]. The simultaneous observation of hypoxic and metabolic disturbances in neurons far from the ischemic lesion such as changes in cytosolic tinctorial properties, DNA and RNA changes in cytoplasm and nuclei, and reduction of endoplasmic reticulum confirm the presence of active transneuronal degeneration [12, 15].

Already 6–8 h after the onset of stroke reactive changes of neutrophils in small vessels ensue. They are caused by microglial activation as well as by acute increase in the synthesis of pro-inflammatory cytokines and cell adhesion molecules. The characteristic features are the adhesion of neutrophils to endothelium in small vessels, their permeation through the BBB, and infiltration of the ischemized brain tissue [20, 22, 26, 57, 58]. The duration and dynamics of the neutrophilic reaction varies with the peculiarities of the ischemic process. Following constant experimental occlusion of the middle cerebral artery in rats, the maximal neutrophilic infiltration appears within 48–96 h after the stroke onset [20, 58, 59], and then the number of neutrophils in brain tissue begins to decrease.

Macrophages begin to extravasate into ischemized neural tissue at the end of the 1st day of experimental stroke; however, the process of macrophagic infiltration in the ischemized area becomes maximally pronounced by the fifth to seventh day from the stroke onset (see Section 8.2).

The changes in endothelial cells in small vessels begin from the first minutes of acute focal ischemia. Approximately 30 min later one can see endothelial swelling; an hour later increased permeability of membranes ensues; 6 h later features of necrosis in some cells appear, and by the second or third day one can see proliferation of endothelial as well as smooth muscle cells in the vessel walls [31]. At the same time, cerebral ischemia activates such processes as angiogenesis and neovascularization. By the fifth to seventh day of experimental stroke the formation of new capillaries [20] is observed in the peri-infarction area. It is induced by special angiogenic growth factors and by cell adhesion molecules. Krupinsky *et al.* [33, 34] showed that neovascularization is a beneficial sign that predicts improvement of brain tissue recovery and the survival of penumbral neurons.

REFERENCES

1. Akulinin, V. A., *et al.*, 1993, *Morphology.* **104** (7–8): 17 (in Russian).
2. Andriezen, W. L., 1993, *Anatomic and Physiologic.* **10**: 532–540.

3. Ashmarin, I. P., Stukalov, P. V., 1996, *Neurochemistry*. Institute of Biological and Medical Chemistry of Russian Scientific Academy Publishing House, Moscow (in Russian).

4. Banati, R. B., Gehrmann, S., Schubert, P., Kreutsberg, G. W., 1993, *Glia*. **7**: 111–118.

5. Barinaga, M., 1997, *Science*. **276**: 196–198.

6. Barres, B. A., 1991, *J Neurosci*. **11**: 3685–3694.

7. Beck, T., Lutz, B., Thole, U., Wree, A., 1993, *Brain Res*. **605** (2): 280–286.

8. Bittar, P. G., Charnay, Y., Pellerin, L., *et al.*, 1996, *J Cerebr Blood Flow Metab*. **16**: 1079–1089.

9. Bogolepov, N. N., 1975, *Ultrastructure of Synapses under Normal Condition and in Pathology*. Meditsina, Moscow (in Russian).

10. Bogolepov, N. N., 1979, *Structure in Hypoxia*. Moscow (in Russian).

11. Bogolepov, N. N., 1996, Synaptic mechanisms of compensatory and reparation processes in hypoxia and ishaemia. In *Compensatory and Adaptive Mechanisms in Brain under Normal Condition, in Pathology and in Experiments,* Tumen. pp. 112–113 (in Russian).

12. Bogolepov, N. N., Burd, G. S., 1981, *Abst VII All-Union Symp. of Neurologists and Psychiatrists*. Moscow. **2**: pp. 32–35 (in Russian).

13. Bogolepov, N. N., Dovedova, E. L., Gershtein, L. M., 1997, The influence of experimental hypoxia on oxidant and protein metabolism in rat brains. In *Hypoxia: Mechanisms, Adaptation and Correction.* LME, Moscow, pp. 34–35 (in Russian).

14. Brierley, J. B., Graham, D. I., 1984, In *Greenfield's Neuropathology* (Adams, J. H., Corsellis, J. A. N., Duchen, L. W., eds.) London, pp. 125–207.

15. Burd, G. S., 1983, *Respiratory Deficiency in Patients with Cerebrovascular Diseases*. Doctoral dissertation. Moscow, 355 (in Russian).

16. Charriaut–Marlangue, C., Margaill, I., Plotkine, M., *et al.*, 1995, *J Cerebr Blood Flow Metab*. **15**: 385–388.

17. Charriaut–Marlangue, C., Margaill, I., Represa, A., *et al.*, 1996, *J Cerebr Blood Flow Metab*. **16**: 186–194.

18. Chen, H., Chopp, M., Schultz, L., *et al.*, 1993, *J Neurol Sci*. **118**: 109–106.

19. Chopp, M., Li, Y., 1996, *Acta Neurochir*. **66**: 21–26.

20. Clark, R. K., Lee, E. V., Fish, C. J., *et al.*, 1993, *Brain Res Bull*. **31**: 565–572.

21. Flyorov, M. A., 1996, Biochemical peculiarities and interaction between neurons and glia. In *Neurochemistry* (Ashmarin, I. P., Stukalov, P. V., eds.) Moscow, pp. 193–200 (in Russian).

22. Garcia, J. H., Liu, K. F., Ho, K. L., 1995, *Stroke*. **26**: 636–642.

23. Garcia, J. H., Liu, K. F., Fisher, M., Tatlisumak, T., 1996, *Cerebrovasc Dis*. **6**: 180.

24. Garcia, J. H., Yoshida, Y., Chen, H., *et al.*, 1993, *Am J Pathol*. **142**: 623–635.

25. Garcia, J. H., Liu, K. F., Yoshida, Y., *et al.*, 1994, *Am J Pathol*. **144**: 188–199.

26. Garcia, J. H., Liu, K. F., Yoshida, Y., *et al.*, 1994, *Am J Pathol*. **145**: 728–740.

27. Giulian, D., 1997, Reactive microglia and ischemic injury. In *Primer on Cerebrovascular Diseases* (Welsh, M., Caplan, L., Siesjo, B., Weir, B., Reis, D. J., eds.) Academic, CA, San Diego, pp. 117–124.

28. Giulian, D., Corpuz, M., Chapman, S., *et al.*, 1993, *J Neurosci Res*. **36**: 681–693.

29. Graeber, M. D., Streit, W. J., 1990, *Brain Pathol*. **1**: 2–5.

30. Gusev, E. I., Bogolepov, N. N., Burd, G. S., 1979, *Cerebrovascular Disorders*. Moscow (in Russian).

31. Iadecola, C., 1999, Mechanisms of cerebral ischemic damage. In *Cerebral Ischemia* (Walz, W., ed.) Humana Press, Totowa, New Jersey, pp. 3–33.

32. Kogure, K., Yamasaki, Y., Malsuo, Y., *et al.*, 1996, *Acta Neurochir*. **66**: 40–43.

33. Krupinski, J., Kaluza, J., Kumar, P., *et al.*, 1993, *Lancet*. **342**: 742.
34. Krupinski, J., Kaluza, J., Kumar, P., *et al.*, 1994, *Stroke*. **25**: 1794–1798.
35. Li, Y., Chopp, M., Jiang, N., *et al.*, 1995, *J Cerebr Blood Flow Metab*. **15**: 389–397.
36. Linnik, M. D., Zobrist, R. H., Hatfield, M. D., 1993, *Stroke*. **24**: 2002–2009.
37. Magistretti, P. J., 1997, Coupling of cerebral blood flow and metabolism. In *Primer of Cerebrovascular Diseases* (Welch, K. M. A., Caplan, L. R., Reis, D. J., eds.) Academic Press, San Diego, pp. 70–75.
38. Magistretti, P. J., Pellerin, L., 1997, The central role of astrocytes in brain energy metabolism. In *Neuroscience, Neurology and Health*. WHO, Geneva, pp. 53–64.
39. Marchetti, B., Gallo, F., Romeo, C., *et al.*, 1996, *Ann NY Acad Sci*. **784**: 209–236.
40. McGeer, E. G., McGeer, P. L., 1994, Neurodegeneration and the immune system. In *Neurodegenerative Diseases* (Calne, D. B., ed.) Brace & World, Harcourt, Saunders, Philadelphia, pp. 277–299.
41. McGeer, P. L., Itagaki, S., Togo, H., McGeer, E. G., 1987, *Neurosci Lett*. **9**: 195–200.
42. McGeer, P. L., Kawamala, T., Walker, D. G., *et al.*, 1993, *Glia*. **7**: 84–92.
43. Morioka, T., Kalehua, A. N., Streit, W. J., 1993, *J Comp Neurol*. **327**: 123–132.
44. Pulsinelli, W. A., 1995, *Am Sci Med*. **2**: 16–25.
45. Ransom, B. R., Kettenmann, H., 1995, *Neuroglia*. Oxford University Press, Oxford, 1445.
46. Rinner, W. A., Bauer, J., Schmidts, M., el al., 1995, *Glia*. **14**: 257–266.
47. Sadoul, R., Dubois–Dauphin, M., Fernandel, P. A., *et al.*, 1996, *Adv Neurol*. **71**: 419–424.
48. Sala, L., 1981, *Zur Feineren Anatomic des Grossen Seepferdefusses*. **2**:18–45.
49. Schurr, A., West, C. A., Rigor, B. M., 1988, *Science*. **240**: 1326–1328.
50. Semchenco, V. V., Stepanov, S. S., 1987, *Arch Anat Histol Embryol*. **93** (11): 43–48 (in Russian).
51. Shmidt-Kastner, R., Freund, T. F., 1991, *J Neurosci*. **40**: 599–636.
52. Tagaya, M., Liu, K. F., Copeland, B., *et al.*, 1997, *Stroke*. **28**: 1245–1254.
53. Tsacopoulos, M., Magistretti, P. J., 1996, *J Neurosci*. **16**: 877–885.
54. Winfree, A., 1993, *SFI Studies in the Sciences of Complexity*. Reading, MA, Addision–Wesley, pp. 207–298.
55. Wood, P. L., 1994, *Life Sci*. **55** (9): 666–668.
56. Yamamoto, K., et. al., 1992, *J Neurochem*. **58**: 1100–1117.
57. Zhang, F., White, J. G., Iadecola C., 1994, *J Cerebr Blood Flow Metab*. **14**: 217–226.
58. Zhang, R. L., Chopp, M., Chen, H., *et al.*, 1994, *J Neural Sci*. **125**: 3–10.
59. Zhang, R. L., Chopp, M., Li, Y., *et al.*, 1994, *Neurology*. **44**: 1747–1751.

Chapter 3

Energy Failure Induced by Brain Ischemia

Decrease in brain perfusion is accompanied by decrease in oxygen delivery to brain tissue, where oxygen is involved in aerobic generation of energy as a substrate for cytochrome oxidase, the terminal enzyme of the respiratory chain. The resulting state of hypoxia is the consequence of a complicated multi-stage process involving sequential changes in the properties of the mitochondrial enzyme complexes [27].

Lukyanova *et al.* [19–37] elucidated sequential stages of the development of hypoxic changes in brain tissue (Fig. 3.1).

In *the first (compensatory) stage* the nicotinamide-dinucleotide (NAD/NADH)-dependent oxidative pathway is inactivated. This is accompanied by functional enhancement of the succinate oxidase pathway. The functional changes of the respiratory chain start in its substrate portion, i.e. in mitochondrial enzyme complex I. The earliest response to hypoxic state is the enhancement of NADH-dependent oxidative pathways as well of their contribution to overall respiration [4, 8, 15, 16, 36]. This is accompanied by an increase in maximal activity of rotenone-sensitive NADH-cytochrome-*c*-oxidoreductase [8]. The subsequent enhancement of electron flow from NAD-dependent substrates may be the reason for increase in the extent of reduction of cytochrome-oxidase during the first stages of ischemia, which correlates with an increase in ATP content [4, 32] and increase in functional activity of neurons [36, 37]. As the hypoxic state is aggravated, activation of the mitochondrial enzyme complex I gives place to its inhibition: 1) oxidation of NAD-dependent substrates decreases along with the coupled oxidative phosphorylation, and the sensitivity of respiration to the specific inhibitors of the NAD-dependent portion of the respiratory chain decreases [4, 20–24, 26]; 2) respiratory carriers of complex I such as pyridine nucleotides and flavins undergo reduction, reflecting disturbances in transfer of reduced equivalents via this portion of the respiratory chain

31

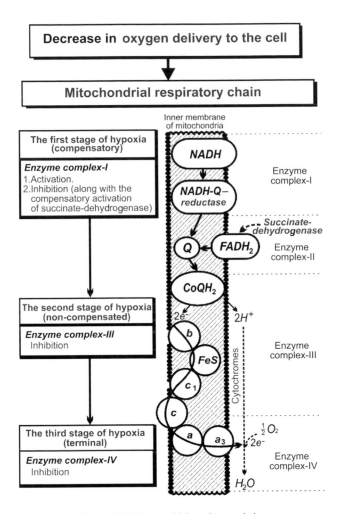

Figure 3.1. Mitochondrial respiratory chain.

[15, 28, 34]. Following these events, the cell looses its capacity to oxidize various energy substrates even when they are present in the medium. Thus, already at the early stages of hypoxia "substrate deprivation" develops, which is especially typical for ischemia [27]. Those substances shunting transfer of electrons within the NADH-coenzyme Q (CoQ) portion are still able to recover respiration as well as the redox state of respiratory carriers that is suggestive of reversibility of the cytochrome portion of the damage during this period [27, 40]. Dudchenko [6] showed that the activity of the other enzymes in the respiratory chain remain stable when the activity of rotenone-sensitive NADH-cytochrome-*c*-oxidoreductase decreases.

Dysfunction of mitochondrial enzyme complex I in the first stage of hypoxia does not lead to significant changes in intracellular ATP concentration and of the cellular functional activity [2, 4, 22, 25, 50]. This might be due to the activation of compensatory metabolic pathways that preserve energy producing function of the cytochrome portion of the chain and the capacity to conduct oxidative phosphorylation [13, 14, 38].

The escalation of oxidative failure leads to superimposition of disturbances typical for the first stage of hypoxia and the inhibition of electron-transfer function of the respiratory chain in the cytochrome *b*–*c* region. That is the essence of *the second (non-compensated) stage.* Although the ability to generate ATP by means of mitochondrial enzyme complex IV is still preserved, the activity of cytochrome-oxidase decreases and the delivery of electrons from the substrate portion (either NAD-dependent or succinate-oxidase) significantly decreases or ceases completely [15, 23, 26, 27, 32]. Depletion of ATP follows. A linear dependence between ATP concentration and partial oxygen pressure ensues.

Complete inactivation of cytochrome oxidase is observed under conditions close to anoxic. This means the cessation of both respiration and oxidative phosphorylation. This is *the third, i.e. terminal, stage* of hypoxia.

The stages of hypoxia correlate with phased changes in ATP content and pivotal energy-dependent processes in cells [2]. Only at the last stages of oxygen deprivation the level of energy deficit becomes sufficient to trigger basic mechanisms responsible for failure of cell vital functions and death (see Section 4.1). The increase in adenosine monophosphate (AMP) concentration is accompanied by activation of a protein kinase system, which is regarded as an additional cause of the destruction of cellular membranes [44].

Glycolysis does not prevent the decrease in ATP level during the last stages of oxygen deprivation [27]. However, much experimental evidence has been obtained that the increase in glucose uptake by astrocytes following glutamate-induced activation of Na^+/K^+-ATPase [39, 48] mediates glucose to be metabolized via the glycolytic pathway to lactate. Released lactate is transformed in neurons to pyruvate in order to be used as an adequate energy substrate [49] (see Section 4.1.2).

Experimental studies [3, 5, 7, 8] in animal models of acute cerebral ischemia revealed the variable extent of creatine phosphate content, ATP, and adenosine diphosphate (ADP) decrease in brain tissue along with the increase in inorganic phosphate and lactate content [42]. It was shown that during several minutes after the onset of acute focal ischemia the deficit of macroergic substances such as ATP and creatine phosphate would ensue in brain tissue [9, 10, 44, 46, 47]. During the first minutes of ischemia in rats less sensitive to hypoxia, only ADP and ATP concentrations decrease significantly, whereas only a short-term change is detected in creatine phosphate level followed by its complete restoration until normal values are reached. The gradient of ADP level decrease under the conditions of low values of oxygen partial pressure is pronounced far more severely in the brain of animals highly sensitive to hypoxia than in brain of the animals less sensitive to hypoxia [4, 20].

The peculiarities of energy changes in brain tissue depend upon the location of the ischemic process as well. In the majority of brain areas reperfusion is accompanied by complete or partial restoration of energy metabolism until normal values are reached: ATP and creatine phosphate concentrations increase; the level of lactate decreases [7, 42]. At the same time, changes of energy metabolism in ischemia-sensitive brain areas such as pyramidal neurons in the CA_1 hippocampal area or neurons in the dorsolateral area of the striatum have a biphasic pattern: following a short-term normalization its secondary inhibition is observed [11, 51]. The development of "post-ischemic hypoxia" leads to significant dysfunction of mitochondria [12, 18] and decrease in nicotinamide-dinucleotide phosphate (NADP) production in synaptosomes. The latter additionally influences the events leading to irreversible damage to brain tissue [45].

Analysis of the correlation between different stages of energy metabolism parameters by nuclear magnetic resonance spectroscopy of ischemized brain tissue revealed significant increase in the overall power of linkage, as well as considerable decrease in dispersion of correlation matrix elements [8]. This suggests a stricter organization of the energy system under conditions of moderate ischemia. Such over-regulation of the energy system may play a compensatory role during the first stages of ischemia. However, according to the "theory of chaos" proposed by Prigogine and Stengers [41], more organized systems are less stable, and, under certain conditions, more vulnerable. Thus, the decreased plasticity of the energy system in the case of prolonged ischemia becomes one of those factors promoting the formation of cerebral infarction as well as those determining the survival of neurons [3, 17].

Neurons undergoing severe ischemia (CBF < 10–15 ml/100 g brain tissue per minute) and hence developing severe dysfunction of energy metabolism fail to support the ion gradients of membranes due to lack of energy delivery to the Na^+/K^+-ATPase enzyme system [1, 9]. As mentioned in Chapter 1, the velocity of anoxic depolarization development of neuronal membranes depends upon the severity and the duration of ischemia and determines cell death. Obviously, the energy deficit is a chief mechanism of neuronal death in the ischemic core. Mild ischemia in the penumbral area initiates the development of complex biochemical transformations supported by the genomic reaction and by molecular consequences of ischemic events— "switching on" the genes of early response along with the secondary expression of genes coding cytokines, adhesion molecules, other pro-inflammatory and trophic factors, as well as the apoptotic genes (see Chapter 7).

Energy deficit and lactate acidosis appear to be the triggers of the cascade of pathological biochemical reactions taking place in all CNS cell pools and leading to the formation of brain infarction via two basic pathways— necrosis and apoptosis.

REFERENCES

1. Astrup, J., Symon, L., Branston, N. M., *et al.*, 1977, *Stroke.* **8**: 51–57.
2. Belousiva, V. V., Dudchenko, A. M., Lukyanova, L. D., 1992, *Bull Exp Biol Med.* **114** (12): 588–590 (in Russian).
3. Bruhn, H., Frahm, V., Gynsell, M. L., *et al.*, 1989, *Magn Reson Med.* **9** (1): 126–131.
4. Chernobayeva, G. N., 1984, *Peculiarties of Oxidative Metabolism Regulation in Animal Brain with Different Individual Sensitivity to Hypoxia.* Candidate's dissertation. Moscow, 211 (in Russian).
5. Cheung, J. Y., Bonventre, J. V., Malis, C. D., *et al.*, 1986, *N Engl J Med.* **314**: 1670–1676.
6. Dubchenko, A. M., 1976, *Activity of Mitochondrial Enzymes and Content of Energy Metabolites in Cerebral Cortex of Rats Possessing Different Sensitivity to Hypoxia.* Doctoral dissertation. Moscow, 289 (in Russian).
7. Faden, A. I., *et al.*, 1990, *J Pharm Exp Ther.* **255** (2): 451–458.
8. Gannushkina, I. V., Baranchikova, M. V., Semyonova, N. A., *et al.*, 1989, *J Neurol Psychiatr.* 51–66 (in Russian).
9. Hansen, A. J., 1985, *Revieves.* **65**: 101–148.
10. Hansson, E., 1985, *Physiol Rev.* **65**: 101–148.
11. Higuchi, T., Fernandez, E. J., Maudsley, A. A., *et al.*, 1996, *Neurosurgery.* **38** (1): 121–129.
12. Ishii, H., Stanimirovic, D. B., Chang, C. J., *et al.*, 1993, *Neurochem Res.* **18** (11): 1193–1201.
13. Khazanov, V. A., Panina, O. N., Kobzeva, Ye. A., *et al.*, 1989, In *Pharmacological Correction of Hypoxic States.* Moscow, pp. 71–79 (in Russian).
14. Kondrashova, M. N., 1989, In *Pharmacological Correction of Hypoxic States.* Moscow, pp. 51–66 (in Russian).
15. Korneyev, A. A., 1985, *The Study of Certain Oxygen-Dependent Events on the Isolated Contracting Heart in Hypoxia.* Candidate's dissertation. Moscow, 191 (in Russian).
16. Korneyev, A. A., Lukyanova, L. D., 1987, *Pathophysiology.* **3**: 3–56 (in Russian).
17. Lemasters, J. J., Diguiseppi, J., Nieminen, A. L., *et al.*, 1987, *Nature.* **325**: 78–81.
18. Linn, F., Paschen, W., Ophoff, B. G., Hossmann, K. A., 1987, *Exp Neurol.* **96** (2): 321–333.

19. Lukyanova, L. D., 1983, In *Models of Pathological Events Used for Research of Biologically Active Agents*. Moscow, pp. 94–95 (in Russian).
20. Lukyanova, L. D., 1984, In *Pharmacological Correction of Several Oxygen-Dependent Pathological Events*. Moscow, pp. 67–68 (in Russian).
21. Lukyanova, L. D., 1984, In *Physiological Problems of Adaptation*. Tartu, pp. 128–131 (in Russian).
22. Lukyanova, L. D., 1987, In *Molecular Mechanisms and Regulation of Energy Metabolism*. Pushchino, pp. 153–161 (in Russian).
23. Lukjyanova, L. D., 1988, *Neurochem Int*. **13**: 146.
24. Lukyanova, L. D., 1989, In *Pharmacological Correction of Hypoxic States*. Moscow, pp. 5–44 (in Russian).
25. Lukyanova, L. D., 1991, *Advances in Science and Technology. Pharmacology and Therapeutic Agents*. Moscow, **27**: 5–25 (in Russian).
26. Lukjyanova, L. D., 1996, In *Adaptation Biology and Medicine* (Shanina, B. K., Takeda, N., Ganguly, N. K., eds.) New Delhi. **1**: 261–279 (in Russian).
27. Lukyanova, L. D., 1997, *Bull Exp Biol Med*. **9**: 244–254 (in Russian).
28. Lukyanova, L. D., Balmukhanov, B. S., Ugolev, A. T., 1982, In *Oxygen-Dependent Events in Cell and Cellular Functions*, 301.
29. Lukyanova, L. D., Chernobayeva, G. N., Romanova, V. Ye., 1989, *Bull Exp Biol Med*. **107** (4): 431–433 (in Russian).
30. Lukyanova, L. D., Chernobayeva, G. N., Romanova, V. Ye., 1991, *Bull Exp Biol Med*. **112** (7): 49–51 (in Russian).
31. Lukyanova, L. D., Dudchenko, A. M., Belousova, V. V., 1994, *Bull Exp Biol Med*. **118**: 576–581 (in Russian).
32. Lukyanova, L. D., Dudchenko, A. M., Chernobayeva, G. N., 1999, The role of bioenergy metabolism in the formation of long-term adaptation mechanisms. In *Hypoxia: Mechanisms, Adaptation and Correction*. Moscow, 92 (in Russian).
33. Lukyanova, L. D., Kharadurov, S. V., Ugolev, A. T., *et al*., 1978, In *The Role of Mitochondrial Events in Temporal Arrangements of Vital Activity*. Pushchino, pp. 14–20 (in Russian).
34. Lukyanova, L. D., Korobkov, A. V., 1981, In *Physiological and Clinical Problems of Hypoxia, Hypodynamics and Hyperthermia*. Moscow, pp. 73–76 (in Russian).
35. Lukyanova, L. D., Romanova, V. Ye., 1987, In *Free Radicals and Biostabilizers*. Sophia, 85.
36. Lukyanova, L. D., Vlasova, I. G., 1989, *Bull Exp Biol Med*. **108** (9): 266–269 (in Russian).
37. Lukyanova, L. D., Vlasova, I. G., 1991, *Advances in Science and Technology. Pharmacology and Therapeutic Agents*. Moscow. **27**: 164–176 (in Russian).
38. Okon, Ye. B., Babsky, A. M., Kondrashova, M. N., 1988, In *Pharmacological Correction of Hypoxic States*. Moscow, pp. 45–46 (in Russian).
39. Pellerin, L., Magistretti, P. J., 1994, *Proc Natl Acad Sci USA*. **91**: 10625–10629.
40. Popova, O. A., Zamula, S. V., *et al*., 1989, In *Pharmacological Correction of Hypoxic States*. Moscow, pp. 155–159 (in Russian).
41. Prigogine, I., Stengers, I., 1986, In *Order from Chaos*. Moscow (Russian translation).
42. Pulsinelli, W. A, Sarokin, A., Buchan, A., 1993, *Progr Brain Res*. **96** (-HD-): 125–135.
43. Rothman, S. M., Olney, J. M., 1986, *Ann Neurol*. **19** (2): 105–111.
44. Scheinberg, P., 1991, *Neurology*. **41**: 1867–1873.
45. Shiraishi, K., Sharp, F. R., Simon, R. P., 1989, *J Cerebr Flow Metab*. **9**: 765–773.
46. Siesjo, B. K., 1988, *Ann NY Acad Sci*. **522**: 638–661.

47. Siesjo, B. K., Bengtsson, F., 1989, *J Cerebr Blood Flow Metab*. **9**: 127–140.
48. Takahashi, S., Driscoll, B. F., Law, M. J., Sokoloff, L., 1995, *Proc Natl Acad Sci USA*. **92**: 4616–4620.
49. Tsacopoulos, M., Magistretti, P. J., 1996, *J Neurosci*. **16**: 877–885.
50. Vlasova, I. G., Kutsov, G. M., Lomakin, Yu. V., *et al.*, 1999, The influence of hypoxia on neurons belonging to different brain structures in outliving slices. In *Hypoxia: Mechanisms, Adaptation and Correction*. Moscow, 14 (in Russian).
51. Welsh, F. A., O'Connor, M. J., Marcy, V. R., *et al.*, 1982, *Stroke*. **1**: 234–242.

Chapter 4

The Glutamate–Calcium Cascade

When glutamate is released from presynaptic terminals, it diffuses to the postsynaptic terminals where it binds to glutamate receptors, allowing Na^+ influx that depolarizes the membrane. This opens Ca^{2+} channels allowing Ca^{2+} to enter the postsynaptic cell. Under ischemic conditions, this process becomes excessive—the "glutamate–calcium cascade".

Fast reactions of the glutamate–calcium cascade underlie the formation of a focal ischemic necrotic lesion. These reactions develop in the first minutes and hours after a vascular event and are the borders of the therapeutic window period [32, 144, 145, 166, 167].

The glutamate–calcium cascade can be divided into three major stages: the induction (triggering) stage, the amplification stage (when the damaging potential is enhancing), and the expression stage (comprising terminal reactions of the cascade directly resulting in cell death) (Plate 1; Scheme 4.1).

4.1. The induction stage: energy-dependent ion pump failure and glutamate excitotoxicity

Decrease in ATP content in the ischemized brain area and compensatory activation of anaerobic glycolysis in response to hypoxia cause inorganic phosphate content as well as lactate production and H^+ generation to increase. These events mediate metabolic acidosis (see Chapter 5). Depletion of neuronal ATP inactivates the Na^+/K^+-ATPase enzyme system, which is required for energy-dependent ion transport [31, 88]. *Dysfunction of active ion transport* results in passive efflux of K^+ from cells and influx of Ca^{2+}.

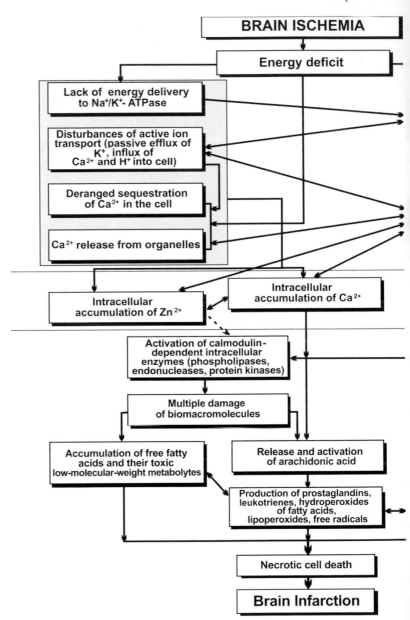

Scheme 4.1. Basic mechanisms of the glutamate–calcium cascade.

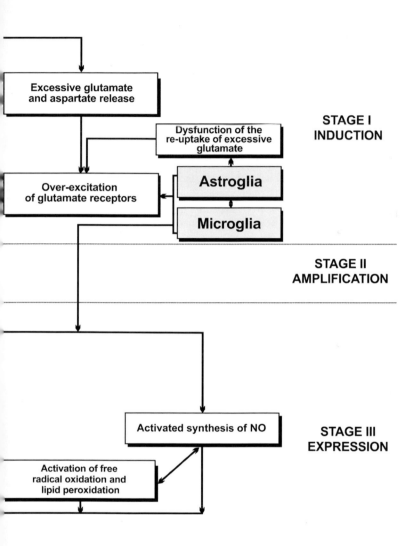

Excessive glutamate
and aspartate release

**STAGE I
INDUCTION**

Dysfunction of the
re-uptake of excessive
glutamate

Over-excitation
of glutamate receptors

Astroglia

Microglia

**STAGE II
AMPLIFICATION**

Activated synthesis of NO

**STAGE III
EXPRESSION**

Activation of free
radical oxidation and
lipid peroxidation

This depolarizes the cell membranes. H^+ continues to accumulate in cells. Along with the primary passive influx of Ca^{2+} via potential-dependent (potential-regulated) calcium channels, the sequestration of calcium in mitochondria and endoplasmic reticulum is disrupted due to oxygen and ATP deficit [24, 248]. Besides, H^+ and Ca^{2+} can compete for similar binding sites and the accumulation of acidic moieties in the cell may cause a considerable amount of Ca^{2+} release from organelles [58, 118, 213, 214]. Intracellular accumulation of Ca^{2+} in brain ischemia causes stress of mitochondria with the uncoupling of oxidative phosphorylation and escalation of catabolic processes [39, 170, 171, 198]. After this Ca^{2+} binds to the intracellular receptor calmodulin and becomes activated and in turn leads to activation of calmodulin-dependent protein kinases, lipases, and endonucleases, with subsequent DNA fragmentation and cell death [10, 170].

Thus, even at the very early steps of the pathological biochemical cascade triggered by deficit of macroergic substances *intracellular calcium accumulation* begins. This process appears to be one of the pivotal mechanisms in the destructive process underlying necrotic neuronal death [9, 128, 150, 157, 209, 212].

In the initial formulation of the hypothesis of necrotic neuronal death [209], it was stated that Ca^{2+} was delivered to cells via potential-dependent calcium channels, which are abundant in apical dendrites of neurons. However, later it was shown that blocking of these channels with dihydropyridinic antagonists never provided significant anti-ischemic defense and, moreover, never considerably blocked the influx of Ca^{2+} into neurons [72, 198, 212]. McDermott *et al.* [131] showed that agonist-dependent calcium channels present a considerably more important pathway for calcium influx, especially that controlled by receptors that are activated by excitatory amino acid neurotransmitters, glutamate and aspartate.

Olney [166] was the first to propose the hypothesis of **"excitotoxic neuronal death"**. His experiments showed that glutamate and aspartate have cytotoxic properties. Whenever they interacted with overly excited postsynaptic receptors damage to neuronal dendrites and soma occurred; however, this was not accompanied by axonal damage. The hyperexcitation of glutamate receptors was the result of an increased release of excitatory neurotransmitter amino acids into the extracellular space, with insufficiency of their reuptake mechanisms and utilization by astroglia, accumulating excessive concentrations of neurotransmitters in the synaptic cleft. Olney suggested that this theory of excitotoxicity could explain neuronal death in various brain diseases including ischemic stroke.

Since the beginning of the 1980s the theory of excitotoxicity has been further developed in several experimental works that confirmed the crucial role of amino acid neurotransmitters in the pathogenesis of ischemic stroke [57, 73, 144, 162]. The glutamate–calcium cascade is thought to be switched on by the excessive release of glutamate and aspartate from terminals of ischemized neurons into the extracellular space [85, 91]. The release of neurotransmitter amino acids is observed within 10–30 min from the onset of acute focal ischemia due to disregulation of active ion transport and depolarization of presynaptic membranes; their concentrations return to previous levels 30–40 min after blood flow recovery [12, 14, 52, 53, 58].

Obrenovitch *et al.* [163] showed that in rats the significant release of glutamate occurred only in rather severe ischemia—in the area of the ischemic core. It is noteworthy, however, that the elevation of glutamate level in the extracellular space is a biphasic process. During the first phase (10 min after the occlusion of the middle cerebral artery in rats) glutamate content increases to 30 mM, whereas a few minutes later it starts to increase progressively reaching the peak value of 80–85 mM after 55–60 min and then remains on this level for not less than 3 h [235]. The early component of the glutamate kinetics is not detected in the absence of Ca^{2+} in the medium [235]. This is suggestive of vesicular mode of glutamate release and, hence, for the level of remaining ATP being sufficient to support this exocytosis-mediated release. The second phase appeared to be relatively independent of Ca^{2+} concentration. This suggests a role of insufficiency of glutamate reuptake mechanisms in elevation of its extracellular level.

4.1.1. Cerebrospinal fluid levels of neurotransmitter amino acids in patients with acute ischemic stroke

The importance of changes in extra- and intracellular neurotransmitter amino acid concentrations in the development of acute ischemic stroke demonstrated using experimental models led to interest in studies of the levels of excitatory (glutamate, aspartate) and inhibitory (GABA, glycine) amino acid neurotransmitters in the cerebrospinal fluid (CSF) of patients with acute ischemic stroke [77, 78, 83, 84, 217]. The goal of these studies was to provide further objective evidence of the severity of the ischemic process, to prognosticate the course and outcome of the stroke, and to obtain more precise data on the mechanisms of brain tissue damage.

In our study done in the Department of Neurology and Neurosurgery of the Russian State Medical University 110 patients (50 male, 60 female, mean age 63.2 years) who had been admitted to the Intensive Stroke Unit within the first day after the onset of stroke in the carotid artery territory were examined. The major etiological factor was atherosclerosis (in 87% of cases) alone or combined with arterial hypertension (62%) and diabetes mellitus (8%). In 87 (79.1%) patients stroke exhibited atherothrombotic origin, in 20

(18.2%) it followed cardiac embolism, and in 3 (2.7%) stroke occurred on the background of transmural myocardial infarction, i.e. was of hemodynamic origin. There were no patients with lacunar (microcirculation) strokes in this group.

From the time of admission patients received complex and maximally unified treatment aimed at correcting disturbances in the respiratory and cardiovascular systems, acid–base and water–electrolyte balance (as required), and controlling edema (as required) to improve brain tissue perfusion; treatment included hemodilution, antiaggregants, angioprotectors, and antocoagulant therapy when needed (under laboratory control). Neuroprotective agents were not administered.

The severity of patients' condition and levels of neurological deficit were assessed using three complementary clinical scales: the Scandinavian scale (SSS) [197], the Orgogozo scale (OSS) [169], and an original scale (OS) developed by us [76]. The SSS, OSS, and OS are all graded 10 neurological items; for the SSS, OSS, and OS the maximum score (absence of neurological deficit) is 66, 100 and 49, respectively, and the minimum score is 0. Based on OSS scores patients were divided into several groups corresponding to severity of ischemic stroke; a score ≥65 corresponded to mild stroke, from 64 to 41 to moderate stroke, from 40 to 26 to severe stroke, and ≤25 to extremely severe stroke. The initial division was maintained throughout the trial; the SSS and OS score assessment and other measures were considered within the initial OSS score groups.

Neurotransmitter amino acid contents in cerebrospinal fluid were measured on days 1 and 3 after the stroke onset using high performance liquid chromatography (HPLC) with electrochemical detection following Pearson [175].

CSF was diluted 1:20 with 0.1 N solution of $HClO_4$ and then centrifuged for 10 min at 10,000 rpm. For determination of amino acids, 0.02 mg/ml L-homoserine internal standard in 0.2 μl NaOH and 25 μl of *o*-phthalaldehyde–sulfite reagent in 0.1 M borate buffer (pH 9.5) were added to 50 μl of supernatant for 20 min (20°C). The resulting samples were applied to a reverse-phase column (3 × 150 mm, C-18) and eluted with 0.01 M phosphate buffer pH 5.6 containing 0.05 mM EDTA and 5% methanol. A sample of L-homoserine (0.02 mg/ml) in 0.1 M borate buffer (50 μl) pH 9.5 was used as internal standard. The flow rate was 1 ml/min with electrochemical detection of glutamate and glycine using an LC-4B carbon electrode (Bioanalytical Systems Inc., USA). All reagents used were supplied by Sigma (USA) and Fluka (Switzerland). Amino acid contents in CSF in μmol per liter were calculated using a calibration curve: $S_{amino\ acid}/S_o$ (where $S_{amino\ acid}$ is the area of the amino acid peak and S_o is the area of the internal standard peak on the chromatogram).

Parameters were compared with normal levels obtained from peridural anesthesia in a control group consisting of 27 patients (13 male, 14 female, mean age 58.8 years) with orthopedic pathology without CNS lesions, systemic diseases, or thermoregulatory disturbances. Considering published data on the variability in neurotransmitter amino acid contents in cerebrospinal fluid, experimental errors were monitored by duplicating the control samples and measuring aliquots in parallel with samples from patients in the acute phase of ischemic stroke. Amino acid levels in controls were 1.22 ± 0.21 µM for aspartate, 4.48 ± 0.74 µM for glutamate, 15.15 ± 1.4 µM for glycine, and 0.97 ± 0.34 µM for GABA. There were no significant differences between men and women in the control group ($p > 0.05$).

The data were analyzed statistically using parametric (Fisher's F-test) and non-parametric (the Mann–Whitney U test and the Spearman rank correlation coefficient r) methods. Differences between values were recognized as significant at the 5% level in all comparisons.

Comparative analysis of the concentrations of the four neurotransmitter amino acids demonstrated significant increases in the cerebrospinal fluid of patients during the acute period of ischemic stroke ($p < 0.05$) as compared with controls (Fig. 4.1).

Considering the therapeutic window concept, there was special interest in studying the dynamics of neurotransmitter amino acid concentrations during the first day of illness. This established that levels of excitatory amino acids (aspartate and glutamate) increased dramatically within the first 6 h: aspartate increased 65-fold ($p < 0.004$) and glutamate 8-fold ($p < 0.024$) compared with controls; these parameters showed a positive correlation ($r = 0.4$; $p < 0.02$). The concentrations of inhibitory neurotransmitter amino acids (GABA and glycine) remained at the control level during this period. The peak increases in the concentrations of inhibitory neurotransmitter amino acids were shifted to later times (18–24 h): GABA levels increased 1.5-fold ($p < 0.05$) and glycine 3-fold ($p < 0.05$) compared with controls; this may be evidence for a late activation of compensatory mechanisms of protective inhibition. Under normal conditions there is known to be a stable equilibrium between the activities of the glutamatergic and GABAergic neurotransmitter systems [144, 183, 209]. The divergence in the dynamics of changes in the levels of excitatory and inhibitory amino acid neurotransmitters is evidence that the development of acute cerebral ischemia involves not only the phenomenon of "excitotoxicity", but also the formation of an imbalance between the excitatory and inhibitory mechanisms with signs of insufficiency of protective inhibition in the first hours from the stroke onset (the therapeutic window period) [217].

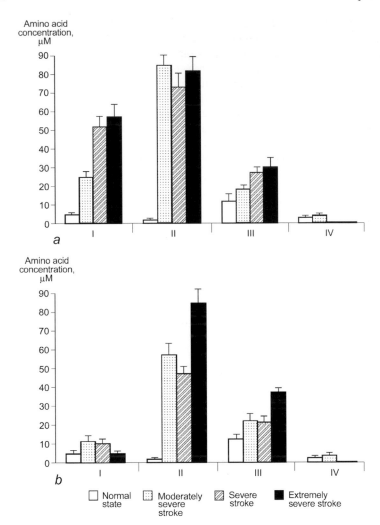

Figure 4.1. Concentrations of neurotransmitter amino acids in CSF of patients with carotid ischemic stroke on days 1 (a) and 3 (b) of illness: I – glutamate; II – aspartate; III – glycine; IV – GABA.

Overall evaluation of integral measures of amino acid concentrations on day 1 of stroke should take into consideration the positive correlation between the sizes of the increases in glutamate and aspartate ($r = 0.4$; $p < 0.02$) and between glutamate and glycine ($r = 0.56$; $p < 0.00009$), which may reflect the double role of glycine in ischemia—along with its inhibitory action, it is involved in processes activating newly formed glutamate N-methyl-D-aspartate (NMDA) receptors as a co-agonist [78, 83]. Comparison of amino acid concentrations in the first day of stroke for cases of differing initial severity revealed a significant predominance of the glutamate level ($p < 0.02$–0.008) and glycine ($p < 0.2$–0.04) in patients with severe and extremely severe illness as compared to patients with strokes of moderate severity. There were no significant correlations between the integral measures of these amino acids in patients with severe (total clinical score by the Orgogozo scale and the original scale 26–40 and 30–35, respectively) and extremely severe (total clinical score by the Orgogozo scale and the original scale lower than 25 and 30, respectively) neurological deficit ($p > 0.05$; see Fig. 4.1). GABA concentrations showed an inverse correlation with the initial severity of neurological deficit and were significantly higher in patients with moderately severe disease as compared with those with severe disease ($p < 0.02$) and extremely severe disease ($p < 0.00005$) (Fig. 4.2).

It is noteworthy that significant differences between the contents of neurotransmitter amino acids in patients with different pathogenic variants of stroke were not found ($p > 0.05$).

These studies demonstrated that the dynamics of neurotransmitter amino acid concentrations also depended on the severity of ischemic stroke and were of prognostic value.

Thus, patients in moderately severe condition had maximum glutamate concentrations during the first 6 h of stroke, these being 18 times higher than in controls ($p < 0.05$), followed by a 3-fold reduction by 12 h as compared with the level during the first 6 h ($p < 0.05$), with essentially no further change to day 3, i.e. the concentration remained moderately elevated (see Fig. 4.2). In patients with severe stroke there was a progressive increase in the glutamate level during the first day (there was a 21-fold elevation at 24 h compared with controls), and it was only by day 3 that a tendency for the concentration to decrease to a moderately elevated level, close to that in patients with moderately severe stroke, was seen ($p > 0.05$). It is interesting to note that patients with extremely severe stroke had progressive increases in the glutamate concentration only to 12–18 h, after which there was a sharp decrease; in some cases, glutamate became undetectable in the cerebrospinal fluid, this lasting to day 3 of illness (see Fig. 4.2). This phenomenon of early "normalization" in the glutamate level in CSF in the most severe forms of ischemic stroke could be associated with disruption of energy-dependent intracellular synthesis from products of the Krebs cycle due to progressive energy deficit. This suggestion is consistent with experimental data on decreases in the intracellular glutamate concentration under conditions of increasing global and focal brain ischemia [206].

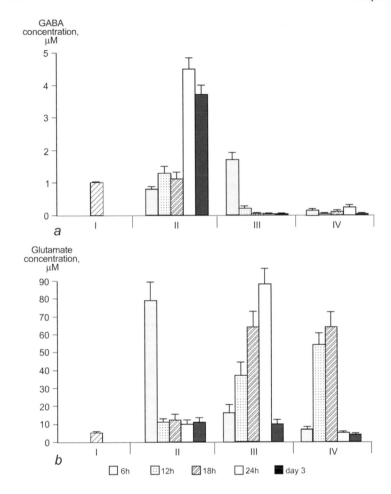

Figure 4.2. Dynamics of the concentrations of GABA (a) and glutamate (b) in CSF of patients with carotid ischemic stroke: I – normal state; II – moderately severe stroke (OSS > 40); III – severe stroke (OSS 25–39); IV – extremely severe stroke (OSS < 25).

   The aspartate concentration in patients with moderately severe and severe stroke increased progressively by the end of the first day of illness; by day 3 the aspartate level tended to decrease, though the level remained elevated as compared with controls. In extremely severe cases the CSF aspartate concentration continued to increase during the first days of illness and peaked on day 3.

The GABA level in patients with moderately severe stroke began to increase 12–18 h from the onset of disease and increased significantly by the end of day 1 (there was a 4.7-fold increase over control at 24 h; $p < 0.05$), which reflected sequential activation of inhibitory protective mechanisms (see Fig. 4.2). On day 3 of stroke, patients with a moderate neurological deficit persisted with high concentrations of GABA (3.7-fold increase compared with control; $p < 0.05$), which appears to be evidence of the importance of this protective mechanism throughout the period during which infarction-related changes develop. In patients with severe stroke, GABA levels increased only in the first 6 h from the onset of illness (1.8-fold compared with controls; $p < 0.05$); by 12 h the GABA level had decreased sharply (9-fold compared with the level at 6 h and 5-fold compared with controls), and GABA was subsequently not detected in the cerebrospinal fluid (the concentration was below the resolving ability of the method). Among patients with extremely severe stroke, the GABA concentration decreased from the first hours of illness to day 3, as compared with controls, which reflected the failure of protective inhibitory mechanisms. The inadequacy of inhibitory GABAergic protective mechanisms in conditions of increased release of excitatory amino acids may have resulted in the high severity of this contingent of patients.

So, the significance of *imbalance between the excitatory and inhibitory neurotransmitter systems* and of insufficiency of natural protective mechanisms in the pathogenesis of ischemic stroke was clinically verified. This is consistent with experimental data [54, 58, 62, 158]. Findings obtained from animal models of brain ischemia suggest that release of glutamate and aspartate into the extracellular space is accompanied by considerable decrease in their intracellular concentration. Along with this, intraneuronal GABA level significantly elevates despite the excessive release of GABA into the extracellular space. Under conditions of acute focal ischemia, GABA accumulates in neurons because of acidosis-mediated activation of glutamate-decarboxylase, which is a GABA-synthesizing enzyme, as well as of inhibition of GABA-transaminase, which is a GABA-degrading enzyme. Thus, experimental works demonstrated the compensatory role of additional synthesis and limited degradation of the natural inhibitory neurotransmitter GABA, which has a protective role.

Unlike the situation with GABA, concentrations of the other inhibitory neurotransmitter, glycine, were elevated in all examined patients during the first 12 h from the onset of stroke, especially in patients with severe and extremely severe neurological deficit (increases were by factors of 1.5–2 compared with control, $p < 0.05$). Increases in the glycine concentration persisted through three days regardless of the initial severity of disease (see Fig. 4.1).

Differences in the responses of the two inhibitory neurotransmitter systems (the GABAergic and glycinergic systems) to acute cerebral ischemia may be associated with differences in the topographic distributions of their receptor zones in brain tissue. Morphological and radionuclide studies have demonstrated that GABA receptors are mainly located in hemispheric structures of the brain (the temporal and frontal areas of the cortex, the hippocampus, and the amygdaloid and hypothalamic nuclei) and the midbrain (the substantia nigra, the periaqueductal gray matter), while glycine receptors are mainly located in brainstem and spinal cord structures [87, 183]. The inclusion in the study only of patients with hemispheric localizations of strokes with no significant morphological changes in the brainstem and cerebellar structures may have resulted in a predominance of dysfunction in the GABAergic neurotransmitter system in this cohort.

Retrospective analysis of the dynamics of biochemical parameters as related to the course of illness and its outcome revealed the most important prognostic criteria. During the first day of stroke, the extent of increase in glutamate concentration was prognostically significant: higher than 31.3 ± 5.5 μM levels were prognostic of a more severe and progressive course of disease (the differences compared with the group with a regressive course were significant at $p < 0.002$). Besides, the high level of glutamate during the first day of stroke predicted the formation of more extensive lesion by days 5–7 ($r = 0.45; p = 0.04$) measured morphometrically by MRI [28].

The most favorable prognostic sign on the first day was, according to our data, an early increase in the GABA concentration, indicative of a regressive course of stroke and good recovery of lost functions. Conversely, rapid and significant decreases in GABA concentrations or undetectable GABA levels were prognostic of an unfavorable course of stroke, with grave disability as the outcome.

The dynamics of amino acid levels by day 3 of stroke also had significant prognostic value. The extent of the increase in the glycine concentration in the cerebrospinal fluid by day 3 showed a strong inverse correlation with the increase in the total clinical points score by days 5 ($r = -0.56; p < 0.009$) and 21 ($r = -0.51; p < 0.03$) of illness. The most important unfavorable prognostic sign was undetectable level of glutamate and GABA in the cerebrospinal fluid on day 3 of stroke, with persisting increases in the aspartate concentration (there was an inverse correlation with the increase in the total clinical points score at the end of the acute period of stroke, with $r = -0.45; p < 0.048$, and a direct correlation with the frequency of lethal outcomes, with $r = 0.65; p < 0.02$).

The findings were later confirmed by Serena *et al.* [205] who showed that in patients with progressive lacunar stroke in the first 48 h from the onset of illness significant increase in the glutamate concentration (253 ± 70

µM compared with 123 ± 73 µM in patients with non-progressive stroke; $p <$ 0.001), as well as significant decrease in the GABA concentration ($140 \pm$ 63 nM compared with 411 ± 97 nM in patients with non-progressive stroke; $p < 0.001$) were measured in peripheral blood. It was shown that glutamate concentrations of more than 200 µM and GABA levels less than 240 µM had predictive value for neurological deterioration in 67 and 84% of cases, respectively. The excitotoxic index (glutamate × glycine/GABA) of more than 106 also had a prognostic value of 85%. So, these findings confirmed our suggestion that an imbalance between glutamate and GABA concentrations may have a role in the progression of ischemic stroke.

It is noteworthy that the levels of non-neurotransmitter amino acids did not change significantly under conditions of ischemia [58].

Thus, the clinical and biochemical studies [77, 78, 83, 84, 217] supported the role of the excitotoxicity phenomenon in the mechanisms underlying the development of acute focal cerebral ischemia forming a brain infarction. There were significant increases in the concentrations of excitatory amino acid neurotransmitters (glutamate and aspartate) in the cerebrospinal fluid in patients with acute hemispheric ischemic strokes from the first hours of illness with no dependency on vascular origin of strokes (atherothrombosis, cardiac embolism, or depression of systemic hemodynamics). The extent and duration of these increases were of prognostic value for determining the course and outcome of the stroke. Along with excitotoxicity, the pathogenesis of ischemic stroke may also involve a deficiency of protective inhibitory mechanisms, mainly due to the GABAergic system, allowing for the hemispheric localization of ischemia in the patients studied here. The degree of imbalance between the excitatory and inhibitory neurotransmitters systems, assessed in terms of the levels of the corresponding amino acids in the cerebrospinal fluid, is to a considerable extent responsible for determining the severity of the clinical manifestations of ischemic stroke and the potential of recovery processes.

The important biochemical criteria established here allow objective assessment of the severity of acute focal brain ischemia and prognosis of the course and outcomes of stroke and can be used in clinical settings. These clinical–biochemical relationships allow the strategy for neuroprotective treatment to be determined in the first hours and days of ischemic stroke.

4.1.2. Regulation of concentration of excitatory amino acids in the synaptic cleft

During recent years the important role of glial cells, especially astrocytes, has been shown in energy metabolism and functioning of the glutamate system [3, 134, 194]. Astrocytes should not be regarded as passive elements of the nervous system, but as cells intensely supplying neurons with energy substrates in response to increase in synaptic activity. Astrocytic processes encircle synaptic contacts and can detect the increase in synaptic neurotransmitter concentration (Fig. 4.3). Such functions of astrocytes as glutamate binding have been thoroughly studied [7]. The action of glutamate on postsynaptic neurons is swiftly determined by the high-affinity binding system located on the astrocytic processes that envelop a synaptic contact [192]. The transition of glutamate from the synaptic cleft is mediated via specific transport systems, two of which are glia-specific (GLUT-1 and GLAST transporters). The third transport system (EAAC-1) located exclusively in neurons is never involved in the transport of synaptic glutamate [192]. Transport of glutamate into astrocytes is driven by the electrochemical gradient of sodium ions.

Experimental and imaging techniques have shown that under the conditions of glutamate-induced cortical activation a local increase in glucose uptake occurs in the astrocytes [6, 132, 133]. This effect is not connected with the receptor-mediator mechanism as it cannot be stopped or imitated by glutamate-receptor antagonists or agonists [176]. Pellerin and Magistretti [176] and Takahashi *et al.* [109] suggest that the activation of Na^+/K^+-ATPase following massive intracellular influx of sodium ions plays a pivotal role in this process. Astrocytes metabolize the additional volume of glucose via the glycolytic pathway to lactate, which, on release, is transformed to pyruvate in neurons to be then included in the tricarboxylic acid cycle and, hence, to become an adequate energy substrate. For one molecule of glutamate taken up with three sodium ions, one molecule of glucose enters astrocytes, two molecules of ATP are produced through glycolysis, and two molecules of lactate are released. Within astrocytes, one ATP molecule is spent on "switching off" the Na^+/K^+ pump, whereas the other provides the energy needed to convert glutamate into glutamine by glutamine synthase.

There are many experimental findings showing that lactate can be used by neurons as an adequate energy substrate [11, 199, 234]. Microdialysis studies in rodents *in vivo* and H^1-MR-spectroscopy (MRS) analysis documented an increase in lactate content during physiological glutamate activation [66, 67, 180]. The possibility of selective distribution of lactate dehydrogenase (LDH) isoenzymes between lactate-producing and lactate-consuming cells has been proven. Thus, in human hippocampus and cerebral cortex the immunoreactivity to LDH_5 isoenzyme bound to tissues actively producing lactate is limited to the astrocytic population, whereas neurons react only to antibodies to LDH_1 isoenzyme present in lactate-consuming tissues [11]. These findings agree with the hypothesis of glutamate-induced glycolysis, an "astrocyte–neuron lactate shuttle" (see Fig. 4.3).

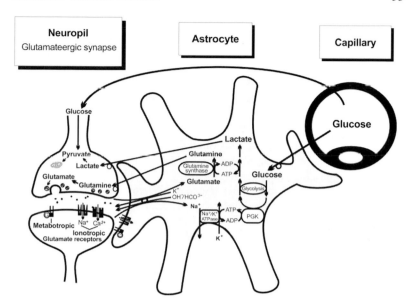

Figure 4.3. Schematic model of relationships between intraparenchymal capillaries, astrocytes and the neuropil, as well as of interaction between neuronal activity and glucose utilization (modified from Magistretti, Pellerin, 1997 [134]). Astrocytic processes surround capillaries (end-feet) and ensheath synapses. Receptors and binding sites for neurotransmitters are located on astrocytes. These features allow astrocytes to sense synaptic activity and to regulate consumption and metabolism of energy substrates originating from the circulation. The action of glutamate is terminated by a glutamate-binding system primarily located in astrocytes. Astrocytes remove the excess of glutamate from the synaptic cleft and transform it into glutamine, which then is returned to neurons. Glutamate is co-transported with sodium ion, leading to an increase in the intracellular concentration of sodium ions and promoting activation of Na^+/K^+-ATPase which in turn stimulates glycolysis. Lactate is produced as a result of glutamate-induced glycolysis in astrocytes, and then it is released and taken up by neurons to serve as an adequate energy substrate. Direct glucose uptake into neurons under basal conditions is also shown by the bow-shaped arrow above.

PET activation studies in humans showed an uncoupling between glucose utilization and oxygen consumption [65]. Experimental research demonstrated that the efficacy of glutamate consumption by synaptosomes was also first of all determined by glucose concentration, not by the level of hypoxia [178, 182].

Thus, the important task of functional collaboration between neurons and astrocytes is the regulation of adequate energy flow during neuronal activation. Glutamate-induced glucose consumption and lactate production in astrocytes appear to be a simple and direct mechanism to provide neuronal functioning.

Temperature, pH, and sodium and potassium ion contents in the ischemized brain area are additional factors that influence extracellular glutamate and aspartate levels and, henceforth, on the extent of excitotoxicity. The increase in extracellular glutamate level in acute focal cerebral ischemia, according to experimental findings [153, 223], is in direct proportion with temperature in the ischemic lesion. One of the main effects of moderate hypothermia is a significant decrease in excitatory amino acid neurotransmitter concentrations in the penumbral area encircling a focal infarction, which inhibits the expansion of the damaged area.

Metabolic acidosis that accompanies the glutamate–calcium cascade reactions inhibits the reuptake of glutamate from the synaptic cleft. When the pH decreases to 5.8, astrocytic transport systems for glutamate become significantly if not completely inhibited [181, 222].

According to experimental data [161], the elevation of extracellular potassium ion level in brain ischemia to 15–60 μM enhances the expansion of neuronal damage. Hirota *et al.* [96] and Zablocka and Damanska-Janik [248] point out that urgent elevation of potassium ion level in the extracellular space leads to excessive release of glutamate and aspartate into the synaptic cleft, to simultaneous inhibition of its reuptake, and to swelling of astrocytes. Although the release of aspartate is slightly delayed compared with that of glutamate, later its level continues to increase with no dependence on extracellular sodium ion concentration.

Mainprize *et al.* [135] emphasize that the destabilization of membranes by high concentrations of extracellular potassium ions leads to the extracellular efflux of glutamate, aspartate, GABA, and glycine in approximately equal concentrations. However, the resulting elevation of extracellular glutamate content on the background of potassium-induced depolarization during brain ischemia does not depend on the level of potassium ions but is connected with the vesicular efflux modulated by NO.

The extracellular concentration of sodium ions does not significantly influence the release of glutamate and aspartate but changes the sodium-dependent outflow of glutamate from the synaptic cleft into the cell [69]. The application of sodium-channel blockers in brain ischemia leads to 50% decrease in ischemic damage versus control cases of the experimental model of ischemic stroke. The authors explain this as a hampering of depolarization of neuronal membranes due to restriction of sodium-dependent glutamate transport and intracellular calcium accumulation. Blocking of endogenous glutamate release out of presynaptic terminals is a third possible reason.

Thus, in the normal state we have relative constancy of extracellular glutamate and aspartate concentrations. Their elevation switch on compensatory mechanisms: the reuptake of their excessive amounts by neurons and astrocytes from the extracellular space; presynaptic inhibition of neurotransmitter release; metabolic consumption, etc. [187]. Under conditions of ischemia there is dysfunction of the highly selective system of glutamate and aspartate transport from synaptic cleft to astroglia [7, 34, 52] and the system of pathways responsible for mediator formation changes [173]. These changes cause the absolute concentration and the exposure interval of glutamate and aspartate in the synaptic cleft to exceed admissible limits, and the process of neuronal membrane depolarization becomes irreversible.

4.1.3. Glutamate receptors and mechanisms of brain ischemia

The accumulation of high concentrations of excitatory neurotransmitter amino acids causes over-excitation of glutamate receptors, which constitute a heterogeneous population. In 1982 Davies and Watkins classified glutamate receptors into 2 classes: NMDA and non-NMDA [45]. Further studies applying modern electrophysiological techniques, radio-ligand binding, and molecular biology showed that there exist families of ionotropic and metabotropic glutamate receptors in the mammalian CNS.

Ionotropic glutamate receptors regulate the permeability of ion channels and are classified by their sensitivity to the action of N-methyl-D-aspartate (NMDA), alpha-amino-3-(3-hydroxy-5-methyl-4-isoxazol)-propionic acid (AMPA), kainic acid (K), and L-2-amino-4-phosphobutyric acid (L-AP$_4$), the most selective ligands of these receptor types. The names of these substances were assigned to the respective types of receptors: NMDA, AMPA, K, and L-AP$_4$ [40, 42, 206, 219].

Ionotropic receptors play a crucial role in excitatory neurotransmission in the CNS that is the most important for triggering of the pathological biochemical cascade in acute focal cerebral ischemia. The over-excitation of **NMDA receptors** (Fig. 4.4) by glutamate leads to the "shock" opening of calcium channels and, hence, to massive influx of calcium ions into neurons with sudden elevation of their concentration up to the threshold value [32, 123, 149].

Studies of how various agents influence NMDA receptors have shown that the postsynaptic NMDA receptor is a supramolecular complex that includes several sites of regulation: a site for the specific mediator (L-glutaminic acid) binding; a site for the specific co-agonist (glycine) binding, and allosteric modulator sites located on the membrane (polyaminic) and inside the ion channel attended by the receptor (sites binding bivalent cations and a "phencyclidine" site for binding of noncompetitive antagonists). That is why along with the term "NMDA receptor" several authors use the term "NMDA receptor ionophore complex", emphasizing the complexity of its

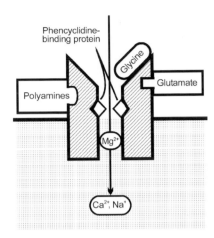

Figure 4.4. Glutamate NMDA receptor.

structure and the multitude of possibilities to regulate its functions [35, 42, 206, 225].

In our collaborative study done in the Department of Neurology and Neurosurgery of the Russian State Medical University together with the Institute of the Human Brain of the Russian Academy of Sciences [78, 80, 82] we for the first time showed the **increase in the level of autoantibodies to phencyclidine-binding protein of NMDA receptors** in blood serum of patients with acute ischemic stroke (Russian patent # 2123705 "New diagnostic facility in acute cerebral ischemia", 1995/1998).

We conducted clinical and immunobiochemical study of 70 patients (30 males and 40 females, average age 65.3 years). The study was held using the standard solid-phase enzyme-linked immunosorbent assay (ELISA) modified by the Institute of the Human Brain [40, 41]. Patients were admitted to the Intensive Stroke Unit within the first 2–6 h from the onset of stroke in the carotid artery territory. All patients received multi-dimensional and at the same time maximally standardized treatment (see Section 4.1.1). Neuroprotective agents were not administered.

Serum content of autoantibodies to phencyclidine-binding membrane protein, isolated from human brain tissue, was measured longitudinally: 3, 6, 9, 12, and 24 h from the stroke onset and on the third and the fifth day of stroke. We used an experimental lot of diagnostic kits manufactured by the Laboratory of Neurobiology of the Institute of the Human Brain. The normal value of the level of autoantibodies was established after examination of 200 healthy volunteers (average age 63.7 years) and comprised 0.3–1.5 ng/ml, which obviously reflects the presence of a background level of glutamate receptor renewal during general reparation processes in the human body [40, 41].

In all patients admitted to the clinic with moderate or severe neurological deficit, significant increase of the level of autoantibodies to phencyclidine-binding protein of NMDA receptors were detected already within 2–3 h of stroke ($p < 0.01$) (Table 4.1; Fig. 4.5). This increase persisted for 3 to 5 days. The hypothetical reason for this increase is damage to receptors by their over-excitation caused by excessive synaptic glutamate or the destruction of newly formed structurally immature receptors.

The increase in the level of these autoantibodies depended on the severity of ischemic stroke. In severe strokes (total clinical score by the Orgogozo scale and the original scale was 26–40 and 30–35, respectively) the level was significantly higher than in those with moderate deficit (total clinical score by the Orgogozo scale and the original scale more than 40 and 36, respectively); the most prominent differences ensued 9–12 h after the stroke onset ($p = 0.01$). In patients with moderate neurological deficit the level of autoantibodies tended to decrease by the end of the 1st day of stroke. By the 5th day in cases of stroke improvement we registered its normalization. In patients with severe neurological deficit the level of autoantibodies remained elevated for a longer time: when stroke improved it tended to decrease by the 3rd–5th day after stroke, when stroke progressed high values were registered during the entire period of observation (see Table 4.1, Fig. 4.5).

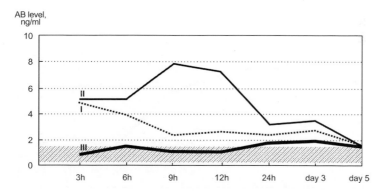

Figure 4.5. Level of autoantibodies (AB) to phencyclidine-binding protein of glutamate NMDA receptors in blood serum in relation to baseline neurological deficit. Here and in Table 4.1: I – moderately severe stroke (OSS > 40); II – severe stroke (OSS 25–39); III – extremely severe stroke (OSS < 25). Shaded area corresponds to normal range.

Table 4.1. Level of autoantibodies to phencyclidine-binding protein of glutamate NMDA receptors (ng/ml) in blood serum in relation to baseline neurological deficit

Group of patients	TCS on day 1	Time after stroke onset						
		3 h	6 h	9 h	12 h	24 h	day 3	day 5
I	38.6±0.4	4.86±0.8	3.95±0.7	2.54±0.3	2.72±0.3	2.45±0.3	2.8±0.5	1.63±0.1
II	32.7±0.5	5.04±0.9	5.10±0.7	7.90±1.2	7.30±1.5	3.20±0.6	3.50±0.5	1.67±0.3
III	26.3±0.4	0.9±0.2	1.56±0.2	1.26±0.6	1.23±0.2	1.77±0.4	2.00±0.7	1.65±0.5
p_{I-II}	0.01	>0.05	>0.05	0.05	0.05	>0.05	>0.05	>0.05
p_{I-III}	0.001	0.01	0.01	0.05	0.05	>0.05	>0.05	>0.05
p_{II-III}	0.01	0.01	0.01	0.001	0.001	0.05	>0.05	>0.05

Note: TCS – total clinical score by the Original Scale.

The retrospective study of autoantibodies level versus outcome of stroke and the degree of recovery by the end of the acute period (the 21st day) allowed the derivation of a prognostic significance of its baseline values (during the 1st day) and its dynamics as well (see Table 4.2, Fig. 4.6). In patients with prominent disability and severe neurological deficit (Barthel Index of 0–45) we measured the highest values of the level within the first 9 h after the stroke onset (7.02–7.5 ng/ml), which significantly exceeded the values obtained from patients with substantial recovery (Barthel Index of 75–100) ($p < 0.01$) and moderate disability (Barthel Index of 50–70) ($p < 0.01$) (see Table 4.2, Fig. 4.6). In the most propitious cases (mild strokes) the complete recovery was preceded by a substantial decrease in autoantibody level by the end of the 1st day.

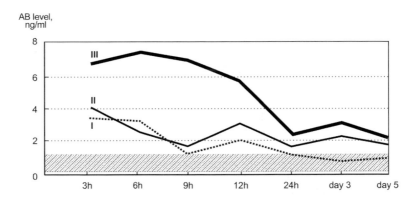

Figure 4.6. Level of autoantibodies (AB) to phencyclidine-binding protein of glutamate NMDA receptors in blood serum in relation to recovery on day 21. Here and in Table 4.2: I – good functional recovery (BI ≥ 75); II – moderately severe disability (BI 50–70); III – severe disability (BI < 50). BI – Barthel Index.

Table 4.2. Level of autoantibodies to phencyclidine-binding protein of glutamate NMDA receptors (ng/ml) in blood serum in relation to recovery on day 21

Group of patients	TCS on day 21	Time after stroke onset						
		3 h	6 h	9 h	12 h	24 h	day 3	day 5
I	46.0±0.5	3.60±0.6	3.36±0.6	1.48±0.3	2.25±0.3	1.60±0.1	1.15±0.5	1.26±0.3
II	42.0±0.4	4.15±0.6	2.87±0.6	1.90±0.2	3.21±0.4	1.99±0.4	2.60±0.4	2.13±0.2
III	34.7±0.6	6.83±0.9	7.50±1.3	7.02±1.4	5.83±1.4	2.68±0.4	3.36±0.3	2.60±0.3
p_{I-II}	0.01	>0.05	>0.05	>0.05	0.05	>0.05	0.05	0.05
p_{I-III}	0.001	0.01	0.01	0.01	0.05	0.05	0.05	0.05
p_{II-III}	0.01	0.05	<0.01	<0.01	0.05	>0.05	>0.05	>0.05

Note: TCS – total clinical score by the Original Scale.

Thus, the significant increase in the level of the specific autoantibodies to phencyclidine-binding protein of NMDA receptors, directly depending on the severity of ischemic stroke, confirms the importance of changes in the glutamate system in the development of brain ischemia and reflects the immune response to the destruction and the renewal of glutamate receptors. No correlation between the level of autoantibodies and glutamate, aspartate, and glycine concentration in CSF was found.

Significant and stable elevation of the serum level of autoantibodies in the first 3 h after stroke may reflect the preceding sensitization of brain tissue to fragments of NMDA receptors due to the background pre-stroke changes in neuronal membranes and in blood–brain barrier, i.e. the preparedness of brain tissue to form infarction, which may be due to chronic ischemic changes in brain and by the progressive course of ischemic brain disease (see Chapter 9).

Of utmost interest were findings acquired from patients with the most severe variants of carotid ischemic stroke (subtotal ischemia of a cerebral hemisphere) that were admitted to the clinic with clouding of consciousness, signs of brain edema, autonomic and trophic disturbances (total clinical score by the Orgogozo scale and the original scale was lower than 25 and 30, respectively). In contrast to the others, these patients had non-elevated levels of autoantibodies to phencyclidine-binding protein of NMDA receptors. There was a slight but non-significant trend ($p > 0.05$) to its increase within the first 6 h after the vascular event. It is noteworthy that in fatal cases the level of autoantibodies tended to decrease versus baseline level down to the lower limit of normal value (Fig. 4.7). The last value acquired before the patient's death was the lowest; its average value was 0.9 ± 0.15 ng/ml. In contrast, in patients who survived to the end of the 1st day of stroke there was a tendency towards slight increase in the level of autoantibodies. The differences in average levels in the deceased and the survivors were significant on the 3rd–5th days after stroke ($p < 0.05$).

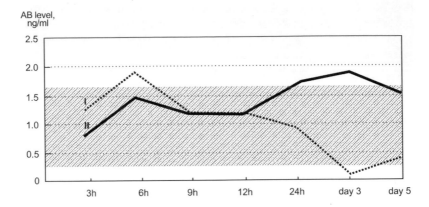

Figure 4.7. Level of autoantibodies (AB) to phencyclidine-binding protein of glutamate NMDA receptors in blood serum of patients with extremely severe neurological deficit. I – deceased; II – survived.

The absence of autoimmune activation phenomenon in patients with extremely severe stroke may be due to the acute immune deficiency caused by the stressing ischemia. The decrease of the level of autoantibodies to subnormal values may be regarded as the most unfavorable predictive criterion usually preceding the patient's death.

The role of the glycine site of the NMDA receptor is peculiar. Glycine at concentration of 0.1 μM enhances responses of NMDA receptor; moreover, this effect cannot be blocked by strychnine, whereas strychnine itself is a blocker of intrinsic glycine receptors. Glycine itself does not lead to the excitatory response but only increases the frequency of the channel opening, never influencing the current amplitude when NMDA agonists act [42, 105].

In recent years evidence has appeared indicating that other bivalent cations besides calcium take part in the ischemic brain damage. Researchers have analyzed the selective neuronal vulnerability in experimental transient global ischemia and described zinc-mediated excitotoxicity. Zinc exists in cells as a component of various metal-dependent enzymes and transcription factors. It also acts as a CNS neurotransmitter and neuromodulator [68]. Considerable concentrations of zinc ions were found in presynaptic vesicles and the synaptic cleft right after the stimulation of neurons. It was showed that zinc ions may influence potential-dependent ion channels and depolarization of neuronal membranes [2, 33], as well as the neurotoxic change induced by NMDA agonists, but would not influence similar processes induced by kainate and quisqualate [32]. This can be interpreted as the subject of modulating effects of zinc ions towards NMDA receptor complex being allosterically connected with the mediator-binding site [190].

The concept of zinc-mediated excitotoxicity was first formulated by Tonder *et al.* [231]. It was shown that in transient global ischemia intra-axonal reserves of zinc ions in hippocampal neurons become exhausted, whereas abnormally high concentrations of zinc ions are registered in CA_3 neuron soma, which are the most sensitive to ischemia. It was also demonstrated that zinc ions are transported from synaptic terminals into postsynaptic neurons selectively vulnerable during transient global ischemia. Such transport occurs diffusely in different brain areas and closely precedes neuronal degeneration [33]. Such transport of zinc ions as well as the neuronal degeneration were halted under conditions of craniocerebral hypothermia [104].

Choi and Koh [33] proved that in cerebral ischemia zinc can play a role similar to glutamate and may lead to neuronal death. In experiments made on cell cultures, high zinc concentrations killed central neurons, especially depolarized ones [237]. Sensi *et al.* [203] and Kim *et al.* [113, 114] showed that zinc being in excess penetrates into cells similarly to calcium and may cause their necrosis and apoptosis depending on the exposure intensity. The permeation of zinc ions into cells is through calcium channels of any kind (through potential-dependent ones in exchange for intracellular sodium ions; through agonist-dependent channels attending NMDA receptors; through calcium-permeable AMPA receptors) [203]. Excess of intracellular zinc leads to decrease in intracellular ATP concentration and can significantly disturb glycolysis with subsequent accumulation of its intermediate products and dihydroxyacetone phosphate [124]. The rate of zinc-mediated cell death can be decreased by injection of pyruvate, which under normal condition is produced from metabolic reactions depressing the excess of zinc.

The mechanism of high-affinity permeability of the NMDA receptor channel for calcium ions differs from that for the potential-dependent calcium channels and is determined mainly by magnesium ion concentrations, which exert a non-competitive potential-dependent mode of inhibition of synaptic excitation [4, 249].

According to Lee *et al.* [124], there are several factors acting on the ischemized brain area that are potent in inhibiting the function of NMDA receptors and decreasing NMDA-mediated excitotoxicity *in vivo* versus the conditions *in vitro*. They are accumulation of lactate, which increases the concentration of extracellular hydrogen ions [249], the influence of magnesium ions, free radicals [2], and dephosphorylation of NMDA receptors due to low intracellular ATP concentration [230], and the direct inhibition of their function by calmodulin [251].

In recent years reports have appeared on neurotoxic properties of glutamate connected with activation of **AMPA and kainate receptors**, which leads to changes in permeability of the postsynaptic membrane for monovalent ions (sodium and potassium), to enhancement of sodium influx, and to short-term depolarization of the postsynaptic membrane [200]; this in turn leads to enhancement of calcium influx into cells via both agonist-dependent and potential-dependent channels [20, 21, 50, 125, 207]. The flow of sodium ions is accompanied by an influx of water and chlorine ions, which can cause swelling of apical dendrites and neuronal lysis (the theory of osmotic damage to neurons).

The relative contribution of AMPA/kainate excitotoxicity can increase due to the elevation of extracellular proton (hydrogen ion) concentration, which delays the recovery of intracellular calcium homeostasis, and, possibly, can depress NMDA excitotoxicity [143]. Another contribution of AMPA and kainate receptors to brain damage has been discovered. It can also be connected with the increase in expression of Zn^{2+}-permeable AMPA receptors in response to ischemia [97].

Lee *et al.* [124] showed that NMDA excitotoxicity is a predominant mechanism of glutamate-induced excitotoxicity for certain cell pools. At the same time, inhibitory GABA-containing cortical neurons and oligodendrocytes are characterized by the increased expression of calcium-permeable AMPA receptors. That highly predisposes them to AMPA-mediated death. The death rate of oligodendrocyte culture decreases when AMPA antagonist is injected [106, 142, 143, 244].

Metabotropic glutamate receptors, connected with G-protein and activating intracellular metabolism, play an important role in regulation of calcium flow caused by NMDA activation in brain ischemia [174]. They also fulfill their modulator function, inducing long-term changes in cell functioning [27, 129, 199, 241]. These receptors do not influence ion channel functioning but stimulate the formation of two intracellular messengers, diacylglycerol (DAG) and inositol-1,4,5-triphosphate (IP_3), which take part in further reactions of the ischemic cascade.

4.2.   The amplification stage: intracellular accumulation of calcium ions, spreading glutamate release, spreading depression waves

The stage of amplification (enhancing of a deleterious potential) is mainly connected with *the continuing elevation of the intracellular calcium ion concentration* [32, 198, 212]. Investigations of selective vulnerability of certain neuronal pools in ischemia provided evidence that calcium-mediated excitotoxicity plays an important role in the formation of a brain infarction. It was shown that under similar conditions of both cerebral blood flow reduction and energy deficit certain cell pools (for instance, CA_1 hippocampal neurons, medium-size neurons in the dorsolateral area of the striatum) exhibit increased sensitivity to ischemia [19, 73, 183]. Neuropathology studies connected the high vulnerability of these cells in ischemia with the high density of calcium channels on their neuronal surface that produces "hot spots" of calcium accumulation in dendrites. Rothman and Olney [191] and Meldrum [144] showed that the accumulation of calcium ions is due to the high density of agonist-dependent calcium channels controlled by NMDA receptors. The ischemia-induced extracellular accumulation of glutamate and aspartate appears to be the crucial factor for the selective vulnerability in ischemia [53, 104, 239]. It is noteworthy that neurons highly sensitive to ischemia possess many GABAergic terminals in their soma along with the powerful glutamate–aspartate excitatory entrance located within the "dendritic complex" [209], and under the normal conditions the mechanisms of glutamate-mediated excitement and GABA-mediated inhibition do always exist in constant equilibrium. This confirms our clinical and biochemical results concerning the importance of the imbalance between the excitatory and the inhibitory neurotransmitter systems, which imposes limitations on timely activation of the protective inhibitory mechanisms in the brain.

Continued elevation of intracellular calcium ion concentration is partially caused by the inositol-1,4,5-triphosphate-mediated induction of calcium release from its intracellular deposits [147]. Also, additional calcium ions continue to be delivered to cells through potential-dependent calcium channels.

Linde *et al.* [127] showed that accumulation of intracellular calcium ions is promoted by the migration of blood serum proteins into brain tissue through the altered blood–brain barrier. It was found that the presence of albumin under conditions of excessive glutamate release significantly enlarges the area of ischemic damage due to increasing excitotoxicity by means of prolonged opening of agonist-dependent calcium channels of NMDA receptors [55].

The elevation of intracellular calcium ion content accompanied by an increase in DAG concentration alters activities of the enzymes that modify membrane proteins including glutamate receptors. As a result the sensitivity of neurons to excitatory signals increases. And the "vicious circle" closes: increased excitability provides the accumulation of calcium ions to continue and the release of glutamate from terminals to increase (see Plate 1).

According to experimental findings by Siesjo and Bengtsson [212], one strongly depolarized cell in brain areas where NMDA-containing neurons lie tight induces such high glutamate release that adjacent cells become excited.

Lasting and prominent elevation of glutamate concentration is usually not registered in the penumbral area. This may indicate that excitotoxic events in the penumbra are not connected with the increased synthesis of excitatory amino acids. Excessive amounts of glutamate may be delivered to the penumbra from the ischemic core area. They can damage adjacent neurons, inducing further release of the neurotransmitter [162]. Thus, the **self-spreading glutamate release** from neuron to neuron ("domino" mechanism), from the core of the ischemized territory towards its periphery, is a possible cause of excitotoxicity and neuronal damage in the penumbral area [212].

The other reason for elevation of extracellular glutamate level in the penumbral area is **spreading depression** (or "peri-infarction area depolarization") [88, 156]. The phenomenon of spreading depression has become well-known since it was described in 1944. Transient disturbance of ion gradients of brain cell membranes underlies this process. One can compare this disturbance to a wave slowly distributing throughout CNS tissues [102] according to their cellular architectonics. The presence of astrocytes and neurons are required for the wave distribution [101, 218]; the depression wave neither distributes along white matter pathways nor via commissural tissue where there are no neurons.

The spreading depression constitutes a concentric-shaped wave of near-total membrane depolarization. Its velocity is about 1.5–1.7 mm/min. It is accompanied by a negative shift of the extracellular potential to 30 mV amplitude [23]. The extracellular space contracts by 50% due to swelling of astrocytes and dendrites. Under conditions of the spreading depression the extracellular potassium ion concentration increases to 30–80 mM, and at the same time there is an intracellular increase in sodium, chlorine, and calcium ion contents [7]. Anion channels permeable to bicarbonate and water molecules open as well [177]. The bicarbonate shift leads to the development of acidosis in neurons and extracellular space; oppositely, in astrocytes alkalosis develops. One or two minutes after the wave of depression passes potassium content begins to reach baseline level, the ion gradient starts to normalize, and repolarization of cell membranes ensues meaning that the spreading depression is terminated [102].

Passage of the depression wave is accompanied by dilation of blood vessels with and by increase of local blood supply. Also, mitochondrial oxidation and lactate production increases [130], but reverse changes then follow [122]. These phase changes may possibly reflect that brain tissue involved in spreading depression causes additional energy expenses.

The mechanism of the spreading depression needs further study. It was shown that the wave of depression may spread even when the axonal conduction and the synaptic transmission are completely blocked [93]. Glutamate antagonists decrease the extent of the spreading depression and can change its features, but fail to inhibit it completely [229]. At the same time, blockage of growth-associated phosphoproteins (GAP) may avert the wave of spreading depression in all tissues [102]. It was demonstrated in experiment that astrocytes in cell cultures and brain slices promote the self-spreading of a calcium wave through their GAP junctions with velocity similar to that seen in the spreading depression *in vitro* [37, 43]. Moreover, such calcium waves, spreading throughout the astrocytic syncytium in cell cultures can elevate the cytosolic level of calcium ions in neurons growing on top of an astrocytic layer [154]. This elevation was attenuated by blockers of GAP junctions. These observations confirm a tight glial–neuronal relationship and the important role of astrocytes during passage of such a wave of spreading depression [102].

Largo *et al.* [121] emphasized the importance of the neuronal GAP junctions in the expanding of depression. According to their data, energy deficit in glial cells makes neurons more susceptible to the spreading depression. The neuronal GAP junctions, closed under quiescent conditions, open when the wave of depression comes.

Since 1977 much evidence has been continuing to suggest that the phenomenon of the spreading depression plays a role in enlargement of the infarction zone in focal ischemia. Branston *et al.* [16] were the first to demonstrate that the waves of depression begin to originate after transient focal ischemia from the peri-infarction area located between the ischemic core and the penumbra. The elevation of potassium ion concentration in the necrotic tissue due to dysfunction of active ion transport appears to be the crucial mechanism triggering the waves of depression, which repeat approximately each 10 min [156]. During the spreading depression one can register a discrepancy between the energy expenses of depolarized tissue and its energy demands. That is a factor which additionally damages the penumbra and slowly enlarges the necrotic core area [5]. The role of spreading depression in the extension of the infarction area is confirmed by the derived correlation between the volume of necrotic zone and the number of depression waves [146]. Each wave enlarges the volume of the infarction zone by approximately 23%.

At the same time, the spreading depression neither damages healthy tissues nor involves them into the penumbral area. Moreover, the question is discussed whether the action of the depression waves is protective for the peripheral areas of ischemized brain tissue due to the stimulation of the synthesis of trophic factors [115].

So, the studies of recent years have demonstrated the role of metabolic glutamate–calcium changes and of membrane depolarization spreading from the core to penumbral periphery in the development of ischemic brain damage. The firm linkage between the changes in active ionic transport, membrane depolarization, and glutamate release is suggestive of concomitance and tight interrelation between the processes of the spreading glutamate release and the spreading depression.

4.3. The expression stage: calcium-induced activation of intracellular enzymes, oxidative stress and abundant NO production, and accumulation of low molecular weight cytotoxic substances

As a result of the amplification stage, conditions for the third stage of the glutamate–calcium cascade are prepared. This is the expression stage, during which irreversible changes leading to cell death occur.

Excessive accumulation of intracellular calcium and its transformation to an active form bound to the intracellular receptor calmodulin cause ***activation of calmodulin-dependent intracellular enzymes: phospholipases, protein kinases, and endonucleases*** [10, 170]. Triggering of the cascade of enzymatic reactions leads to multiple damage to biomacromolecules and finally to cell death [166].

The most injuring is the destruction of phospholipids in the external cell membrane and in membranes of intracellular organelles. Several minutes after the onset of ischemia the release of palmitate and docosonoate from membranes occurs. This causes the destruction of phosphatidylcholine and phosphatidylethanolamine [1]. It was shown [139, 245] that during 30-min experimental ischemia about 16% of membrane-associated phosphatidyl-ethanolamine is decomposed and about 37% of free arachidonic acid is released.

Arachidonic acid metabolism is connected with the formation of prostaglandins, thromboxanes, hydroxylated and hydroperoxidized fatty acids, leukotrienes, lipoperoxides, and reactive oxygen species [29, 242], i.e. it significantly intensifies the processes of free radical oxidation and lipid peroxidation [172, 198, 221, 250, 252]. Rapid enhancement of oxidative processes along with the deficiency of the antioxidant system leads to the development of ***oxidative stress***, which is one of the universal mechanisms of tissue damage in the human body.

The development of oxidative stress in the CNS is especially dangerous because oxidative metabolism is very intense in the human brain. Although brain constitutes only 2% of body weight, it consumes 50% of all oxygen used by the body. The intensity of oxygen consumption by neurons many dozens of times exceeds the demands of other cells and tissues (350–450 µl O_2/g per min versus 70–90 µl for heart, 1.6–2.4 µl for skeletal muscles, 9–24 µl for phagocytosing leukocytes) [250]. The additional factors for the development of oxidative stress in brain tissue are high content of lipids (about 50% of brain solid substance), the unsaturated bonds of which are substrates for lipid peroxidation, and high content of ascorbate (100-fold exceeding those in peripheral blood), which participates in nonenzymatic processes of lipid peroxidation as a pro-oxidant. The activity of enzymatic antioxidant systems (catalase, glutathione peroxidase) in brain is significantly lower than in other tissues, which also increases the risk of the development of oxidative stress.

A free radical is a part of a molecule with an unpaired electron on one of the orbitals. Superoxide anion ($O_2^{\bullet-}$) and hydroxyl radical ($^\bullet OH$) are oxygen-containing free radicals which contain oxygen with an unpaired electron.

Under physiological conditions oxygen radicals are generated in all cells, being a link in aerobic metabolism. Complete reduction of an oxygen molecule to water requires 4 electrons: during the transfer of the first electron superoxide radical is formed, and during the transfer of the second electron hydrogen peroxide is formed; the most toxic and reactive hydroxyl radical ensues after the transfer of the third electron, and the fourth transfer leads to the formation of water. Low molecular weight membrane- or cytosol-associated substances become acceptors of the free radicals. Under normal conditions antioxidants such as superoxide dismutase (SOD), glutathione peroxidase, catalase, ceruloplasmin, vitamins A, E, C, and K, flavonoids, and coumarins are active.

Under conditions of blood flow decrease and the development of focal brain ischemia damaged mitochondrial phosphorylation becomes a source of reactive oxygen intermediates [15] (see Chapter 3). Decrease in oxygen delivery to neurons and increase in the reduction level of respiratory chain components stimulate single-electron reduction of oxygen with the formation of free radicals as well as of oxidants having non-radical origin (hydrogen peroxide and hypochlorite), because oxygen in the reduced state easily reacts with intermediate components of the respiratory chain [30].

Basic oxygen intermediates ($O_2^{\bullet-}$, superoxide radical; $HO_2^{\bullet-}$, perhydroxyl radical; H_2O_2, hydrogen peroxide; $^\bullet OH$, hydroxyl radical) are generated in the sequence of Haber–Weiss reactions [30]:

1. $O_2 + e^- \rightarrow O_2^{\bullet-}$

2. $O_2^{\bullet-} + H^+ \rightarrow HO_2^{\bullet}$

3. $O_2^{\bullet-} + O_2^{\bullet-} \xrightarrow{+2H^+} H_2O_2 + O_2$

4. $2\,H_2O_2 \rightarrow 2\,H_2O + O_2$

5. $O_2^{\bullet-} + H_2O_2 \rightarrow {}^{\bullet}OH + OH^- + O_2$

6. $O_2^{\bullet-} + NO^{\bullet} \rightarrow ONOO^-$

7. $ONOO^- + H^+ \rightarrow ONOOH$

8. $ONOOH \rightarrow {}^{\bullet}OH + NO_2^{\bullet}$

Superoxide dismutase (SOD) catalyzes the third reaction at physiological pH values with high and constant rate (2×10^9 liter/M per sec), producing H_2O_2 which is then split to H_2O and O_2 by catalase in mammalian cells or in brain tissue by glutathione peroxidase at the expense of decreased glutathione (the fourth reaction). Oxidized glutathione can again be transformed to glutathione by glutathione reductase in the presence of nicotinamide-adenine dinucleotide phosphate (NADP).

Hydroxyl radicals (${}^{\bullet}OH$) are the most active oxidants: they initiate lipid peroxidation, oxidize proteins, and damage cellular DNA. Superoxide radicals ($O_2^{\bullet-}$) are much less reactive, but have a longer half-life period and can form ${}^{\bullet}OH$ in the 5th reaction of the Haber–Weiss sequence, which takes place only in the presence of Fe^{2+} (the Fenton reaction).

Another mechanism for hydroxyl radical generation is represented by the 6th reaction between superoxide anion and nitric oxide radical ($NO^{\bullet}$), the product of which is peroxynitrite ($ONOO^-$) [8, 46, 151]. The rate of this reaction is extremely high (6.7×10^9 liter/M per sec), close to the rate constant for diffusion [30]. At neutral pH, $ONOO^-$ degrades instantly to ${}^{\bullet}OH$ and nitrogen dioxide ($NO_2^{\bullet}$) (see the eighth reaction).

In addition to the mitochondrial respiratory chain, free radical oxygen intermediates are also generated in enzymatic reactions, in autooxidation of monoamines, and in synthesis of prostaglandins and leukotrienes. As the majority of these reactions are calcium-dependent, their considerable activation ensues during ischemia, which is accompanied by elevation of intracellular calcium ion concentration.

Highly reactive oxygen species oxidize biomacromolecules and initiate chain reactions of lipid peroxidation in membrane lipids [25, 86, 172, 246] as well as direct oxidative damage to nucleic acids and proteins [26, 60, 94, 100].

Hydroperoxides generated in lipid peroxidation are unstable. As their decay goes on, various secondary and terminal products of lipid peroxidation appear. These substances are highly toxic (diene conjugates, Schiff bases, etc.). They cause damage to membranes and cellular structures [252]. As a result, cross-links between biopolymers appear, uncoupling of oxidative phosphorylation and swelling of mitochondria are observed, inactivation of thiol enzymes taking part in respiration and glycolysis occurs, and further destruction of membrane lipid compounds goes on [250].

Many fatty acids are released, and this activates the arachidonic cycle, and its secondary products accumulate. These are eicosanoids which enhance aggregation of blood cells and constriction of vessels [32, 221]. Eicosanoids and platelet activation factor (PAF) cause additional disturbances of microcirculation which aggravate the ischemic process.

The elevation of free arachidonic acid concentration leads to additional inhibition of mitochondrial respiration [95] and causes direct cytotoxic action as well. Due to its amphiphilic properties, arachidonic acid conjugates with the cytosolic membrane and incorporates in membrane lipids [109]. Fatty acids (including arachidonic acid) can aggregate into micelles, which have detergent properties that allow them to solubilize cytosolic membranes [109]. Alcohols which appear after the oxidation of free fatty acids as well as ketones and aldehydes also damage neuronal membranes [198].

Oxidative stress reactions are tightly connected with energy metabolism and glutamate excitotoxicity and form closed vicious circles of pathological interrelations. Oxygen intermediates and lipid peroxidation products inhibit the activity of the Na^+/K^+-ATPase enzyme system, aggravating changes in energy-dependent ion transport [198]; they inhibit glutamine synthase, the astroglial enzyme, and disrupt the reuptake and metabolism of glutamate, mediating its accumulation in the synaptic cleft [165]. Altering the structure of receptor proteins, especially G-proteins, free radicals can change their functions, causing either inhibition or activation under different conditions. The opening of potential-dependent and agonist-dependent ion channels leads to fast influx of calcium and sodium ions into neurons, which aggravates all the reactions of the glutamate–calcium cascade described earlier and causes the specific activation of calcium-dependent calpain-I and NO-synthase.

Under conditions of increased ATP catabolism and accumulation of xanthine and other products of purine metabolism, the activation of calcium-dependent calpain-I leads to transformation of xanthine dehydrogenase to xanthine oxidase. During the reperfusion, when oxygen comes into ischemized tissue, xanthine is oxidized and huge amounts of superoxide anion ($O_2{}^{\bullet-}$) accumulate [250]. The activation of NO-synthase causes excessive production of $NO^{\bullet}$ radical. As a result, the conditions when peroxynitrite ($ONOO^-$) is formed ensue (see the 6th reaction), which stimulates generation of hydroxyl radicals and mediates nitration of tyrosine residues in proteins.

Free radicals induce reactions between proteins and other cellular components leading to fragmentation of protein molecules and their dysfunction [26]. At the same time, under conditions of free radical oxygen species attack, membrane-associated proteins suffer not only from direct chemical *de novo* modification, but also from the disturbances in their "lipid environment" due to lipid peroxidation [48, 112, 152].

Migration of phagocytes into the ischemized area leads to accumulation of myeloperoxidase, which having its substrate hydroperoxide in proximity is able to produce hypochlorite anion. The direct targets of hypochlorite anion, which is a strong oxidant, are thiol compounds, SH-groups, as well as lipids and proteins containing NH_2-groups. The intermediate products of these transformations are toxic.

De novo modified membrane-associated lipids may induce apoptosis (see Chapter 10). Their action can be connected with changes of cytosolic calcium content that mediates free radical oxygen species effects on various signal transmission systems as well as with their direct influence on gene expression, particularly of those participating in programmed cell death [49]. It was hypothesized that mechanisms connected with the damaging action of reactive oxygen species are fixed in evolution and the cell uses them to conduct the apoptotic program. The genes involved in regulation of the redox state in the cell, of balance in generators of reactive oxygen species, and of antioxidant support have already been described. Expression of the "protective" *bcl-2* gene averts apoptosis caused by several pro-oxidant inducers.

Increasing attention is now being given to the participation of **nitric oxide free radical ($NO^{\bullet}$)** in mechanisms of acute brain ischemia. This compound appears to be a multi-functional physiological regulator [17, 148]. $NO^{\bullet}$ is considered the first representative of a new family of signal molecules possessing neurotransmitter properties [185, 201], which in contrast to conventional neurotransmitters is not supplied in the synaptic vesicles of neuronal terminals, but is released into the synaptic cleft via the mechanism of free diffusion, not via exocytosis. The $NO^{\bullet}$ molecule is synthesized in response to the physiological demand by the enzyme called NO-synthase (NOS) from its metabolic precursor—the amino acid L-arginine. The property of $NO^{\bullet}$ to cause a biological effect is determined to a considerable extent by the small size of the molecule, its highly reactive nature, and its property of being distributed via diffusion throughout tissues, including neural tissues [187]. This is why NO• is called a retrograde messenger [46, 49].

The calcium–calmodulin dependence of tissue isoform of NOS increases NO synthesis at elevated intracellular calcium ion level. That is why any process leading to accumulation of calcium ions in the cells (be that energy deficit, glutamate excitotoxicity, or oxidative stress) is accompanied by elevation of NO• concentration [185, 201].

It was shown that NO• produces a modulating effect on synaptic transduction [148] and on the functional state of glutamate NMDA receptors [35]. Activating soluble heme-containing guanylate cyclase and ADP-ribosyl transferase, NO• is involved in intracellular calcium content regulation [196] and is involved in pH regulation under conditions of brain ischemia [188].

The excitation of NMDA receptors leads to activation of neuronal NOS and to enhanced release of NO•. At present there is no unequivocal opinion about the role of nitric oxide in mechanisms of glutamate excitotoxicity. According to Sorokina *et al.* [220], Haspekov *et al.* [90], Wilcox [240], and Wright [243], NO• takes part in neuronal damage. Its toxic action involves induced disturbances of mitochondrial oxidative phosphorylation and of ribonucleotide reductase metabolism, and the formation of such free radical compounds as peroxynitrite anion, which blocks several neuronal receptors, can change activity of SOD and aggravate free radical damage leading to cell death. Peroxynitrite is able to inhibit tyrosine kinases incorporated in the active site of receptors to neurotrophins, increasing insufficiency of brain neurotrophic support [110]. The mechanism of NO• toxicity also includes S-nitrosylation of cellular matrix by means of transfer of NO to a critical cysteine-SH group of proteins. One candidate molecular pathway of damaging action of NO• and peroxynitrite is direct DNA damage leading to activation of the nuclear enzyme, poly(ADP-ribose) polymerase (PARP), which catalyzes attachment of ADP ribose units from NAD to nuclear proteins following DNA damage. Excessive activation of PARP can deplete NAD and ATP, which is consumed in regeneration of NAD, leading to cell death by energy depletion [56, 75, 136].

Nevertheless, Ravindran *et al.* [184] and Takizawa *et al.* [226] suggest that NO• can protect neurons from cytotoxic effects of glutamate by means of activation of soluble heme-containing guanylate cyclase, of cyclic guanosine monophosphate (cGMP) level elevation, and by influencing ferrous iron-containing components of the mitochondrial respiratory chain.

The investigations of Yun *et al.* [247] and Gonzalez-Zulueta *et al.* [74] showed that NO is a key link between NMDA-mediated increase in cytoplasmic calcium and activity-dependent long-term changes such as differentiation, survival, and synaptic plasticity. The transmission of the glutamate signal to the nucleus, which is ultimately important for long-lasting neuronal responses, is realized through NO generation by calcium-dependent neuronal NOS [195]. It has been demonstrated that NO activates Ras (p21) with following induction of the extracellular signal-regulated kinases (Erks) cascade, which is a critical mechanism for the development of tolerance to oxygen–glucose deprivation in neurons [74].

Newly formed nitrosonium ion (NO^+), which binds to a regulatory center of NMDA receptors, decreases their excitability and sensitivity to their agonists. Retrograde blocking of NMDA receptors by NO^+ is also possible [138].

Experimental works by Kamii *et al.* [108], who selectively inhibited neuronal $NO^•$, revealed a decrease in ischemic damage. However, inhibition of both neuronal and endothelial nitric oxide caused blood flow decrease due to spasm of brain vessels.

Many authors [13, 86, 137, 227, 248] have shown that in acute cerebral ischemia $NO^•$ improves cerebral blood supply by means of vasodilatation, decrease of platelet aggregation and mural adhesion of neutrophils, inhibition NMDA receptor activity, and decrease in the excitotoxic effect of glutamate. However, when reperfusion begins, deleterious effects of $NO^•$ start to dominate, aggravating the destruction of dying cells. Thus, an ambiguous nature of $NO^•$ is manifest. Probably, NO as well as antioxidants fulfill modulatory functions by manifesting the peculiarities of their effects, which depend on the conditions of the environment and on their intracellular concentration. It is known that glutathione, ascorbate, and carnosine, being prominent inhibitors of free radical oxidation at high concentrations, can exhibit pro-oxidant effects when their concentrations change.

4.3.1. CSF levels of thiobarbituric acid reactive substances, SOD, cGMP, and N-acetylneuraminic acid in patients with ischemic stroke

To objectify the significance of oxidant stress reactions in the development of acute focal ischemia in humans, we, together with colleagues from the Institute of Pharmacology of the Russian Academy of Medical Sciences (K. S. Raevsky, V. G. Bashkatova, A. I. Lysko), conducted *a study of secondary lipid peroxidation products (thiobarbituric acid reactive substances or TBARS) and the antioxidant enzyme superoxide dismutase (SOD)* in CSF of 50 patients (27 males and 23 females; average age 65.1 years). Patients were admitted to the Intensive Stroke Unit within the first 2–6 h after the onset of stroke in the carotid artery territory and received complex and at the same time maximally standardized treatment without neuroprotective agents (see Section 4.1.1).

The leading etiological factor of stroke was atherosclerosis (92%) with concomitant arterial hypertension (74%) and diabetes mellitus (6%). In 37 (74%) patients stroke exhibited atherothrombotic origin, in 12 (24%) it followed cardiac embolism, and in 1 (2%) stroke occurred on the background of total atrio-ventricular blockade, was of hemodynamic origin. There were no patients with lacunar strokes in this group.

The study of TBARS was conducted by spectrophotometric assay according to Ohkawa *et al.* [164]. Cooled 10× isotonic sodium solution was added to CSF sample. Control sample contained 200 μl of isotonic sodium chloride solution. Then 0.2 μl of 45% dodecyl sodium sulfate solution, 1.5 ml of 20% acetic acid, and 8% thiobarbituric acid solutions were sequentially added to each sample. Then the volume of each sample was increased to 4 ml with distilled water. The reaction mixture was incubated on a water bath at 95°C for 60 min. The samples were cooled and centrifuged for 10 min at 4000 rpm using a K-70 centrifuge. The red pigment produced was extracted with *n*-butanol–pyridine mixture and measured in absorbance at 532 nm using an Aminco (USA) spectrophotometer.

SOD activity in CSF was measured by method by Beauchamp and Fridovich (1971). Nitro tetrazolium blue was used as a detector of superoxide anion generated in the xanthine–xanthine oxidase system. Nitro tetrazolium blue competes for superoxide anion with superoxide dismutase present in the medium and the reduction of the anion occurs as observed by the formation of a peak in time with absorption spectrum at 560 nm. The rate of increase depends on the concentration of SOD present in the medium.

TBARS concentration and SOD activity in CSF were measured when patients were admitted to clinic (within the first 2–6 h after stroke onset) and on day 3. Values were compared to normals acquired during peridural anesthesia from a control group comprising 20 individuals with chronic orthopedic pathology, no neurological or systemic disease, and no thermal regulation disorders. Control values of TBARS concentration and SOD activity were 2.01 ± 0.17 μM and 1.3 ± 0.4 U/liter, respectively.

All patients with ischemic stroke were found to exhibit elevated level of TBARS even within the first 3–6 h after stroke ($p < 0.01$; Table 4.3, Fig. 4.8). No correlation was found between TBARS concentration on day 1 and severity, location, or pathogenic variant of stroke development. By day 3 the elevation of TBARS level was found to dominate significantly in patients with severe neurological deficit (total clinical scores by the Orgogozo and the original scales were 26–40 and 30–35, respectively) versus patients with moderate deficit (total clinical score more than 40 and 36, respectively; $p < 0.05$).

Clinical and biochemical analysis revealed reverse correlation between the dynamics of total clinical score, which shows the extent of recovery, and CSF concentration of TBARS ($r = -0.45$; $p = 0.04$). In progressive stroke higher concentration of TBARS was seen by day 3 along with the aggravation of neurological deficit (5.18 ± 0.42 versus 4.03 ± 0.14 μM in patients with regressive course of the disease; $p < 0.05$; see Table 4.3 and Fig. 4.8).

Table 4.3. Concentrations of TBARS (μM) in CSF of patients with carotid ischemic stroke

Group of patients	Time after stroke onset		Dynamics, %
	6 h	day 3	
Neurological deficit on admission:			
moderate (OSS > 40)	3.04 ± 0.30	4.56 ± 0.44 *	49.7
severe (OSS < 39)	3.27 ± 0.38	5.10 ± 0.64 **	56.0
p	0.04	0.02	
Course of illness:			
regressive	3.14 ± 0.31	4.03 ± 0.14 *	28.34
progressive	3.54 ± 0.36	5.18 ± 0.42 **	46.33
p	0.006	0.001	
Functional recovery:			
good (BI ≥ 75)	3.03 ± 0.44	2.36 ± 0.65 **	−22.11
severe disability (BI < 50)	3.67 ± 0.47	5.47 ± 0.55 **	49.05
p	0.01	< 0.001	

Note: Significance of the differences versus the first examination: * $p < 0.05$, ** $p < 0.001$. OSS – total clinical score by the Orgogozo Stroke Scale. BI – Barthel Index.

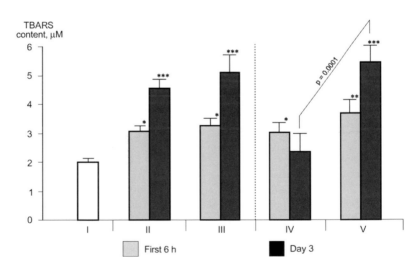

Figure 4.8. Concentrations of TBARS in CSF of patients with carotid ischemic stroke. I – control; II – moderately severe stroke (OSS > 40); III – severe stroke (OSS < 39); IV – good functional recovery (BI ≥ 75); V – severe disability (BI < 50). Significance of the differences versus controls: * $p < 0.01$, ** $p < 0.001$, *** $p < 0.0001$.

The most significant was the correlation between TBARS level and the stroke outcome by the end of the acute period (by day 21). In cases with good recovery (Barthel Index 75–100) we found no elevation but decrease in TBARS concentration by 22.1% by day 3 (see Table 4.3 and Fig. 4.8). In cases of severe disability (Barthel Index 0–45) the increase in TBARS concentration was maximal (by 49.1%, to 5.47 ± 0.55 µM; $p < 0.01$ versus values obtained from patients with good recovery).

The comparison of TBARS concentration in CSF with the ischemic lesion volume (on MR images) on day 1 showed no correlation. However, we found direct dependence between the volume of infarction formed by days 5–7 after the stroke onset and the dynamics of TBARS concentration within the period by day 3 ($r = 0.45$; $p = 0.05$): maximal accumulation of TBARS concentration (up to 5.47 ± 0.55 µM) correlated with the largest infarction volume (29.7 ± 1.6 cm^3). When TBARS level decreased the volume of ischemic lesion decreased as well.

Thus, the study of secondary products of lipid peroxidation reacting with thiobarbituric acid supported data about the role of lipid peroxidation in the development of acute focal ischemia and the formation of brain infarction; we found a correlation between the extent of TBARS elevation and the severity of ischemic process as well with its course and outcomes. Sequential elevation of TBARS content within the first days of stroke and the maximal predictive value of its extent on the 3rd day confirms the remaining activity of oxidant stress reactions beyond the therapeutic window, which may reflect their role in the "up-formation" of brain infarction.

The study of SOD revealed the increase in its activity on day 1 after the stroke onset in all cases, especially in patients with severe and extremely severe neurological deficit ($p < 0.001$–0.006 in comparison with mild and moderately severe strokes; see Table 4.4 and Fig. 4.9). A tendency toward reverse correlation between SOD activity and total clinical score by the Orgogozo scale was found ($r = -0.32$; $p = 0.07$; Fig. 4.10). There was no dependence between SOD activity and stroke location and its pathogenetic variant.

The dynamics of SOD activity by day 3 directly correlated with clinical improvement and the extent of neurological deficit decrease assessed by the Orgogozo scale by day 21 after the stroke onset ($r = 0.47$; $p = 0.04$): the better was recovery the higher was the SOD activity (Fig. 4.11). In patients with regressive course of the disease having had moderate deficit on admission (total clinical score by the Orgogozo scale more than 40) SOD activity elevated by day 3 by 68.5%; progressive course of the disease in patients with severe neurological deficit on admission was accompanied by its decrease by 42.2% (see Table 4.4 and Fig. 4.9).

Table 4.4. SOD activity (U/liter) in CSF of patients with carotid ischemic stroke

Group of patients	Time after stroke onset		Dynamics, %
	6 h	day 3	
Neurological deficit on admission:			
moderate (OSS > 40)	2.45 ± 0.24	3.44 ± 0.15 **	40.4
severe (OSS 25–39)	3.72 ± 0.66	4.18 ± 0.45 *	12.4
extremely severe (OSS < 25)	3.82 ± 0.45	1.75 ± 0.42 **	−55.0
p_{I-II}	0.006	0.005	
p_{I-III}	0.001	0.0001	
p_{II-III}	> 0.05	0.0001	
Course of illness:			
regressive	2.38 ± 0.64	4.01 ± 0.64 **	68.58
progressive	3.74 ± 0.63	2.16 ± 0.76 **	−42.2
p	0.001	0.0006	
Functional recovery:			
good (BI ≥ 75)	2.35 ± 0.42	4.37 ± 0.47 **	86.0
severe disability (BI < 50)	3.48 ± 0.57	2.32 ± 0.61 **	−33.3
p	0.001	0.0008	

Note: Significance of the differences versus the first examination: * $p < 0.05$, ** $p < 0.001$. OSS – total clinical score by the Orgogozo Stroke Scale. BI – Barthel Index.

The retrospective analysis of SOD activity dynamics in relation to clinical outcome by day 21 after the stroke onset also revealed that maximal elevation of SOD activity (by 86%) predicted good recovery, whereas in cases with severe disability we found decrease of its activity by 33.3% (see Table 4.4 and Fig. 4.9).

Comparing SOD activity and TBARS concentration, we revealed a tendency to inverse correlation between them ($r = −0.36$; $p = 0.06$; Fig. 4.12).

Thus, the study demonstrated the activation of the superoxide dismutase system in the acute period of ischemic stroke. We found a direct correlation between the elevation of SOD activity and the extent of recovery, as well as a tendency to reverse correlation between the dynamics of SOD activity and TBARS concentration that obviously determines the compensatory "protective" function of SOD and its role in the restriction of ischemic cascade mechanisms. The decrease of SOD activity by day 3 in severe course of stroke shows the importance of quick exhaustion of natural antioxidant systems in the damaging potential of oxidative stress.

By now the mechanisms of intracellular regulation via "second messenger" systems have been demonstrated for almost all receptors, including glutamate receptors. It appears that the regulation of glutamate receptors is mediated via cGMP and calcium ions [17, 70, 179]. The significance of cGMP metabolism in glutamate neurotransmission is

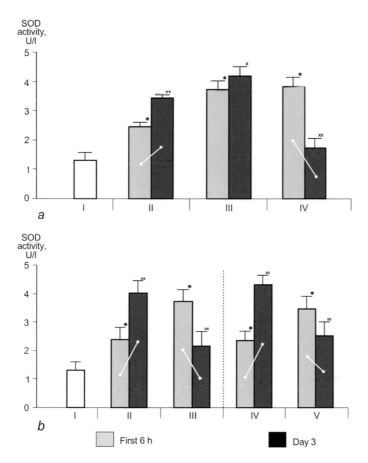

Figure 4.9. SOD activity in CSF of patients with carotid ischemic stroke in relation to neurological deficit (a), course of the disease, and the extent of functional recovery on day 21 (b). a: I – control; II – moderately severe stroke; III – severe stroke; IV – extremely severe stroke; b: I – controls; II – regressive course of illness; III – progressive course of illness; IV – good functional recovery; V – severe disability. Linear histogram shows the dynamics of SOD activity. Significance of the differences versus controls: * $p < 0.0001$. Significance of the differences versus the first examination: $^x p < 0.05$, $^{xx} p < 0.001$.

emphasized by the possibility of extra-synaptic localization of glutamate receptors due to detection of glutamate-sensitive guanylate cyclase, involved in the system of active transport of glutamate [70].

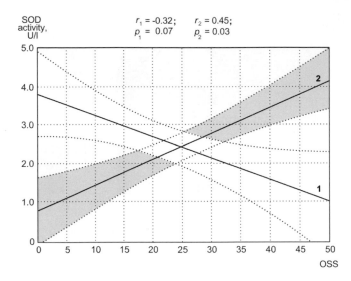

Figure 4.10. Correlation between SOD activity in CSF on day 1 and total clinical score by the Orgogozo Stroke Scale (OSS) on day 1 (1) and day 3 (2) in patients with carotid ischemic stroke.

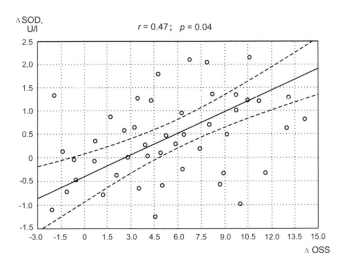

Figure 4.11. Correlation between the dynamics of SOD activity in CSF on day 3 and the increase in total clinical score (by the Orgogozo Stroke Scale) on day 21 after carotid ischemic stroke.

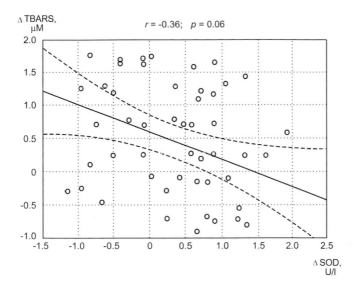

Figure 4.12. Correlation between the dynamics of TBARS concentration and SOD activity in CSF of patients with carotid ischemic stroke.

The interaction between NO and guanylate cyclase leads to activation of cGMP production [46] which is involved in 1) the integrity of electrical and metabolic membrane-associated events; 2) the regulation of calcium flow via the agonist-dependent channels of NMDA receptors and of intracellular calcium content; 3) the activation of cGMP-dependent protein kinases which catalyze phosphorylation of many protein substrates [46, 179, 220].

Together with colleagues from the Laboratory of Membranology of the Children's Health Scientific Center of the Russian Academy of Medical Sciences (V. G. Pinelis *et al.*), we conducted *a study of cGMP* in CSF taken from 50 patients with acute carotid ischemic stroke (see above) by means of standard solid-phase enzyme-linked immunosorbent essay (ELISA). The average value obtained from control individuals was (15.47 ± 8.81 nM) taken as the normal value.

Already by 3–6 h after stroke onset all examined patients exhibited significant elevation of cGMP concentration (Table 4.5; Fig. 4.13). It was far more pronounced in patients with severe (by 186.1%) and extremely severe (by 189.9%) deficit versus patients with moderate neurological deficit (by 66.8%; $p < 0.01$).

By day 3 the level of cGMP continued to increase (on average by 165–167%) reaching in patients with severe and extremely severe deficit a peak value of 116–130 nM.

Table 4.5. Concentrations of cGMP (nM) in CSF of patients with carotid ischemic stroke

Group of patients	Time after stroke onset		Dynamics, %
	6 h	day 3	
Neurological deficit on admission:			
moderate (OSS > 40)	25.8 ± 6.70	68.12 ± 8.46 *	163.9
severe (OSS 25–39)	44.26 ± 12.47	116.42 ± 17.40 *	163.0
extremely severe (OSS < 25)	43.38 ± 14.15	130.25 ± 28.17 *	169.2
p_{I-II}	0.001	< 0.0001	
p_{I-III}	< 0.001	< 0.0001	
p_{II-III}	> 0.05	> 0.05	
Course of illness:			
regressive	29.62 ± 4.28	43.36 ± 5.72 *	46.4
progressive	48.61 ± 7.96	125.83 ± 12.86 *	158.9
p	< 0.001	< 0.0001	
Functional recovery:			
good (BI ≥ 75)	27.56 ± 4.81	45.18 ± 6.73 *	63.9
severe disability (BI ≤ 45)	59.16 ± 7.91	136.21 ± 14.73 *	130.2
p	< 0.001	< 0.0001	

Note: Significance of the differences versus the first examination: * $p < 0.001$. OSS – total clinical score by the Orgogozo Stroke Scale. BI – Barthel Index.

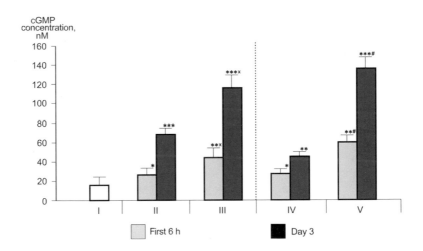

Figure 4.13. Concentrations of cGMP in CSF of patients with carotid ischemic stroke. I – control; II – moderately severe stroke (OSS > 40); III – severe stroke (OSS < 39); IV – good functional recovery (BI ≥ 75); V – severe disability (BI < 50). Significance of the differences versus controls: * $p < 0.01$, ** $p < 0.001$, *** $p < 0.0001$. Significance of the differences between patients with moderately severe versus severe stroke: [x] $p < 0.001$. Significance of the differences between patients with good functional recovery versus severe disability: [#] $p < 0.001$.

The extent of cGMP concentration increase exhibited reverse correlation with the dynamics of total clinical score assessed by the Orgogozo scale ($r = -0.57$; $p = 0.03$; Fig. 4.14), stroke outcome assessed by the Barthel Index on day 21 ($r = -0.45$; $p = 0.04$), and positively correlated with the increase in TBARS concentration ($r = 0.76$; $p = 0.01$; Fig. 4.15). In the most propitious cases with marked improvement and good recovery by the end of the acute period of stroke (by day 21) cGMP concentration in CSF increased by day 3 by 46.4–63.9%; progressive course of disease with severe disability in outcome was associated with the increase in its level by 158–167% ($p < 0.001$).

As in the study of TBARS, we found a tendency for correlation between the increase in cGMP concentration by day 3 and the infarction volume by days 5–7 after the stroke onset ($r = -0.47$; $p = 0.08$).

Close correlation between cGMP and TBARS changes, stroke severity, and infarction size confirms active participation of glutamate neurotransmission disturbances and oxidative stress reactions including those which are connected with excessive NO synthesis in mechanisms of acute focal brain ischemia leading to transformation of reversible metabolic changes into a permanent neuropathological lesion.

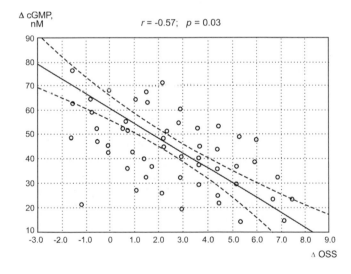

Figure 4.14. Correlation between the dynamics of cGMP concentration in CSF on day 3 and the increase in total clinical score (by the Orgogozo Stroke Scale) on day 21 after carotid ischemic stroke.

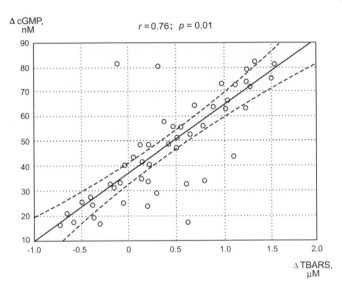

Figure 4.15. Correlation between the dynamics of concentrations of cGMP and TBARS in CSF on day 3 in patients with carotid ischemic stroke.

The dynamics of an infarction lesion can be objectified and assessed morphometrically (by brain CT and MRI). However, of the utmost interest is to verify the necrotic component of infarction. The characteristic morphological feature of necrosis is destruction of neuronal membranes. So, we chose N-acetylneuraminic acid as a biochemical marker of necrotic damage.

N-acetylneuraminic acid (NANA) is a ganglioside of glycosphingolipid class which is one of the basic components of neuronal membranes, especially synaptic (61%), and involved in all basic neuronal functions and inter-neuronal interactions (synaptogenesis, regulation of ionic microenvironment, membrane-associated electrogenesis, mediation, etc.) [215]. The elevation of NANA level in biological media reflects the destruction of membranous structures of brain tissue [179].

Together with colleagues from the Laboratory of Membranology of the Children's Health Scientific Center of the Russian Academy of Medical Sciences (V. G. Pinelis *et al.*), we studied *NANA* in CSF taken from 50 patients with acute carotid ischemic stroke (see above). The average value obtained from control individuals (45.9 ± 3.9 mg/liter) was taken as the normal value.

When plasma glycoproteids are heated with trichloroacetic acid, sialic acids are cleaved and then are hydrolyzed with formation of free neuraminic and acetic acids. In the presence of copper ions and neuraminic acid, resorcinol produces a blue staining.

To 0.5 ml of 5% trichloroacetic acid solution 0.5 ml of CSF was added and then the sample placed for 7 min onto a boiling water bath for hydrolysis. Then the mixture was cooled and filtered through paper filter. To 0.5 ml of transparent filtrate 0.5 ml of water and 1 ml of resorcinol reagent were added. The tubes containing the mixture were closed by glass stoppers and then placed again on a boiling water bath for 15 min. After cooling 3 ml of extracting compound were added, then the mixture was shaken and kept for 15 min to reach phase stratification. After the tinctured water moved to the upper layer, it was aspirated away and photometry was performed at 575–590 nm wavelength in a cuvette with 0.5 cm optical pathlength versus a blank experiment which was performed with 1 ml of water added to 1 ml of resorcinol reagent (the other ingredients except CSF were the same as in the main experiment). To construct the calibration curve we took 1 ml of each calibrating solution prepared according to the table, added 1 ml of resorcinol reagent, and processed them in similar way to the main experiment. Calculations were made using the calibration curve.

Within the first hours after stroke onset all patients exhibited increase in NANA concentration in CSF which was most pronounced in patients with severe neurological deficit (by 198–204% versus controls, $p < 0.0001$, and by 46.8–49.9% versus patients with moderate deficit, $p < 0.01$). The highest values of NANA content were obtained from patients with large ischemic lesion (more than 25 cm^3) as determined by MRI (it exceeded controls by more than 200%, $p < 0.001$); in patients with small lesions (less than 5 cm^3) NANA content was significantly lower (i.e. it exceeded controls by 90–105%; $p < 0.05$).

By day 3 of stroke mixed changes of NANA concentration were found: in patients with moderate deficit on admission its content decreased by 28.6%, whereas in patients with severe and extremely severe deficit it increased by 5.0 and 31.7%, respectively (Fig. 4.16).

By the end of the acute period of stroke an inverse correlation was revealed between the dynamics of NANA concentration and the dynamics of total clinical score assessed by the Orgogozo scale ($r = -0.51$; $p = 0.03$) and stroke outcome assessed by the Barthel Index ($r = -0.45$; $p = 0.05$). In patients with regressive course of the disease and good functional recovery we registered a decrease in NANA level by 14–21% unrelated to the total clinical score on admission. In mild strokes with complete functional recovery we saw more prominent normalization of NANA concentration (down to the level of 8–10% above normal value). In contrast to this, in progressive strokes with severe disability we observed elevation of NANA concentration by 28–37% (Fig. 4.17).

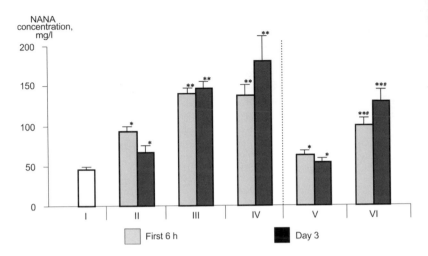

Figure 4.16. Concentrations of NANA in CSF of patients with carotid ischemic stroke. I – control; II – moderately severe stroke; III – severe stroke; IV – extremely severe stroke; V – good functional recovery; VI – severe disability. Significance of the differences versus controls: * $p < 0.001$, ** $p < 0.0001$. Significance of the differences between patients with good functional recovery versus severe disability: # $p < 0.0001$.

The strongest correlation was found between NANA concentration on day 1 and day 3 and the infarction volume on day 5 ($r = 0.71$; $p = 0.01$ and $r = 0.87$; $p = 0.001$, respectively; Fig. 4.16).

The revealed elevation of NANA concentration within day 1 after stroke onset correlating with the stroke severity and infarction volume reflects the intensity of neuronal membrane disruption which accompanies necrotic damage. It is interesting that in mild strokes with small lesion sizes a tendency to decrease or relative normalization of NANA content (within 10%) was found by day 3. Such NANA dynamics may reflect the termination of necrotic events. In severe strokes with the formation of vast infarction we found prolonged elevation of NANA concentration (for not less than 3 days), which possibly indicated prolongation of necrotic processes, and, i.e. suggested that the glutamate–calcium cascade reactions continued.

So, the reactions of **the glutamate–calcium cascade** represent the most acute response of brain tissue to CBF decrease and development of focal ischemia. Their result is the formation of focal necrosis in the ischemized brain area [32, 144, 145, 166, 167].

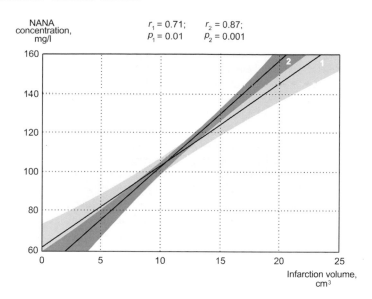

Figure 4.17. Correlation between NANA concentrations in CSF on day 1 (1) and day 3 (2) and the infarction volume on day 5 after carotid ischemic stroke.

The maximal activity of the glutamate–calcium cascade lies within the first minutes and hours after ischemic stroke onset (the period of the therapeutic window). At the same time deleterious events underlying the terminal stage of "expression" (free radical reactions and lipid peroxidation, abundant generation of NO, eicosanoids, platelet activation factor, etc.) remain important also in the more distant period especially when the size of ischemized brain area is large. Interacting with other delayed consequences of brain ischemia (see Chapter 6) they will be involved in the "up-formation" of brain infarction within 3–5 days after the vascular event and also induce generalized post-ischemic changes of the neuro–immune–endocrine system.

It should be noted that the division of the glutamate–calcium cascade into different stages is artificial and of only relative value; it to a far greater extent reflects the necessity to systematize such sophisticated and voluminous data. The tight connection and interrelation between all events taking place in the cascade are obvious. Apparently, a very short interval after being "triggered" the cascade begins to work as a united metabolic "common stock" where reasons and consequences frequently exchange their places and support each other. The complexity of analysis is due to the multifaceted character of metabolic regulators (cyclic nucleotides, mono-

and bivalent metal cations, free radicals and antioxidants, etc.), the effects of which are rather varied and frequently controversial being related to peculiarities of the extra- and intracellular environment.

The clinical and biochemical studies performed in patients with carotid ischemic stroke failed to reveal significant differences in metabolic changes of neurotransmitter amino acids, autoantibody production to NMDA receptors, prominence of lipid peroxidation, SOD activity, and cGMP and NANA concentrations in relation to the character of vascular event that caused the disorder of cerebral blood supply (atherothrombosis, cardiac embolism, or systemic blood supply breakdown). This additionally highlights the universal character of mechanisms underlying brain ischemia. Peculiarities of the discussed metabolic changes were first of all determined by the volume and the location of the ischemic lesion and correlated with clinical severity and stroke outcome by the end of the acute period of the disease.

REFERENCES

1. Abe, K., Kogure, K., Yamamoto, H., *et al.*, 1987, *J Neurochem.* **48**: 503–509.
2. Aizenman, E., Lipton, S. A., Loring, R. H., 1989, *Neuron.* **2**: 1257–1263.
3. Andriezen, W. L., 1893, On a system of fibre-like cells surrounding the blood vessels of the brain of man and mammals, and its physiological significance. *Int Monatsschrift Anatomic and Physiologic.* **10**: 532–540.
4. Ascher, P., Nowak, L., 1988, *J Physiol.* **399**: 247–266.
5. Back, T., Kohno, K., Hossmann, K.-A., 1994, *J Cerebr Blood Flow Metab.* **14**: 12–19.
6. Barinaga, M., 1997, *Science.* **276**: 196–198.
7. Barres, B. A., 1991, *J Neurosci.* II: 3685–3694.
8. Beckman, J. S., Beckman, T. W., Chen, J., *et al.*, 1990, *Proc Natl Acad Sci USA.* **87**: 1620–1624.
9. Bennett, M. R., Huxlin, K. R., 1996, *Gen Pharmacol.* **27**: 407–419.
10. Berridge, M. J., 1985, *Triangi.* **3** (4): 79–90.
11. Bittar, P. G., Charnay, Y., Pellerin, L., *et al.*, 1996, *Acta Neural Scand Suppl.* **76**: 940.
12. Bogolepov, N. N., Burd, G. S., 1981, *Abst VII All-Union Symp Neurologists and Psychiatrists*, Moscow. **2**: 32–35 (in Russian).
13. Boldyrev, A. A., 1994, *Trends Neurosci.* **17**: 468.
14. Bosley, T. M., Woodhans, P. L., Gordon, R. D., *et al.*, 1983, *J Neurochem.* **40** (1): 189–201.
15. Boveris, A., Chance, B., 1973, *Biochem J.* **134**: 707–716.
16. Branston, N. M., Strong, A. J., Symon, L., 1977, *J Neurol Sci.* **32**: 305–321.
17. Bredt, D. S., Snyder, S. H., 1989, *Proc Natl Acad Sci USA*, 9030–9033.
18. Bredt, D. S., Glatt, C. E., Hwang, P. M., *et al.*, 1991, *Neuron.* **44** (7): 615–624.
19. Brierley, L., 1976, *Neuropathology*, 43–85.
20. Buchan, A. M., Li, H., Sunghee, C., *et al.*, 1991, *Neurosci Lett.* **132**: 255–258.
21. Buchan, A. M., Xue, D., Huang, Z.-G., *et al.*, 1991, *Neuroreport.* **2**: 473–476.
22. Buchan, A. M., Lesiuk, H., Bames, K. A., *et al.*, 1993, *Stroke.* **24**: 148–152.

23. Bures, J., Buresova, O., Krivanek, J., 1974, *The Mechanisms and Applications of Leao's Spreading Depression of Electroencephalographic Activity*. Academic, New York.
24. Busa, W. B., Nuccitelli, R., 1984, *Am J Physiol*. **246**: 409–438.
25. Carney, J. M., Carney, A. M., 1994, *Life Sci*. **55**: 2097.
26. Carney, J. M., Starke-Reed, P. E., Oliver, C. N., *et al.*, 1991, *Proc Natl Acad Sci USA*. **88**: 3633–3636.
27. Carmell, J., Curtis, A. R., Kemp, J. A., *et al.*, 1993, *Neurosci Lett*. **153**: 107–110.
28. Castillo, J., Davalos, A., Noya, M., 1997, *Lancet*. **349**: 79–83.
29. Chan, P. H., 1988, in *Cellular Antioxidant Defense Mechanisms* (Chow, C. K., ed.) CRC Press, Boca Raton FL, pp. 89–109.
30. Chan, P. K., 1999, *Cerebral Ischemia* (Walz, W., ed.) Humana Press, Totowa, New Jersey, pp. 105–125.
31. Choi, D. W., 1990, *J Neurosci*. **10**: 2493–2501.
32. Choi, D. W., 1993, *Proc Natl Acad Sci USA*. **90**: 9741.
33. Choi, D. W., Koh, J. Y., 1998, *Annu Rev Neurosci*. **21**: 347–375.
34. Clements, J. D., 1996, *Trend Neurosci*. **19**: 163–171.
35. Conley, E. C., 1996, *Cell Calcium*. II: 233–239.
36. Cooper, J. R., Bloom, F. E., Roth, R. H., 1996, in *The Biochemical Basis of Neuropharmacology*. University Press, Oxford, New York, pp. 126–458.
37. Cornell-Bell, A. H., Finkbeiner, S. M., Cooper, M. S., *et al.*, 1990, *Science*. **247**: 470–473.
38. Czeh, G., Somjen, G. G., 1990, *Brain Res*. **527**: 224–233.
39. Dalton, D. W., Busto, R., Ginsberg, M. D., 1989, *J Cerebr Blood Flow Metab*. **9**: 812.
40. Dambinova, S. A., 1989, *Glutamate Neuroreceptors*, Leningrad (in Russian).
41. Dambinova, S. A., 1993, *Eur Patent Bull*. 93/69 WO 93/00586.
42. Dambinova, S. A., Kamenskaya, M. A., 1996, Molecular mechanisms of impulse transduction in neuronal membranes. Ion channels and receptors. In *Neurochemistry* (Ashmarin, I. P., Stukalov, P. V., eds.) Moscow, pp. 246–295 (in Russian).
43. Dani, J. W., Chernjavsky, A., Smith, S. J., 1992, *Neuron*. **8**: 429–440.
44. Davalos, A., Naveiro, J., Noya, M., 1996, *Stroke*. **27**: 1060–1065.
45. Davies, J., Watkins, J. T., 1982, *Brain Res*. **235**: 378–386.
46. Dawson, T. M., Dawson, V. L., Snyder, S. H., 1992, *Ann Neurol*. **32**: 297–311.
47. Dawson, D. A., Kusumoto, K., Graham, D. I., *et al.*, 1992, *Neurosci Lett*. **142**: 151–154.
48. Dawson, T. L., Gores, G. I., Nieminen, A. L., *et al.*, 1993, *Am J Physiol*. **264**: 961–967.
49. Dawson, T. M., Steiner, J. P., Dawson, V. L., *et al.*, 1993, *Proc Natl Acad Sci USA*. **90**: 9808–9812.
50. Diemer, N. H., Jorgensen, M. B., Johansen, F. F., *et al.*, 1992, *Acta Neuropathol Scand*. **86**: 45–49.
51. Dietrich, W. D., Feng, Z. C., Leistra, H., *et al.*, 1994, *J Cerebr Blood Flow Metab*. **14**: 20–28.
52. Drejer, J., Larson, O. M., Schousboe, A., 1982, *Exp Brain Res*. **47**: 259–269.
53. Drejer, J., Nielsen, E. G., Honore, T., *et al.*, 1989, *Neurosci Lett*. **98**: 333–338.
54. Duffy, F., Nelson, S., Lowry, O., 1972, *J Neurochem*. **19**: 959–977.
55. Eimerl, S., Schramm, M., 1991, *Neurosci Lett*. **130**: 125–127.
56. Eliasson, M. J., Sampei, K., Mandir, A. S., *et al.*, 1997, *Nat Med*. **3** (10): 1089–1095.
57. Engelsen, B. A., Fosse, V. M., Myrseth, E., *et al.*, 1985, *Neurosci Lett*. **62**: 97–102.
58. Erecinska, M., Silver, I. A., 1997, Loss of neuronal calcium homeostasis, in ischemia. In *Primer on Cerebrovascular Diseases* (Welch, K. M. A., Reis, D., Caplan, L. R., Siesjo, B. K., Weir, B., eds.) Academic, New York, pp. 178–183.

59. Erecinska, M., Nelson, D., Wilson, D. R., *et al.*, 1984, *Brain Res.* **304**: 19–23.
60. Evans, P. V., 1993, *Br Med Bull.* **49**: 5477.
61. Fagg, G. E., Foster, A. C., 1983, *Neurosci.* **9**: 701–719.
62. Felbergrova, J., ljunggren, B., Norberg, K., *et al.*, 1974, *Brain Res.* **90**: 265–279.
63. Fisher, M. A., 1991, *J Cerebrovasc Dis.* **1**: 112–119.
64. Fisher, M. A., Garcia, J. H., 1996, *Neurology.* **47**: 884–888.
65. Fox, P. T., Raichle, M. E., Mintun, M. A., *et al.*, 1988, *Science.* **241**: 462–464.
66. Frahm, J., Kruger, G., Merboldt, K.-D., *et al.*, 1996, *Magn Reson Med.* **35**: 143–148.
67. Fray, A. E., Forsyth, R. J., Boutelle, M. G., *et al.*, 1996, *J Physiol.* **496**: 49–57.
68. Frederickson, C. J., 1989, *Int Rev Neurobiol.* **31**: 145–238.
69. Fujisawa, F. F., Dawson, D., Browne, S. E., *et al.*, 1993, *Brain Res.* **629**: 73–78.
70. Garthwaite, G., Garthwaite, J., 1988, *Neuroscience.* **26**: 321–326.
71. Ginsberg, M. C., 1990, Models of cerebral ischemia in the rodent. In *Cerebral Ischemia and Resuscitation* (Schurr, A., Rigor, B. M., eds.) CRC Press, Boca Raton FL, pp. 1–15.
72. Ginsberg, M. D., 1990, *Cerebrovasc Brain Metab Rev.* **2**: 68–93.
73. Globus, M. Y.-T., Rusto, R., *et al.*, 1991, *J Neurochem.* **57**: 470–478.
74. Gonzalez-Zulueta, M., Feldman, A. B., Klesse, L. J., *et al.*, 2000, *Proc Natl Acad Sci USA.* **97** (1): 436–441.
75. Goto, S., Xue, R., Sugo, N., *et al.*, 2002, *Stroke.* **33** (4): 1101–1106.
76. Gusev, E. I., Skvortsova, V. I., 1990, *Case Record Form for the Examination and Treatment of Patients with Ischemic Stroke*, Moscow, pp. 1–44 (in Russian).
77. Gusev, E. I., Skvortsova, V. I., Burd, G. S., *et al.*, 1997, *J Neurol Sci.* **150**: 81–82.
78. Gusev, E. I., Skvortsova, V. I., Dambinova, S. A., *et al.*, 1996, *Cerebrovasc Dis.* **6**: 44.
79. Gusev, E. I., Skvortsova, V. I., Dambinova, S. A., *et al.*, 2000, *Cerebrovasc Dis.* **10**: 49–60.
80. Gusev, E. I., Skvortsova, V. I., Izykenova, G. A., *et al.*, 1996, *Korsakoff J Neurol Psychiatr.* **5**: 68–72 (in Russian).
81. Gusev, E. I., Skvortsova, V. I., Komissarova, I. A., *et al.*, 1999, *Korsakoff J Neurol Psychiatr.* **2**: 18–26 (in Russian).
82. Gusev, E. I., Skvortsova, V. I., Kovalenko, A. V., Sokolov, M. A., 1999, *Korsakoff J Neurol Psychiatr.* **2**: 65–70 (in Russian).
83. Gusev, E. I., Skvortsova, V. I., Rayevsky, K. S., *et al.*, 1995, Neurotransmitter amino acids in CSF in patients with acute ischemic stroke. In *Functional Studies as a Basis for Creation of Drugs*, Moscow, pp. 133–134 (in Russian).
84. Gusev, E. I., Skvortsova, V. I., Raevsky, K. S., *et al.*, 1997, *Eur J Neurol.* **4** (1): 78.
85. Haddad, G. G., Jiang, C., 1993, *Progr Neurobiol.* **40**: 277–318.
86. Halliwell, B. L., Packer, L., Prilipko, Y., 1992, *Free Radicals in the Brain. Aging, Neurological and Mental Disorders* (Christen, Y., ed.) Berlin, pp. 21–40.
87. Hammerstad, J. F., Murray, J. E., Cutter, R. W. P., 1971, *Brain Res.* 35: 357–367.
88. Hansen, A. J., 1985, *Physiol Rev.* **65**: 101–148.
89. Harms, L., Enchtnja, S., Timm, G., *et al.*, 1992, *J Neurol.* **239** (64): 3–32.
90. Haspekov, L. G., Onufriyev, M. V., Lyzhin, A. A., *et al.*, 1999, The influence of ischemia on the activity of nitric oxide synthase in the organotypic culture of hippocampal tissue. In *Hypoxia: Mechanisms, Correction, Adaptation*, Moscow, 81 (in Russian).
91. Hegstad, E., Berg-Johnsen, J., Haugstad, T. S., *et al.*, 1996, *Acta Neurochir* (Wien). **138** (2): 234–241.
92. Heiss, W. D., Graf, R., 1994, *Curr Opin Neurobiol.* **1**: 11–19.
93. Herreras, O., Somjen, G. G., 1993, *Brain Res.* **610**: 276–282.

94. Higami, Y., Shimokawa, I., Okimoto, T., *et al.*, 1994, *Mut Res.* **316**: 59.
95. Hillered, L., Chan, P. H., 1988, *J Neurosci Res.* **20**: 451–456.
96. Hirota, H., Katayama, Y., Kawamata, T., *et al.*, 1995, *Neurol Res.* **17** (2): 94–96.
97. Hollmann, M., Heinemann, S., 1994, *Annu Rev Neurosci.* **17**: 31–108.
98. Hossmann, K. A., 1994, *Ann Neurol.* **36** (4): 557–565.
99. Hossmann, K. A., 1994, *Brain Pathol.* **4**: 23–36.
100. Huang, T. T., Carlson, E. J., Leadon, S. A., *et al.*, 1992, *FASEB J.* **6**: 903–910.
101. Hull, C. D., van Harreveld, A., 1964, *Am J Physiol.* **207**: 921–924.
102. Irwin, A., Walz, W., 1999, Spreading depression waves as mediators. In *Cerebral Ischemia* (Walz, W., ed.) Humana Press, Totowa, New Jersey, pp. 35–45.
103. Isayama, K., Pilts, L. H., Nishimura, M. C., 1991, *Stroke.* **22**: 1394–1398.
104. Johansen, F. F., Tonder, N., Berg, M., *et al.*, 1993, *Mol Chem Neuropathol.* **18**: 161–172.
105. Johnson, J. W., Ascher, P., 1987, *Nature.* **325**: 329–331.
106. Jonas, P., Racca, C., Sakmann, B., *et al.*, 1994, *Neuron.* **12**: 1281–1289.
107. Jorgensen, M. B., Johansen, F. F., Dicmer, N. H., 1987, *Acta Neuropharmacol.* **73**: 184–194.
108. Kamii, H., Mikawa, S., Murakami, K., *et al.*, 1996, *J Cerebr Blood Flow Metab.* **16**: 1153–1157.
109. Katz, A. M., Messineo, F. C., 1981, *Circ Res.* **48**: 1–16.
110. Kavanaugh, W. M., Williams, L. T., 1994, *Science.* **266**: 1862–1865.
111. Kempski, O., Staub, F., Jansen, M., *et al.*, 1988, *Stroke.* **19**: 385–392.
112. Kiedrowski, L., Costa, E., Wroblewski, J. T., 1991, *J Neurochem.* **58**: 335–341.
113. Kim, E. Y., Koh, J. Y., Kim, Y. H., *et al.*, 1999, *Eur J Neurosci.* **11** (1): 327–334.
114. Kim, Y. H., Kim, E. Y., Gwag, B. J., *et al.*, 1999, *Neurosci.* **89** (1): 175–182.
115. Kitagawa, K., Matsumoto, M., Kuwabara, K., *et al.*, 1991, *Brain Res.* **561**: 203–211.
116. Kitagawa, K., Matsumoto, M., Tagaya, M., *et al.*, 1991, *J Cerebr Blood Flow Metab.* II: 449–452.
117. Kraig, R. P., Nicholson, C., 1978, *Neuroscience.* **3**: 1045–1059.
118. Kristian, T., Siesjo, B. K., 1997, Changes in ionic fluxes during cerebral ischemia. In *Neuroprotective Agents and Cerebral Ischemia* (Green, R., Cross, A. R., eds.) Academic, New York, pp. 27–45.
119. Lascola, C. D., Kraig, R. P., 1997, Astrocyte reaction in global ischemic brain injury. In *Primer on Cerebrovascular Diseases* (Welsh, M., Caplan, L., Siesjo, B., Weir, B., Reis, D. J., eds.) Academic, CA, San Diego, pp. 114–117.
120. Largo, C., Lbarz, J. M., Herreras, O., 1997, *J Neurophysiol.* **78**: 295–307.
121. Largo, C., Tombaugh, G. C., Aitken, P. G., *et al.*, 1997, *J Neurophysiol.* **77**: 9–16.
122. Lauritzen, M., 1987, *Acta Neural Scand Suppl.* **76**: 940.
123. Lazarewitcz, J. W., 1996, *Acta Neurobiol Exp* (Warsz). **56** (1): 299–311.
124. Lee, J.-M., Zipfel, G. J., Choi, D. W., 1999, *Nature.* **399**: 7–14.
125. Li, H., Buchan, A. M., 1993, *J Cerebr Blood Flow Metab.* **13**: 933–939.
126. Lijima, T., Mies, G., Hossmann, K.-A., 1992, *J Cerebr Blood Flow Metab.* **12**: 727–733.
127. Linde, R., Laursen, H., Hansen, A. J., 1996, Is calcium accumulation post-injury an indicator of cell damage? In *Mechanisms of Secondary Brain Damage in Cerebral Ischemia and Trauma.* Springer Wien NY Acta Neurochir. **66**: 15–20.
128. Lipskaya, L. A., 1994, *Cytology.* **36** (3): 303–309 (in Russian).
129. Littmann, L., Glati, B. S., Robinson, M. B., 1993, *Neurochem.* **61**: 586–593.
130. Lothman, E., LaManna, J., Cordingley, G., *et al.*, 1975, *Brain Res.* **88**: 15–36.

131. MacDermott, A. B., Mayer, M. L., Westbrook, G. L., *et al.*, 1986, *Nature*. **321**: 519–522.
132. Magistretti, P. J., 1997, Coupling of cerebral blood flow and metabolism. In *Primer of Cerebrovascular Diseases* (Welch, K. M. A., Caplan, L. R., Reis, D. J., Siesjo, B. K., Weir, B., eds.) Academic Press, San Diego, pp. 70–75.
133. Magistretti, P. J., Pellerin, L., 1996, *Cerebr Cortex*. **6**: 50–61.
134. Magistretti, P. J., Pellerin, L., 1997, The central role of astrocytes in brain energy metabolism. In *Neuroscience, Neurology and Health*. WHO, Geneva, pp. 53–64.
135. Mainprize, T., Shuaib, A., Ljaz, S., *et al.*, 1995, *Neurochem Res*. **20** (8): 957–961.
136. Mandir, A. S., Poitras, M. F., Berliner, A. R., *et al.*, 2000, *J Neurosci*. **20** (21): 8005–8011.
137. Mans, D., Lafleur, M., Westmuse, E., *et al.*, 1992, *Biochem Pharmacol*. **43**: 1761.
138. Manzoni, O., Prezeau, L., Marin, P., *et al.*, 1992, *Neuron*. **8**: 653–662.
139. Marion, J., Wolfe, L. S., 1979, *Biochim Biophys Acta*. **574**: 25–32.
140. Matsumoto, M., Yamamoto, K., Hamburger, H. A., *et al.*, 1987, *Mayo Clinic Proc*. **62**: 460–472.
141. Matsumoto, K., Yamada, K., Kohmura, E., *et al.*, 1994, *Neurol Res*. **16** (6): 460–464.
142. Mature, C., Sanchez-Gomez, M. V., Martinez-Millan, L., Miledi, R., 1997, *Proc Natl Acad Sci USA*. **94**: 8830–8835.
143. McDonald, J. W., Althomsons, S. P., Hyrc, K. L., *et al.*, 1998, *Nature Med*. **4**: 291–297.
144. Meldrum, B. S., 1989, *Cerebrovasc Dis* (New York). 47–60.
145. Meldrum, B. S., Garthwaite, J., 1990, *Trends Pharmacol Sci*. II: 379–387.
146. Mies, G., Lijima, T., Hossmann, K.-A., 1993, *Neuroreport*. **4**: 709–711.
147. Mitani, A., Yanase, H., Sakai, K., *et al.*, 1993, *Brain Res*. **601**: 103–110.
148. Moncada, S., Palmer, R. M., Higgs, E. A., 1991, *Pharmacol Rev*. **43**: 109–142.
149. Morley, P., Hogan, M. J., Hakim, A. M., 1994, *Brain Pathol*. **4**: 37–47.
150. Morley, P., Tauskela, J.-S., Hakim, A. M., 1999, *Cerebral Ischemia* (Walz, W., ed.) Humana Press, Totowa, New Jersey, pp. 69–105.
151. Murphy, S., Simmons, M. L., Agullo, L., 1993, *Trends Neurosci*. **16**: 323–328.
152. Nagafuji, T., Matsui, T., Koide, T., *et al.*, 1992, *Neurosci Lett*. **147**: 159–162.
153. Nakashima, K., Todd, A., 1996, *Anesthesiology*. **85** (1): 161–168.
154. Nedergaard, M., 1994, *Science*. **263**: 1768–1771.
155. Nedergaard, M., Gjedde, A., Diemer, N. H., 1986, *J Cerebr Blood Flow Metab*. **6**: 414–424.
156. Nedergaard, M., Hansen, A. J., 1993, *J Cerebr Blood Flow Metab*. **13**: 568–574.
157. Nicotera, P., Zhivotovsky, B., Orrenius, S., 1994, *Cell Calcium*. **16**: 279–288.
158. Nordstrom, C. H., Siesjo, B.-K., 1978, *Stroke*. **9**: 327–335.
159. Nowak, T. S., Nowak, Jr., Kiessling, M., 1999, Reprogramming of gene expression after ischemia. In *Cerebral Ischemia* (Walz, W., ed.) Humana Press, Totowa, New Jersey, pp. 145–217.
160. Nowak, T. S., Nowak, Jr., Zhou, Q., Valentine, W. J., *et al.*, 1999, Regulation of heat shock genes by ischemia. In *Stress Proteins* (*Handbook of Experimental Pharmacology*) (Latchman, D. S., ed.) Vol. 136, Springer, Heidelberg, Germany.
161. Obrenovitch, T. P., Zilkha, E., 1995, *J Neurophysiol*. **73** (5): 2107–2114.
162. Obrenovitch, T. P., Zilkha, E., 1995, Changes in extracellular glutamate concentration associated with propagating cortical spreading depression. In *Experimental Headache Models in Animal and Man* (Olesen, J., Moskowitz, M. A., eds.) Raven, New York, pp. 113–117.

163. Obrenovitch, T. P., Zilkha, E., Urenjak, J., 1996, *J Cerebr Blood Flow Metab.* **16** (5): 923–931.

164. Ohkawa, H., Ohisi, N., Yagi, K., 1979, *Analyt Biochem.* **95**: 351–355.

165. Oliver, C. N., Starke-Reed, P. E., Stadtman, E. R., *et al.*, 1990, *Proc Natl Acad Sci USA.* **87**: 5144–5147.

166. Olney, J. W. E., 1994, *J Neural Transm Suppl.* **43**: 41–51.

167. Olney, J. W. E., McGeer, J. W., Oiney, P., 1978, *Neurotoxicity of Excitatory Amino Acids: Kainic Acid as Tool in Neurobiology* (McGeer, J. W., ed.) New York, pp. 95–121.

168. Ooboshi, H., Sadoshima, S., Yao, H., *et al.*, 1992, *J Neurochem.* **58** (1): 298–303.

169. Orgogozo, J. M., Dartigues, J. F., 1986, Clinical trials in acute brain infarction: the question of assessment criteria. In *Acute Brain Ischemia: Medical and Surgical Therapy* (Battistini, N., Courbier, R., Fieschi, C., Fiorani, P., Plum, F., eds.) Raven Press, New York, pp. 282–289.

170. Orrenius, S., McCabe, M. S., Nicotera, P., 1992, *Toxicol Lett.* **64**: 357–364.

171. Orrenius, S., McConkey, D. S., Jones, D. P., *et al.*, 1988, *Athas Sci Pharmacol.* **2**: 319–324.

172. Packer, L. L., Packer, L., Prilipko, Y., Cyirsten, Y., 1992, *Free Radicals in the Brain*, Berlin, pp. 1–20.

173. Palaiologos, G. I., Hertz, L., Schousboe, A., 1988, *J Neurochem.* **51**: 317–320.

174. Paschen, W., 1996, *Acta Neurobiol Exp Warsz.* **56** (1): 313–322.

175. Pearson, S. S., Crudek, C., Mercer, K., Reynolds, G. P., 1991, *J Neural Transm.* **86**: 151–157.

176. Pellerin, L., Magistretti, P. J., 1994, *Proc Natl Acad Sci USA.* **91**: 10625–10629.

177. Phillips, J. M., Nicholson, C., 1979, *Brain Res.* **173**: 567–571.

178. Phillis, J. W., Song, D., O'Regan, M. H., 1996, *Neurosci Lett.* **207** (3): 151–154.

179. Pinelis, V. G., Sorokina, E. G., Reutov, V. P., 1997, *Russ Acad Sci Report.* **352** (2): 259–261 (in Russian).

180. Prichard, J., Rothman, D., Novotny, E., *et al.*, 1991, *Med Sci.* **88**: 5829–5831.

181. Puka, M., Lehmann, A., 1994, *J Neurosci Res.* **37** (5): 641–646.

182. Puka, M. J. C., Wiesebfeld-Hallin, Z., 1994, *Anesth Anaid.* **77**: 104–109.

183. Pulsinelli, W., Brierley, J., Plum, F., 1982, *Ann. Neurol.* II: 491–498.

184. Ravindran, J., Shuaib, A., Ljai, S., *et al.*, 1994, *Neurosci Lett.* **176** (2): 209–211.

185. Rayevsky, K. S., 1997, *Bull Exp Biol Med.* **123**: 484–490 (in Russian).

186. Rayevsky, K. S., Bashkatova, V. G., Vitskova, G. Yu., *et al.*, 1998, *Exp Clin Pharmacol.* **61**: 13–16 (in Russian).

187. Rayevsky, K. S., Georgiev, V. P., 1986, *Mediator Amino Acids: Neuropharmacological and Neurochemical Aspects*, Moscow, p. 240 (in Russian).

188. Regli, L., Spent, M. C., Anderson, R. E., *et al.*, 1996, *J Cerebr Blood Flow Metab.* **16** (5): 988–995.

189. Rehncrona, S., Hauge, H. N., Siesj, B. K., 1989, *J Cerebr Blood Flow Metab.* **9**: 65–70.

190. Reynolds, I. J., 1990, *Eur J Pharmacol.* **177**: 215–216.

191. Rothman, S. M., Olney, J. M., 1986, *Ann Neurol.* **19** (2): 105–111.

192. Rothstein, J. D., Martin, L., Levey, A., *et al.*, 1994, *Neuron.* **13**: 713–725.

193. Ruetzler, C., Lohr, J., Alui, A., 1990, *Acta Neurochir Suppl Faults.* **51**: 186–188.

194. Sala, L., 1891, *Zur Feineren Anatomic des Grossen Seepferdefusses. Zeitsclirift fur Wissenschaftliche Zoologie*, Band Lll, Heft. **2**: 18–45.

195. Sasaki, M., Gonzalez-Zulueta, M., Huang, H., *et al.*, 2000, *Proc Natl Acad Sci USA.* **97** (15): 8617–8622.

196. Savelyeva, K. V., Mikoyan, V. D., Vanin, A. F., *et al.*, 1996, *Int Congr Pathophysiologists*, Moscow, p. 25 (in Russian).
197. *Scandinavian Stroke Study Group Stroke*, 1985. **16**: 885–890.
198. Scheinberg, P., 1991, *Neurology*. **41**: 1867–1873.
199. Schoepp, D. D., Conn, P. J., 1993, *Trends Pharmacol Sci*. **14**: 13–20.
200. Schousboe, A., Frandsen, A., Wayl, P., *et al.*, 1994, *Neurotoxicology*. **15**: 477–481.
201. Schuman, E. M., 1995, *Science*. **267**: 1658–1662.
202. Schurr, A., West, C. A., Rigor, B. M., 1988, *Science*. **240**: 1326–1328.
203. Sensi, S. L., Canzoniero, L. M., Yu, S. P., *et al.*, 1997, *J Neurosci*. **17** (24): 9554–9564.
204. Sensi, S. L., Yin, H. Z., Carriedo, S. G., *et al.*, 1999, *Proc Natl Acad Sci USA*. **96**: 2414–2419.
205. Serena, J., Leira, R., Castillo, J., *et al.*, 2001, *Stroke*. **32** (5): 1154–1161.
206. Sergeyev, P. V., Shimanovsky, N. L., Petrov, V. I., 1999, *Receptors*, Moscow-Volgograd, pp. 346–423 (in Russian).
207. Sheardown, M. J., Suzdak, P. D., Nordholm, E., 1993, *Eur J Pharmacol*. **236**: 347–353.
208. Siesjo, B.-K., 1985, In *Progress in Brain Research* (Hossmann, K.-A., Kogure, K., Siesjo, B. K., Welsh, F., eds.) Elsevier, Amsterdam, pp. 121–154.
209. Siesjo, B.-K., 1986, *Eur Neurol*. **25** (1): 45–56.
210. Siesjo, B.-K., 1988, *Ann NY Acad Sci*. **522**: 638–661.
211. Siesjo, B.-K., Agardh, C. C., Bengtsson, F., 1989, *Cerebrovasc Brain Metab Rev*. **1**: 165–211.
212. Siesjo, B.-K., Bengtsson, F., 1989, *J Cerebr Blood Flow Metab*. **9**: 127–140.
213. Silver, A., Erecinska, M., 1990, *J Gen Physiol*. **95**: 837–866.
214. Silver, I. A., Erecinska, M., 1992, *J Cerebr Blood Flow Metab*. **12**: 759–772.
215. Skvortsov, I. A., Karaseva, A. N., Burkova, A. S., *et al.*, 1987, Actual problems of paediatric neurology. Ganglioside functions in nervous system in normal and pathological states. In *Reviewing Report. Medicine and Health Care*, Moscow, p. 65 (in Russian).
216. Skvortsova, V. I., Nasonov, E. L., Zhuravlyova, E. Yu., *et al.*, 1999, *Korsakoff J Neurol Psychiatr*. **5**: 27–32 (in Russian).
217. Skvortsova, V. I., Rayevsky, K. S., Kovalenko, A. V., *et al.*, 1999, *Korsakoff J Neurol Psychiatr*. **2**: 34–39 (in Russian).
218. Somjen, G. G., Aitken, P. G., Czeh, C. L., 1992, *Can J Physiol Pharmacol*. **70**: 248–254.
219. Sommer, B., Seeburg, P. H., 1992, *Trends Pharmacol Sci*. **13**: 291–296.
220. Sorokina, E. G., Pinelis, V. G., Reutov, V. P., *et al.*, 1996, *Int Congr of Pathophysiologists*, Moscow, p. 187 (in Russian).
221. Suslina, Z. A., 1991, *Ischemic Strokes and Prostanoids System* (*Clinical and Biochemical Study*). Doctoral dissertation, Moscow, 331 (in Russian).
222. Swanson, R. A., Chen, J., Graham, S. H., 1994, *J Cerebr Blood Flow Metab*. **14** (1): 1–6.
223. Takagi, K., Ginsberg, M. D., Globus, M. Y., *et al.*, 1993, *J Cerebr Blood Flow Metab*. **13** (4): 575–585.
224. Takahashi, S., Driscoll, B. F., Law, M. J., Sokoloff, L., 1995, *Proc Natl Acad Sci USA*. **92**: 4616–4620.
225. Takahashi, T., Feidmeyer, D., Suzuki, N., *et al.*, 1996, *J Neurosci*. **16**: 4376–4382.
226. Takizawa, S., Fujila, H., Ogawa, S., *et al.*, 1996, *J Cerebr Blood Flow Metab*. **16**: 1075–1078.
227. Tamda, M., O'Neil, P., 1991, *J Clin Soc Perkin Trans*. II: 1681.

228. Tang, C. M., Dichter, M., Morad, M., 1990, *Proc Natl Acad Sci USA*. **87**: 6445–6449.
229. Tegtmeier, F., 1993, Differences between spreading depression and ischemia. In *Migraine: Basic Mechanisms and Treatment* (Lehmenkuhler, A., Grotemeyer, K. H., Tegtmeier, G., eds.) Munich-Schwarzenberg-Urban, pp. 511–532.
230. Tombaugh, G. C., Sapolsky, R. M., 1993, *J Neurochem*. **61**: 793–803.
231. Tonder, N., Johansen, F. F., Frederickson, C. J., *et al.*, 1990, *Neurosci Lett*. **109**: 247–252.
232. Tong, G., Shepherd, D., Jahr, C. E., 1995, *Science*. **267**: 1510–1512.
233. Touzani, O., Young, A. R., Derlon, J. M., *et al.*, 1995, *Stroke*. **26**: 2112–2119.
234. Tsacopoulos, M., Magistretti, P. J., 1996, *J Neurosci*. **16**: 877–885.
235. Wahl, F., Obrenovitch, T. P., Hardy, A. M., *et al.*, 1994, *J Neurochem*. **63** (3): 1003–1011.
236. Warach, S., Gaa, J., Siewert, B., 1995, *Ann Neurol*. **37**: 231–241.
237. Weiss, J. H., Hartley, D. M., Koh, J. Y., Choi, D. W., 1993, *Neuron*. **10**: 43–49.
238. Welch, K. M. A., Windham, J., Knight, R. A., *et al.*, 1995, *Stroke*. **26**: 1983–1989.
239. Wieloch, T., 1985, *Science*. **230**: 681–683.
240. Wilcox, C. S., 1992, *Proc Natl Acad Sci USA*. **89**: 113.
241. Winder, D. G., Smith, T., Conn, P. J., 1993, *J Pharmacol Exp Ther*. **266**: 518–525.
242. Wolfe, L. S., 1982, *J Neurochem*. **38**: 1–14.
243. Wright, C. E., 1992, *Cardiovasc Res*. **26** (1): 48–57.
244. Yin, H. Z., Turetsky, D., Choi, D. W., Weiss, J. H., 1994, *Neurobiol Dis*. **1**: 43–49.
245. Yoshida, S., Inoh, S., Asano, T., *et al.*, 1980, *J Neurosurg*. **53**: 323–331.
246. Yu, K.-L., Yeo, T. T., Dong, K.-W., *et al.*, 1994, *Mol Cell Endocrinol*. **102**: 85–92.
247. Yun, H. Y., Dawson, V. L., Dawson, T. M., 1999, *Diabetes Res Clin Pract*. **45** (2–3): 113-115.
248. Zablocka, B., Domanska-Janik, K., 1996, *Acta Neurobiol Exp Warsz*. **56** (1): 63–70.
249. Zarei, M. M., Dani, J. A., 1994, *J Gen Physiol*. **103**: 231–248.
250. Zavalishin, I. A., Zakharova, M. N., 1996, *Korsakoff J Neurol Psychiatr*. **2**: 111–114 (in Russian).
251. Zhang, S., Ehlers, M. D., Bernhardt, J. P., *et al.*, 1998, *Neuron*. **21**: 443–453.
252. Zhdanov, G. G., Nechayev, V. P., Nozel, M. L., 1989, *J Anesthesiol Resuscit*. **4**: 63–68 (in Russian).
253. Zivin, J.A., Choi, D.W., 1991, *Scientific American*. Mir, Moscow. **9**: 30–31 (Russian translation).

Chapter 5

Metabolic Acidosis and Ischemic Damage

One of the primary responses of brain tissue to decrease in CBF is acidosis. Decrease in ATP content in the ischemized brain area leads to compensatory activation of anaerobic glycolysis and to increased production of lactate and H^+ that causes the development of lactic acidosis. Modest increase in H^+ concentration in early stages of ischemia plays a compensatory and adaptive role as it promotes the improvement of perfusion in the penumbral area [2]. Significant elevation of lactate level within the first hours of ischemic stroke leads to decrease in pH to 6.4–6.7 [11] and appears to be an unfavorable prognostic sign [5, 12, 19, 20].

Experimental works by Siesjo *et al.* [28, 30, 31, 35] have shown that tissue acidosis plays a very important role when selective neuronal necrosis transforms into brain infarction. The suggested interaction between acidosis and the expanding of tissue damage is based on experiments where pre-ischemic hyperglycemia was shown to increase damage caused by transient ischemia due to decrease in intra- and extracellular pH [23, 35]. After acid was injected into brain tissue of experimental animals, focal necrosis developed [16].

In the 1990s data was obtained showing that acidosis in neuronal culture can ameliorate damage to neurons caused by glutamate and anoxia [8, 38]. However, being based only on *in vitro* experiments, these results have not been subsequently confirmed *in vivo*.

Acidosis acts on a multitude of metabolic processes and modulates membrane processes such as passive and active transport of ions [6, 21]. In general, acidosis depresses metabolic reactions and transport of ions. Slight inhibition of calcium flow can in theory be of some protective benefit, which was also demonstrated experimentally *in vitro*, when extracellular acidosis protected neurons from glutamate or effects of anoxia [8, 38]. At the same time, it was proven that the damaging effect of acidosis is enhanced during

95

acute focal brain ischemia as it is directed to energy-compromised tissue, in which the cellular phosphorylation level is decreased and ischemic metabolic cascade has been induced [17, 18, 29, 35, 36]. Intracellular acidosis leads to disturbances in calcium sequestration in mitochondria and endoplasmic reticulum due to competitive relations with H^+ and calcium for binding sites, to accumulation of free calcium ions and the additional activation of pathogenic mechanisms which they trigger: 1) aggravation of oxidant stress, 2) excessive NO synthesis, and 3) activation of the intracellular enzymes [22, 24].

Excessive production of free radicals on the background of acidosis is connected with elevated release in acidic medium of iron, which is a trigger of oxidative mechanisms. It is released from its binding with transferrin-like proteins [31, 35], this intensifying the Haber–Weiss reactions [25]. As a result, damage to mitochondrial membranes occurs quickly and decrease in the activities of cytochrome oxidases $a–a^3$ ensues. Metabolic acidosis and hyperglycemia hampers the recovery of mitochondrial functions in the post-ischemic period [35].

Decrease in pH in extra- and intracellular spaces causes direct cytotoxic effects [9, 27, 31] leading to weakening of cell membranes. Their physical and chemical properties become altered, leading to increased permeability of neurons and endothelium [5]. Swelling of endothelial cells [15] aggravates the disturbances of microcirculation and, following that, post-ischemic hypoperfusion, i.e. the "no-reflow" phenomenon occurs [1].

Several experiments have demonstrated that pre-ischemic hyperglycemia and acidosis causes condensation of nuclear chromatin in neurons [13] and disruption of the sequence of transduction signals when DNA fragmentation is enhanced and changes in mRNA expression ensue [7]. It was found that hyperglycemia decreased or stopped the expression of mRNA for *c-fos*, normally caused by ischemia and reperfusion [7]. Barry *et al.* [3, 4] obtained data on non-neuronal cells suggesting the presence of pH-dependent endonuclease (DNase II), which is activated when pH falls below 6.5.

In culture astrocytes appear to be more sensitive to ischemia than neurons [8, 10]. It was hypothesized that morphological and functional disintegration of glia–neuron contacts is one of the most important mechanisms of the damaging potential of acidosis [33]. Thus, acidosis-induced damage of astrocytes mediates necrotic neuronal death via derangement of glutamate transport from the synaptic cleft and via influence on the level of excitotoxicity [37, 39].

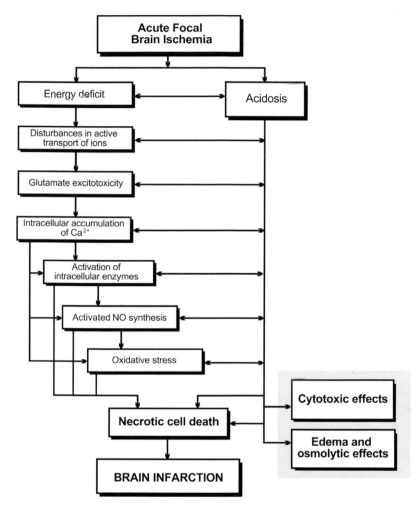

Scheme 5.1. Acidosis-mediated damage in acute focal brain ischemia.

Accumulation of H^+ plays a crucial role in the development of cellular edema. It has been proven that acidosis enhances cell swelling [14, 27, 32, 34]. The initial increase in tissue liquid causes a cytotoxic effect mediating disturbances of energy metabolism [26].

Thus, acidosis-mediated damage is an important component of the process of brain infarction formation (Scheme 5.1).

REFERENCES

1. Ames, A. I., Wright, R.L., Kowada, M., 1968, *Am J Pathol.* **52**: 437–453.
2. Baron, J. C., Frackowiak, R. S. J., Herholi, K., *et al.*, 1989, *J Cerebr Blood Flow Metab.* **9**: 723–742.
3. Barry, M. A., Eastman, A., 1993, *Arch Biochem Biophys.* **300**: 440–450.
4. Barry, M. A., Reynolds, J. E., Eastman, A., 1993, *Cancer Res.* **53**: 2349–2357.
5. Burd, G. S., 1983, *Respiratory Deficiency in Patients with Cerebrovascular Diseases.* Doctoral dissertation, Moscow, 355 (in Russian).
6. Busa, W. B., Nuccitelli, R., 1984, *Am J Physiol.* **246**: 409–438.
7. Combs, D. J., Dempsey, R. J., Donaldson, D., Kindy, M. S., 1992, *J Cerebr Blood Flow Metab.* **12**: 169–172.
8. Giffard, R. G., Monyer, H., Christine, C. W., Choi, D. W., 1990, *Brain Res.* **506**: 339–342.
9. Ginsberg, M. D., 1994, *New Strategies to Prevent Neural Damage from Ischemic Stroke*, Boston, 34.
10. Goldman, S. A., Pulsinelli, W. A., Clarke, W. Y., *et al.*, 1989, *J Cerebr Blood Flow Metab.* **9**: 471–477.
11. Hakim, A. M., Shoubridge, E. A., 1989, *Cerebrovasc Brain Metab Rev.* **1**: 115–132.
12. Harms, L., Enchtnja, S., Timm, G., *et al.*, 1992, *J Neurol.* **239** (64): 3–32.
13. Kalimo, H., Rechncrona, S., Soderfeldt, B., *et al.*, 1981, *J Cerebr Blood Flow Metab.* **1**: 313–327.
14. Kempski, O., Staub, F., Rosen, F. V., *et al.*, 1988, *Neurochem Pathol.* **9**: 109–125.
15. Kozuka, M., Smith, M. L., Siesjo, B. K., 1989, *J Cerebr Blood Flow Metab.* **9**: 478–490.
16. Kraig, R. P., Petito, C. K., Plum, F., Pulsinelli, W. A., 1987, *J Cerebr Blood Flow Metab.* **7**: 379–386.
17. Kristian, T., Gido, G., Siesjo, B. K., 1995, *J Cerebr Blood Flow Metab.* **15**: 78–87.
18. Lascola, C. D., Kraig, R. P., 1997, Astrocyte reaction in global ischemic brain injury. In *Primer on Cerebrovascular Diseases* (Welsh, M., Caplan, L., Siesjo, B., Weir, B., Reis, D. J., eds.) Academic, CA, San Diego, pp. 114–117.
19. Matsumoto, M., Yamamoto, K., Hamburger, H. A., *et al.*, 1987, *Mayo Clinic Proc.* **62**: 460–472.
20. Matsumoto, K., Yamada, K., Kohmura, E., *et al.*, 1994, *Neurol Res.* **16** (6): 460–464.
21. Moody, W., 1984, *Ann Rev Neurosci.* **7**: 257–278.
22. Nakashima, K., Todd, A., 1996, *Anesthesiology.* **85** (1): 161–168.
23. Nedergaard, M., 1988, *Acta Neurologica Scandinavica.* **77**: 1–24.
24. Ooboshi, H., Sadoshima, S., Yao, H., *et al.*, 1992, *J Neurochem.* **58** (1): 298–303.
25. Rehncrona, S., Hauge, H. N., Siesjo, B. K., 1989, *J Cerebr Blood Flow Metab.* **9**: 65–70.
26. Ruetzler, C., Lohr, J., Alui, A., 1990, *Acta Neurochir Suppl Fault.* **51**: 186–188.
27. Scheinberg, P., 1991, *Neurology.* **41**: 1867–1873.
28. Siesjo, B. K., 1985, In *Progress in Brain Research* (Hossmann, K. A., Kogure, K., Siesjo, B. K., Welsh, F., eds.) Elsevier, Amsterdam, pp. 121–154.
29. Siesjo, B. K., 1988, *Crit Care Med.* **16**: 954–963.
30. Siesjo, B. K., Agardh, C. D., Bengtsson, F., 1989, *Cerebrovasc Brain Metab Rev.* **1**: 165–211.
31. Siesjo, B. K., Bengtsson, F., 1989, *J Cerebr Blood Flow Metab.* **9**: 127–140.
32. Siesjo, B. K., Katsura, K., *et al.*, 1993, In *Progress in Brain Research* (Kogure, K., Hossmann, K. A., Siesjo, B. K., eds.) Elsevier, Amsterdam, pp. 23–48.
33. Siesjo, B. K., Lindvall, O., 1996, *Brain Res.* **38**: 139–144.

34. Siesjo, B. J., Siesjo, P., 1996, *Eur J Anesthesiol.* **13**: 247–268.
35. Siesjo, B. K., Katsura, K., Kristian, T., *et al.*, 1996, Molecular mechanisms of acidosis-mediated damage. In *Acta Neurochir.* **66**: 8–14.
36. Silver, I. A., Erecinska, M., 1990, *J Gen Physiol.* 95: 837–866.
37. Stanimirovic, D. B., Ball, R., Durkin, J. P., 1997, *Glia.* **19**: 123–134.
38. Tombaugh, G. C., Sapolsky, R. M., 1993, *J Neurochem.* **61**: 793–803.
39 Vibulsrcth, S., Hefti, F., Ginsberg, M. D., *et al.*, 1987, *Brain Res.* **422**: 303–311.

Chapter 6

Delayed Neuronal Death Following Acute Focal Brain Ischemia

The major part of a brain infarction is formed within the first 3–6 h of the first clinical signs of stroke (Fig. 6.1). However, the temporal evolution of infarction and its "up-formation" are connected with secondary brain cell damage which takes place in hours to days after the initial vascular event.

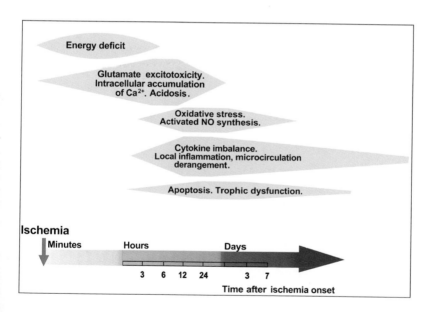

Figure 6.1. Temporal development of processes inducing focal ischemic brain damage.

Some processes that started in the first hours after stroke and underlie the glutamate–calcium cascade (disturbances of ion transport, changes in glutamate and calcium metabolism, free radicals reactions, lipid peroxidation, activated synthesis of NO, eicosanoids, and platelet activation factor, etc.) retain their significance in a more distant period after the onset of stroke, especially in cases of a vast zone of the ischemic lesion. They induce and support other delayed consequences of ischemia: gene expression and following molecular events (see Chapter 7); microglial activation and changes in astroglial cell pool, which lead to development of local inflammation in the ischemic focus and of other immune reactions (see Section 8.1 and Chapter 9); additional disturbances of microcirculation and the blood–brain barrier (see Section 8.2). Programmed cell death plays an important role in the mechanisms of delayed brain damage (see Chapter 10).

All mechanisms of delayed neuronal death not only lead to up-formation of brain infarction, but also cause long-term changes in functional activity of the common neuro–immune–endocrine system and take part in the development of post-stroke encephalopathy.

The duration of the up-formation of brain infarction has not been completely established and is, apparently, individual. Modern immunohistochemical and neuroimaging techniques have demonstrated that penumbral brain tissue can survive for at least 48–72 h after the vascular event or longer. These observations are of the utmost importance as they show that there are no strictly limited borders for the therapeutic window and thus they vindicate the search for new effective methods of neuroprotection.

Chapter 7

Gene Expression and Subsequent Molecular Events in Response to Acute Brain Ischemia

Analysis of pathological biochemical reactions involved in the mechanisms of acute focal brain ischemia suggests that there are tight correlations between all metabolic cascades and that one can observe their time correspondence and mutual regulation via a unified signal system including common systems of cellular messengers.

The signal transduction system is the first which reacts to any damaging stimulus (including ischemia) assuming the function of regulation of all molecular and cellular events taking place in brain tissue. The central link of this signaling chain is the receptor-mediated generation of second intracellular messengers that modulate the activity of such proteins as enzymes, receptors, and ion channels. All these regulatory proteins take part in modulation of events mediating neuronal damage and reparation. The basic mechanism of transduction of extra- and intracellular signals are protein phosphorylation/dephosphorylation by protein kinases and phosphatases, which are directly regulated by receptor activation and indirectly via activated second messengers (Scheme 7.1). The redistribution of protein kinases on cellular membranes caused by ischemia alters phosphorylation of receptors and ion channel proteins. The phosphorylation state influences cell survival and death pathways. Within the first minutes and hours of ischemia, activation of protein kinases stimulates the release of neurotransmitters as well as the activities of ion channels and receptors, which is accompanied by disturbances of active ionic transport, depletion of ATP, and acceleration of neuronal necrosis, especially in the penumbral area. Further phased decrease in the activities of protein kinase leads to the inhibition of intracellular signaling mechanisms and hence to decreased

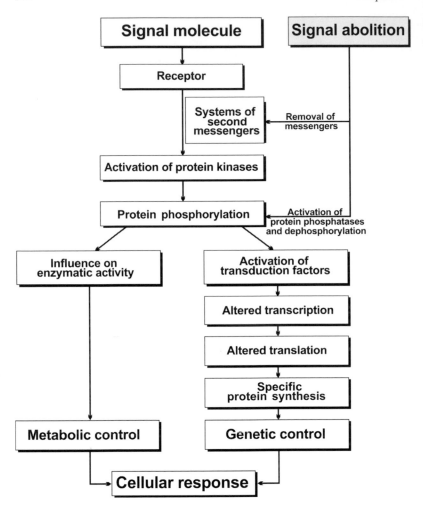

Scheme 7.1. System of intracellular signal transduction (formation of cellular response to external stimulus).

influence of trophic and other life-supporting effects on cells, this being conducive to delayed neuronal death.

Information about changes in the state of membrane-associated structures and receptors, determined by the pattern of redistribution of membrane-associated protein kinases, is transduced by second messengers (adenylate cyclases, protein kinases, phosphatases, etc.) into the cell nucleus, this serving as a signal to trigger the molecular mechanism that embodies the universal algorithm of brain tissue response to any injury (Fig. 7.1) [68, 96]. Thus, acute focal cerebral ischemia activates a complex of genetic programs

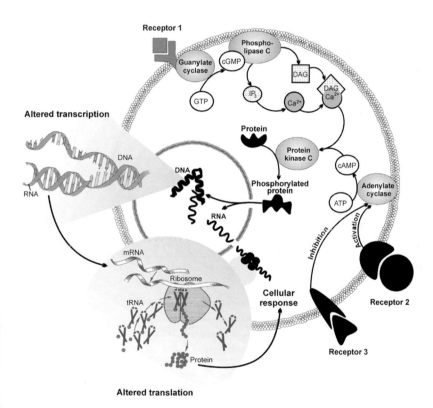

Figure 7.1. Regulation of cell viability by external signals via the system of intracellular signal transduction.

which lead to sequential expression of a large number of genes (Fig. 7.2). Some of them are involved in the immediate early response of the brain to injury, and others are involved in the cellular events determining the development of ischemic damage or coordinating reparative processes [40, 55]. The chronology of gene expression depends on the ischemic model studied and in case of focal ischemia it differs in the central (core) area and the penumbra.

The biochemical events underlying gene expression are transcription (the process of mRNA biosynthesis on DNA template) in the first stage and mRNA translation (the process of protein synthesis based on coding sequence of nucleotides in mRNA) in the second stage. Their effects are determined by the availability of nucleoside triphosphates (ATP and GTP) and by sustained energy metabolism [75]. Thus, it is no surprise that the first response of brain tissue to CBF decrease appears to be the inhibition of mRNA and protein synthesis [109]. To suspend protein synthesis CBF should be maintained within the level of 50–60 ml/100 g brain tissue per min [44, 67, 108]. This significantly exceeds the level required for suspending ATP production [67, 108] (see Chapter 1).

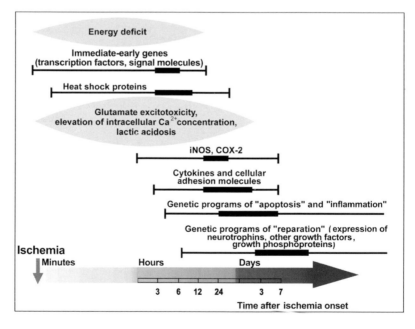

Figure 7.2. Chronology of molecular reactions in brain tissue in response to acute cerebral ischemia.

In an experimental study performed with rats by Kamiya *et al.* [47] it was shown that the critical level of blood supply required for the synthesis of early response gene *c-fos* mRNA in focal ischemia was 25–30 ml/100 g brain tissue per minute, whereas the accumulation of Fos protein requires 35–40 ml/100 g brain tissue per minute in the same animals. This confirms that the maintenance of transcriptional activity closely depends on energy metabolism. Neither mRNA nor proteins can be synthesized in the ischemic core. But in peri-infarction tissue mRNAs may be synthesized, while encoded proteins are not produced.

An experiment involving permanent middle cerebral artery occlusion in mice [30] showed that at the end of 1 h of ischemia, protein synthesis was suppressed in a larger tissue volume than ATP in accordance with the biochemical differentiation between core and penumbra. The final size of a gradually expanding infarction was heralded by the early inhibition of protein synthesis.

When CBF recovers after transient ischemia, mRNA synthesis quickly recovers [65], and the majority of metabolic processes are restored. However, protein synthesis deficiency may persist for many hours of recirculation after ischemia [1, 31, 35]. The extent of protein synthesis recovery reflects the individual vulnerability of neuronal populations and has prognostic value: quick and complete normalization of translational activity predicts neuronal survival, whereas it is slow and incomplete in cell populations such as hippocampal CA_1 neurons whose death is predestined [5, 31, 46, 106].

Despite a conventional opinion that insufficiency of translation mechanisms and, hence, of protein synthesis influences the outcome of ischemia negatively, several findings have appeared that under certain conditions *in vitro* protein deficiency retards the progression of apoptosis [63]. In studies conducted *in vivo*, it was also demonstrated that several inhibitors of protein synthesis possess neuroprotective properties [25, 78, 90]. However, their beneficial effects were possibly connected with the enhancement of transcriptional activity of potentially protective genes [19, 22].

The first and relatively nonspecific genomic response to any damaging stimuli (including ischemia) is the induction of **immediate-early, or early response, genes** (*c-fos, c-jun, krox-20, zif/268*, etc.) [6, 12, 38, 68]. Their triggering occurs already within the first minutes of ischemia in response to changes in ionic homeostasis, resulting from the ischemic depolarization occurring at the infarction border, and to excessive glutamate release and development of oxidative stress [6, 7, 30, 35, 51]. From membrane-associated receptors the signal about cell damage is transduced into the neuronal nucleus, where gene expression begins [68, 96]. The immediate-early genes are only transiently expressed after inducing stimuli due to the short half-life and rapid turnover of the encoded mRNAs and proteins, thus rendering them particularly suitable for regulatory functions [35, 75].

The expression of the majority of immediate-early genes leads to the synthesis of DNA-binding proteins or transcription factors, which act as nuclear third messengers in a mechanism of stimulus–transcription coupling linking neuronal excitation to changes in cellular function by regulating late response of target genes [69, 75]. The activity of transcription factors is regulated by posttranslational modification (e.g., phosphorylation or dephosphorylation) and/or translocation from the cytoplasm into the nucleus [75, 102]. Gene expression is dynamically regulated by ligand-activated transcription factors, e.g., nuclear receptors for steroid hormones [9].

Transcription factors cause the expression of a wide spectrum of target genes [2] via activation of AP1 and CRE promoter elements [3]. Other immediate-early genes encode signal molecules involved in biochemical cascades, such as: protein phosphatases, G-proteins, tissue plasminogen activator, cyclooxygenase, and secreted cytokine-like molecules [68], i.e. we can say that immediate-early genes not only modify the response of a cell to stimulation, but may transmit information from cell to cell [75].

Transcription factors can be subdivided into 2 groups: 1) "leucine zipper" or "basic zipper" ("bzip") proteins from *fos, jun,* and *myc* gene families; 2) "zinc finger" proteins which include KROX-24 (also termed NGFI-A, Zif/268, EGR-1, TIS 8); KROX-20 (EGR-2); EGR-3, NGFI-B (also termed NURR 77 or TIS 1); NGFI-C and NURR 1 [75].

The proteins from the *fos*, *jun*, and *krox* multigene families play a crucial role in the control of the cell cycle, development, growth, and differentiation of cells, and as well determine the destination of differentiated CNS neurons with a defined molecular phenotype [74]. According to recent findings, certain transcription factors are involved in neuronal death or survival programs after excitotoxic injury [34].

NF-kappaB is one of the most important transcription factors, the activation of which leads to expression of a variety of genes. Experimental studies *in vitro* and *in vivo* show that under conditions of brain ischemia, NF-kappaB may avert or initiate cell death. Thus, Irving *et al.* [43] observed NF-kappaB activation in ischemized neurons 3 h after constant occlusion of the middle cerebral artery in rats, and its decrease in dying neurons 6–48 h later. This suggested that inhibition of NF-kappaB activation promotes neuronal death in ischemia. However, Seegers *et al.* [87] found that 7 h after middle cerebral artery occlusion in rats NF-kappaB activation occurred only in dying cells of the penumbra; the expression of HSP70 protein was absent in the same neurons.

NF-kappaB consists of many protein subunits. The better studied subunits are p50 and p65 (Rel A). In the normal state these protein subunits are located in the cytosol combined with protein inhibiters IkappaB-α and IkappaB-β as inactive heterodimers [97]. *In vitro* NF-kappaB activation starts in response to IL-1β, TNF-α, free radicals [17], and activation of glutamate NMDA receptors [43]. Under conditions of cerebral ischemia TNF-α production dramatically increases in the affected neurons [14], whereas in the normal state it is synthesized more by astrocytes and microglial cells [13]. The interaction between TNF-α and cellular receptor p55 leads to activation of sphingomyelinase and to ceramide synthesis, which activates serine-threonine protein kinase. IkappaB is phosphorylated and the proteosome is destroyed [56]. Activated heterodimer moves into the nucleus, where it initiates expression of various genes, including those encoding synthesis of pro-inflammatory cytokines, adhesion molecules, as well as *c-myc* and MnSOD genes.

It was shown in the experimental models of ischemic stroke that the generalized cortical expression of *c-fos* and *zif/268* genes mRNA is quickly enhanced within the first hours after cerebral artery occlusion [35, 103], reaches a peak after 3–8 h, and then returns to baseline levels not later than 24 and 12 h after induced focal ischemia, respectively. Neurons in the ischemic penumbral zone express cellular immediate-early gene-encoded proteins at early post-ischemic time intervals, but synthesis of transcription factors remains stably elevated up to day 4 [100, 101].

Immediate-early gene expression with mRNA and protein induction is measured within a large territory of the total hemisphere homolateral to the occlusion. Several authors have detected it in the ischemic core [3, 105]. Thus, Jacewicz *et al.* [45] found hardly detectable levels of *c-fos* mRNA in the ischemic core, which correlated with the extent of CBF decrease. Also, authors have demonstrated the absence of protein synthesis in the ischemic core [23, 36, 100, 101]. Moreover, the preserved capacity of cells in the ischemized area for *de novo* synthesis of transcription factors encoded by immediate-early genes is regarded to be the most exact and sensitive criterion of cell viability.

Interesting findings were acquired about the elevated expression of immediate-early genes in areas remote from the ischemized area. According to most authors, this phenomenon is connected with the self-spreading peri-infarction depolarization, i.e. with "spreading depression" [29, 72] (see Section 4.2). The identical reactions of immediate-early genes to experimental spreading depression [32–34, 37] and to glutamate NMDA-receptors activation is consistent with that [57, 77].

Heat-shock proteins (HSP), or "stress" proteins, are encoded by the family of evolutionary conserved genes which are expressed in response to various stress influences and which are involved in adaptation mechanisms [28, 64]. Being first discovered in heat-shock in drosophila [82], stress proteins take part in the majority of physiological processes in all living organisms [70] and are component of the unified signaling mechanism [4, 11, 62, 83].

Activation of transcription factors for HSP (heat-shock factors, HSF) occurs via their phosphorylation under conditions of elevation of intracellular calcium concentration [79], free radical reactions [15, 18, 52, 98], lipid peroxidation [16], other events of oxidant stress [21], and activation of protease inhibitors [110] and tyrosine kinases [79]. But the main trigger of HSP synthesis is ATP deficit that accompanies insufficient delivery of glucose and oxygen to brain tissue [10].

There are several classes of HSF. HSF-1 mediates the response to stress, and HSF-2 regulates *hsp*-genes [71, 74, 85, 86]. In cerebral ischemia HSF-1 and HSF-2 synergistically activate gene transcription [28, 66, 92]. They constitute activated trimers, which bind to the regulatory sequences (HSE) in promoter-zones of "stress"-genes, leading to mRNA synthesis [80, 84, 93, 107]. The accumulation of HSP leads to the triggering of an auto-regulation loop which disrupts their further expression [8].

In experimental models of focal brain ischemia produced by middle cerebral artery occlusion it was established that expression of basic stress protein HSP72 is registered within a limited brain area with CBF decrease below 50% of normal value and only in still viable cells [20, 74, 76, 99]. In the ischemic core HSP72 are expressed predominantly in vascular cells, which are more resistant than neurons to ischemia. In the peri-infarction area, where the ischemia is less severe, HSP expression is also registered in glial cells; in the penumbral zone it is also detected in neurons [24, 54]. So, HSP72 are induced in cells that have survived the acute ischemic injury.

Expression of other stress proteins differs in various cell pools. Thus, HSP27 expression was predominantly registered in astrocytes [48, 49] and heme oxygenase-1 expression (HO-1, Hsp32) in microglia.

Recent studies suggest a neuroprotective role of HSP. It was shown that HSP72-positive cells in the penumbral area better survive under ischemic conditions [95]. Rats subjected to prior ischemic stress accompanied by induction of the non-constitutive HSP72 had smaller infarction volume from permanent middle cerebral artery occlusion than control (without prior ischemic stress) animals [91]. The appearance of HSP72 expression shows the border of viable tissue that is promising for therapy [11, 54, 83]. The mechanisms of neuroprotective action of HSP72 have been demonstrated to be connected with its inhibitory influence on the apoptosis induction pathways, including blocking translocation of apoptosis-inducing factor (AIF) from mitochondria to the nucleus and inhibiting activated caspase 3 (see Chapter 10, Scheme 10.1) [89]. Current therapeutic strategies based on protective and reparative influence of exogenous HSP72 on brain tissue as well as on overexpression of endogenous HSP72 are being developed in experimental ischemic models [60, 81, 88]. It was also revealed that HSP can act as a molecular conductor transferring other substances into brain cells [27, 39, 62]. Temporal therapeutic window for HSP72 neuroprotection has been shown to be at least 6 h after ischemia onset in rats [39]. Future strategies aimed at enhancing HSP72 expression or using exogenous HSP72 after stroke onset in a human might be worth pursuing.

The next wave of gene expression involves **genes encoding molecules that take part in all basic pathways of delayed cell death** (see Fig. 7.2). At this stage triggering of pro-inflammatory cytokine (TNF-α, IL-1β, IL-6, MCP-1, CINC) and adhesion molecule (ICAM-1, ELAM-1, P-selectin) synthesis occurs. These compounds induce local inflammation in the ischemic focus and additional disturbances of microcirculation and the blood–brain barrier [53, 58, 59] (see Chapter 8). The synthesis of some enzymes is also triggered. They include inducible NO-synthase (iNOS) and cyclooxygenase-2 (COX-2) that are involved in oxidant stress [26, 41, 42, 73]. Expression of pro-inflammatory cytokines and pro-oxidant enzymes predominantly takes place in infiltrating neutrophils, vascular cells, and microglia. It peaks 12 h after transient occlusion of the middle cerebral artery [41] and determines the corresponding chronology of the delayed consequences of ischemia (see Chapters 8–10).

It is now proven that immediate-early gene expression and HSP synthesis trigger the expression of **genes connected with programmed cell death** (see Chapter 10) [61, 74, 104]. Tight correlation between the triggering of apoptosis and the synthesis of the basic triggering factor for inflammation development, interleukin-1β (IL-1β), reflects the existence of common molecular induction of delayed consequences of ischemia.

Excessive synthesis of pro-inflammatory cytokines, especially IL-1β, directly influences the endocrine system, mediating the involvement of neurohormonal reactions in the common neuro–immune–endocrine stress-performing process (see Chapter 11).

REFERENCES

1. Abe, K., Araki, T., Kogure, K., 1988, *J Neurochem.* **51**: 1470–1476.
2. Akins, P. T., Liu, P. K., Hsu, C. Y., 1996, *Stroke.* **27**: 1682–1687.
3. An, G., Lin, T. N., Liu, J. S., *et al.*, 1993, *Ann Neurol.* **33**: 457–464.
4. Ananthan, J., Goldberg, A. L., Voellmy, R., 1986, *Science.* **232**: 522–524.
5. Araki, T., Kato, H., Inoue, T., Kogure, K., 1990, *Acta Neuropathol.* **79**: 501–505.

6. Ashmarin, I. P., Stukalov, P. V., 1996, *Neurochemistry*. Institute of Biological and Medical Chemistry of Russian Scientific Academy Publishing House, Moscow (in Russian).

7. Ashmarin, I. P., Stukalov, P. V., Eschenko, N. D., 1999, *Brain Biochemistry*. Publishing House of St. Petersburg University, St. Petersburg, 328 (in Russian).

8. Baler, R., Zou, J., Voellmy, R., 1996, *Cell Stress Chaperones*. **1**: 33–39.

9. Beato, M., Herriich, P., Schutz, G., 1995, *Cell*. **83**: 851–857.

10. Benjamin, I. J., Hone, S., Greenberg, M. L., *et al.*, 1992, *J Clin Invest*. **89**: 1685–1689.

11. Blumenfeld, K. S., Welsh, F. A., Harris, V. F., Pesenson, M. A., 1992, *J Cerebr Blood Flow Metab*. **12** (6): 987–995.

12. Bravo, R., 1990, *Cell Growth Diff*. **I**: 305–309.

13. Breder, C., Tsujimoto, M., Terano, Y., *et al.*, 1993, *J Comp Neurol*. **337** (4): 543–567.

14. Bruce, A., Boling, W., Kindy, M. S., *et al.*, 1996, *Nat Med*. **2** (7): 788–794.

15. Bruce, J. L., Price, B. D., Coleman, N., *et al.*, 1993, *Cancer Res*. **53**: 12–15.

16. Cajone, F., Salina, M., Benelli-Zazzera, A., 1989, *Biochem J*. **262**: 977–979.

17. Clemens, J. A., Stephenson, D. T., Dixon, E. P., *et al.*, 1997, *Brain Res Mol*. **48** (2): 187–196.

18. Courgeon, A.-M., Rollet, E., Becker, J., 1988, *Eur J Biochem*. **171**: 163–170.

19. Edwards, D. R., Mahadevan, L. C., 1992, *EMBO J*. **II**: 2415–2424.

20. Fredduzzi, S., Mariucci, G., Tantucci, M., *et al.*, 2001, *Exp Brain Res*. **136** (1): 19–24.

21. Freeman, M. L., Borrelli, M. J., Syed, K., 1995, *J Cell Physiol*. **164**: 356–366.

22. Furukawa, K., Estus, S., Fu, W., *et al.*, 1997, *J Cell Biol*. **136**: 1–13.

23. Gass, P., Spranger, M., Herdegen, T., *et al.*, 1992, *Acta Neuropathol*. **84**: 545–553.

24. Gonzalez, M. F., Shiraishi, K., Hisanaga, K., *et al.*, 1989, *Mol Brain Res*. **6**: 93–100.

25. Goto, K., Ishige, A., Sekiguch, K., *et al.*, 1990, *Brain Res*. **534**: 299–302.

26. Grandati, M., Verrecchia, C., Revaud, M. L., *et al.*, 1997, *J Cerebr Blood Flow Metab*. **17**: 94.

27. Guzhova, I., Kislakova, K., Moskaliova, O., *et al.*, 2001, *Brain Res*. **914**: 1–21.

28. Guzhova, I., Kislakova, K., Moskaliova, O., *et al.*, 2001, *Brain Res*. **914** (1-2): 66–73.

29. Hansen, A. J., 1985, *Physiol Rev*. **65**: 101–148.

30. Hata, R., Maeda, K., Hermann, D., *et al.*, 2000, *J Cerebr Blood Flow Metab*. **20** (2): 306–315.

31. Hata, R., Maeda, K., Hermann, D., *et al.*, 2000, *J Cerebr Blood Flow Metab*. **20** (6): 937–946.

32. Herdegen, T., Brecht, S., Mayer, B., *et al.*, 1993, *J Neurosci*. **13**: 4130–4146.

33. Herdegen, T., Sandkiihier, J., Gass, P., *et al.*, 1993, *J Comp Neurol*. **333**: 271–288.

34. Herdegen, T., Skene, P., Bahr, M., 1997, *Trends Neurosci*. **20**: 227–231.

35. Hermann, D. M., Kilic, E., Hata, R., *et al.*, 2001, *Neuroscience*. **104** (4): 947–955.

36. Herrera, D. G., Robertson, H. A., 1989, *Brain Res*. **503**: 205–213.

37. Herrera, D. G., Robertson, H. A., 1990, *Brain Res*. **510**: 166–170.

38. Herschman, H. R., 1991, *Annu Rev Biochem*. **68**: 281–319.

39. Hoehn, B., Ringer, T. M., Xu, L., *et al.*, 2001, *J Cerebr Blood Flow Metab*. **21** (11): 1303–1309.

40. Iadecola, C., 1999, Mechanisms of cerebral Ischemic damage. In *Cerebral Ischemia* (Walz, W., ed.) Humana Press, Totowa, New Jersey, pp. 3–33.

41. Iadecola, C., Xu, X., Zhang, F., *et al.*, 1995, *J Cerebr Blood Flow Metab*. **15**: 52–59.

42. Iadecola, C., Zhang, P., Casey, R., *et al.*, 1996, *Stroke*. **27**: 1373–1380.

43. Irving, E. A., Hadingham, S. J., Roberts, J., *et al.*, 2000, *Neurosci Lett*. **288** (1): 45–48.

44. Jacewicz, M., Kiessling, M., Pulsinelli, W. A.., 1986, *J Cerebr Blood Flow Metab.* **6**: 263–272.
45. Jacewicz, M., Takeda, Y., Nowak, T. S., Jr., Pulsinelli, W. A., 1993, *J Cerebr Blood Flow Metab.* **13**: 450.
46. Johansen, F. F., Diemer, N. H., 1990, *Acta Neuropathol* (Berl.). **81**: 14–19.
47. Kamiya, T., Jacewicz, M., Pulsinelli, W. A., Nowak, T. S., Jr., 1995, *J Cerebr Blood Flow Metab.* **15**: 1.
48. Kato, H., Kogure, K., Araki, T., *et al.*, 1995, *J Cerebr Blood Flow Metab.* **15**: 60–70.
49. Kato, H., Kogure, K., Liu, Y., *et al.*, 1994, *Brain Res.* **652**: 71–75.
50. Kato, H., Kogure, K., Liu, X.-H., *et al.*, 1995, *Brain Res.* **679**: 1–7.
51. Kiessling, M., Gass, P., 1994, *Brain Pathol.* **4**: 77–83.
52. Kil, H. Y., Zhang, J., Piantadosi, C. A., 1996, *J Cerebr Blood Flow Metab.* **16**: 100–106.
53. Kim, J. S., 1996, *J Neural Sci.* **137**: 69–78.
54. Kinouchi, H., Sharp, P., Koistinaho, J., *et al.*, 1993, *Brain Res.* **619**: 334–338.
55. Koistinaho, J., Hokfelt, T., 1997, *Neuro Report.* **8** (2): 1–8.
56. Kolesnick, R., Golde, D. W., 1994, *Cell.* **77**: 325-328.
57. Lauritzen, M., Hansen, A. J., 1992, *J Cerebr Blood Flow Metab.* **12**: 223–229.
58. Liu, T., Dark, R. K., McDonnell, P. C., *et al.*, 1994, *Stroke.* **25**: 1481–1488.
59. Liu, T., McDonnell, P. C., Young, P. R., *et al.*, 1993, *Stroke.* **24**: 1746–1751.
60. Lu, A., Ran, A., Parmentier-Batteur, S., *et al.*, 2002, *J Neurochem.* **81** (2): 355–364.
61. MacManus, J. P., Linnik, M. D., 1997, *J Cerebr Blood Flow Metab.* **17**: 815–832.
62. Margulis, B., Guzhova, I., 2000, *Cytology.* **42** (4): 323-342 (in Russian).
63. Martin, L. J., *et al.*, 1998, *Brain Res Bull.* **46**: 281–309.
64. Massa, S. M., Swanson, R. A., Sharp, P. R., 1996, *Cerebrovasc Brain Metab Rev.* **8**: 95–158.
65. Matsumoto, K., Yamada, K., Hayakawa, T., 1990, *Neurol Res.* **12**: 45–48.
66. Mestril, R., Ch, S.-H., Sayen, R., Dillmann, W. H., 1994, *Biochem J.* **298**: 561–569.
67. Mies, G., Ishimaru, S., Xie, Y., *et al.*, 1991, *J Cerebr Blood Flow Metab*, 753–761.
68. Morgan, J. I., Curran, T., 1991, *Annu Rev Neurol.* **14**: 421–451.
69. Morgan, J. I., Curran, T., 1995, *The Neuroscientist.* **1**: 68–75.
70. Morimoto, R., Tissieres, A., Georgopoulos, C., 1994, *The Biology of Heat Shock Proteins and Molecular Chaperones.* Cold Spring Harbor Laboratory Press, Plainview, New York.
71. Nakai, A., Morimoto, R., 1993, *Mol Cell Biol.* **13**: 1983–1997.
72. Nedergaard, M., Hansen, A. J., 1993, *J Cerebr Blood Flow Metab.* **13**: 568–574.
73. Nogawa, S., Zhang, F., Ross, M. E., Iadecola, C., 1997, *J Neurosci.* **17**: 2746–2755.
74. Nowak, T. S., Jacewicz, M., 1994, *Brain Pathol.* **4**: 67–76.
75. Nowak, T. S., Jr., Kiessling, M., 1999, Reprogramming of gene expression after ischemia. In *Cerebral Ischemia* (Walz, W., ed.) Humana Press, Totowa, New Jersey, pp. 145–217.
76. Nowak, T. S., Zhou, Q., *et al.*, 1999, Regulation of heat shock genes by ischemia. In *Stress Proteins (Handbook of Experimental Pharmacology)* Vol. 136, Springer, Heidelberg, Germany.
77. Obrenovitch, T. P., Zilkha, E., Urenjak, J., 1996, *J Cerebr Blood Flow Metab.* **160**: 923–931.
78. Papas, S., Crepel, V., Hasboun, D., *et al.*, 1992, *Eur J Neurosci.* **4**: 758–765.
79. Price, B. D., Calderwood, S. K., 1991, *Cell Biol.* **II**: 3365–3368.
80. Rabindran, S. K., Haroun, R. I., Clos, J., 1993, *Science.* **259**: 230–234.
81. Rajdev, S., Hara, K., Kokubo, Y., *et al.*, 2000, *Ann Neurol.* **47** (6): 782–791.
82. Ritossa, F., 1962, *Experientia.* **18**: 571–573.
83. Sanz, O., Estrada, F., Ferrer, I., *et al.*, 1997, *Neurosci.* **80** (1): 221-232.

84. Sarge, K. D., Murphy, S., Morimoto, R. I., 1993, *Mol Cell Biol.* **13**: 1392–1407.
85. Scharf, K.-D., Rose, S., Zott, W., *et al.*, 1990, *EMBO J.* **9**: 4495–4501.
86. Schuetz, T. J., Gallo, G. J., Sheldon, L., *et al.*, 1991, *Proc Natl Acad Sci USA.* **88**: 6911–6915.
87. Seegers, H., Grillon, E., Trioullier, Y., *et al.*, 2000, *Neurosci Lett.* **288** (3): 241-245.
88. Sharp, F. R., 1998, *Ann Neurol.* **44** (4): 581–583.
89. Sharp, F. R., Massa, S. M., Swanson, R. A., 1999, *Trends Neurosci.* **22** (3): 97–99.
90. Shigeno, T., Yamasaki, Y., Kato, G., *et al.*, 1990, *Neurosci Lett.* **120**: 117–119.
91. Simon, R. P., Niiro, M., Gwinn, R., 1993, *Neurosci Lett.* **163** (2): 135-137.
92. Sistonen, L., Sarge, K. D., Morimoto, R. I., 1994, *Mol Cell Biol.* **14**: 2087–2099.
93. Sorger, P. K., Pelham, H. R. B., 1987, *EMBO J.* **6**: 3035–3041.
94. Stabberod, P., Tomasevic, G., Kamme, F., Wieloch, T., 1994, *Abst Soc Neurosci.* **20**: 616.
95. States, B. A., Honkaniemi, J., Weinstein, P. R., Sharp, F. R., 1996, *J Cerebr Blood Flow Metab.* **16**: 1165–1175.
96. Steng, M., Greenberg, M. E., 1990, *Neuron.* **4** (477): 44–85.
97. Stephenson, D., Yin, T., Smalstig, E. B., *et al.*, 2000, *J Cerebr Flow Metab Mar.* **20** (3): 592–603.
98. Suga, S., Nowak, T. S., Jr., 1998, *Neurosci Lett.* **243**: 57–60.
99. Torregrosa, G., Barbera, M. D., Orti, M., *et al.*, 2001, *Hippocampus.* **11** (2): 146–156.
100. Uemura, Y., Kowall, N. W., Beal, M. F., 1991, *Brain Res.* **542**: 343–347.
101. Uemura, Y., Kowall, N. W., Moskowitz, M. A., 1991, *Brain Res.* **552**: 99–105.
102. Vandromme, M., Gauthier-Rouviere, C., Lamb, N., Femandez, A., 1996, *Trends Biol Sci.* **21**: 59–64.
103. Wang, X., Yue, T. L., Barone, F. C., *et al.*, 1994, *Mol Chem Neuropathol.* **23** (2-3): 14–103.
104. Waters, C., 1997, *RBI Neurotransmissions, Newsletter for Neuroscientists.* **XIII** (2): 2–7.
105. Welsh, F. A., Moyer, D. J., Harris, V. A., 1992, *J Cerebr Blood Flow Metab.* **12**: 204–212.
106. Widmann, R., Kuroiwa, T., Bonnekoh, P., Hossmann, K. A., 1991, *J Neurochem.* **56**: 789–796.
107. Wu, C., Wilson, S., Walker, B., *et al.*, 1987, *Science.* **238**: 1247–1253.
108. Xie, Y., Mies, G., Hossmann, K. A., 1989, *Stroke.* **20**: 620–626.
109. Yoshimine, T., Hayakawa, T., Kato, A., *et al.*, 1987, *J Cerebr Blood Flow Metab.* **1**: 387–393.
110. Zhou, M., Wu, X., Ginsberg, H. N., 1996, *J Biol Chem.* **271**: 24769–24775.

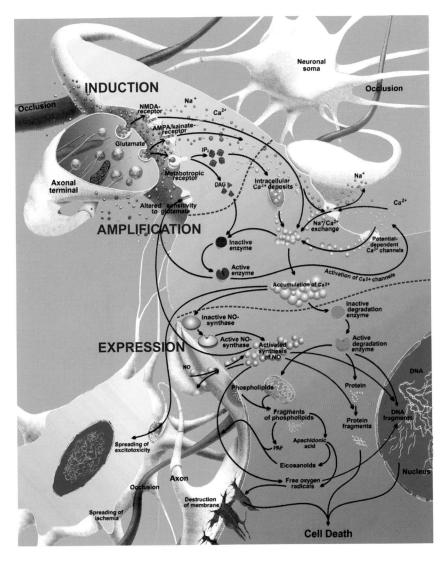

Plate 1. Stages of the glutamate–calcium cascade (modified from J.A.Zivin, D.W.Choi, 1991 [253]).

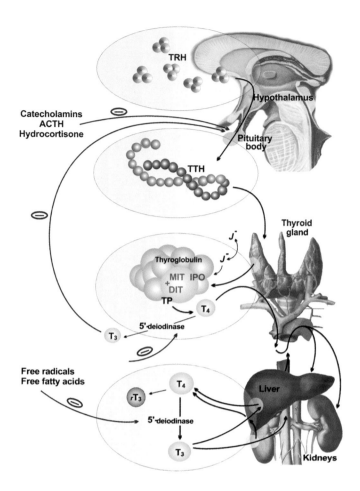

Plate 2. Formation of "low T$_3$-syndrome". Under physiological conditions the only source of free endogenous thyroxine (T$_4$) is the thyroid gland. The formation of T$_4$ in thyroid gland includes oxidation of iodide (I$^-$) delivered into follicles by iodide peroxidase (IPO), generation of inactive T$_4$ precursors – monoiodothyrosine (MIT) and diiodothyrosine (DIT), their following condensation and release into blood by thyroid protease (TP). In contrary, only 20% of triiodothyronine (T$_3$) synthesized in human body are delivered from thyroid gland. The rest of it is generated in extra-thyroid organs (liver and kidneys) via monodeiodination of the external ring of T$_4$ by specific 5'-deiodinase. Simultaneously about 40% of T$_4$ undergo monodeiodination in the 5th site of the internal ring of T$_4$ after which metabolic inactive reversible T$_3$ (rT$_3$) is generated. When free radicals and free fatty acids concentrations are elevated in human organism, for instance, in focal brain ischemia, 5'-deiodinase becomes inhibited, the conversion of T$_4$ to T$_3$ transformation occurs, and low T$_3$-syndrome develops. The rate of rT$_3$ degradation decreases. Along with this, ACTH, hydrocortisone and catecholamines decrease the sensitivity of pituitary to stimulating effects of thyrotropin releasing hormone (TRH) that also promotes low T$_3$-syndrome development.

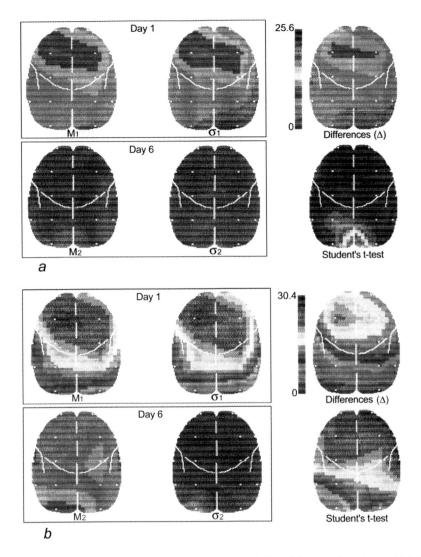

Plate 3. Statistical comparison of averaged EEG maps in δ- and θ-ranges in patients with left hemispheric ischemic stroke before (day 1) and after (day 6) Glycine therapy. a) δ-range; b) θ-range. Here and on Figs. 14.3–14.5, 15.5: M_1 and M_2 – averaged maps; σ_1 and σ_2 – standard deviations.

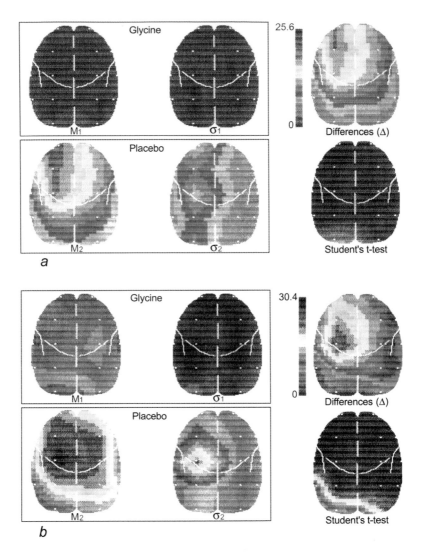

Plate 4. Statistical comparison of averaged EEG maps in δ- and θ-ranges on day 6 in patients with left hemispheric ischemic stroke treated with Glycine or placebo. a) δ-range; b) θ-range.

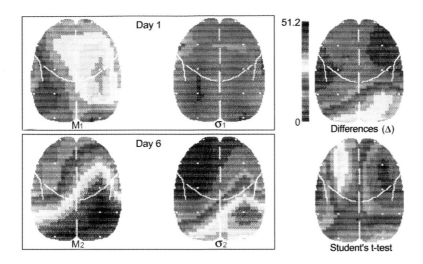

Plate 5. Statistical comparison of averaged EEG maps in α-range in patients with left hemispheric ischemic stroke before (day 1) and after (day 6) Glycine therapy.

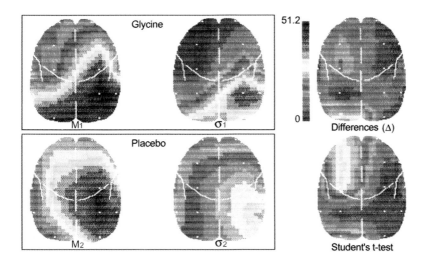

Plate 6. Statistical comparison of averaged EEG maps in α-range on day 6 in patients with left hemispheric ischemic stroke treated with Glycine or placebo.

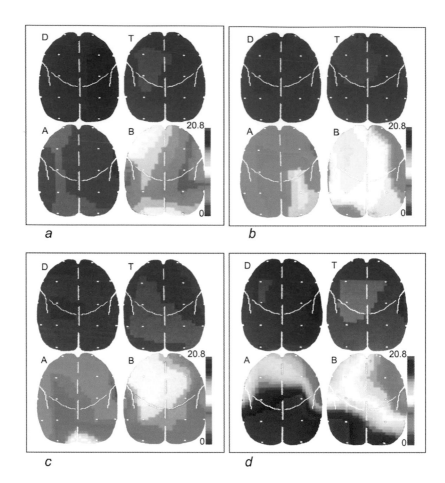

Plate 7. EEG maps of patient K., 49-year-old, treated with Semax (12 mg/day). a) The first 6 h after stroke in the territory of the left middle cerebral artery (before Semax administration): low total energy level of EEG (<20 μV^2/Hz); low power and disordered zonal distribution of α-rhythm; low power of slow frequency ranges, moderate inter-hemispheric asymmetry (IHA) in θ-rhythm, S > D; b) 1.5 h after the first administration of Semax (12 mg): increase in power of α_1-range (7–10 Hz) by 40–53% (versus the baseline level on admission) predominantly in occipital and central zones of "intact hemisphere" with formation of IHA, S > D; no tendency towards an elevation of θ-range power is observed; c) 3 h after the first administration of Semax: steady power of α_1-range at the level exceeding that at baseline by 30–35%; tendency towards IHA regress in α- and θ-rhythms; d) on day 6: increase in total energy level of EEG with dominant frequency in α_1-range (9 Hz); bilateral increase in α-range power with complete regress of IHA in α-rhythm and normalization of its zonal distribution; tendency towards an increase in δ- and θ-rhythms power is absent; IHA in θ-rhythm is preserved on the previous level. Here and on Figs. 15.17, 15.18, 15.20, and 16.4: frequency ranges – δ = 1.4–3.9 Hz; θ = 4–6.9 Hz; α = 7–11.9 Hz; β = 12–31 Hz.

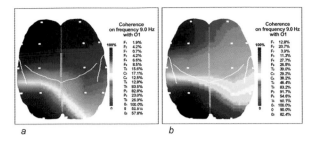

Plate 8. Coherent analysis of dominant α-rhythm (9 Hz) between the occipital derivation of the affected hemisphere (O1) and the other EEG derivations: a) before the first administration of Semax (12 mg); b) 1.5 h after the first administration. Increase in α-waves coherence in the affected hemisphere (with T3, C3, P3, F1, F7 derivations) and between hemispheres (with O2, T6, P4, T4, C4, F8, etc.).

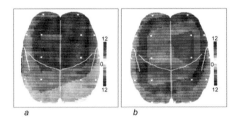

Plate 9. P$_{240}$ SSEP component maps before (a) and after (b) Semax (12 mg/day) treatment. Increase in amplitude, decrease in latency and normalization of zonal distribution of P240 peak.

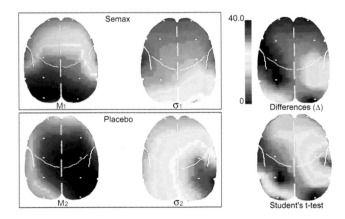

Plate 10. Statistical comparison of the averaged EEG maps in α-range on day 6 after the onset of left hemispheric stroke in patients treated with Semax (12 mg/day) and placebo.

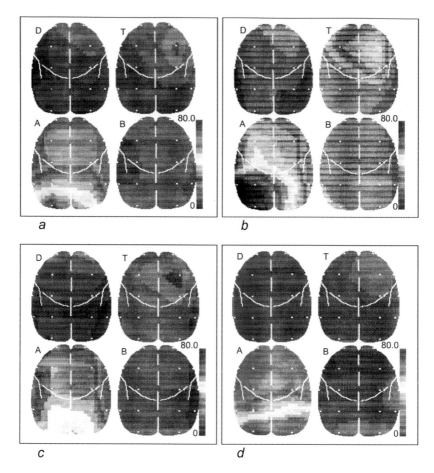

Plate 11. EEG maps of patient B., 67-year-old, on the background of Cerebrolysin treatment: a) the first 12 h after the stroke in the territory of the right middle cerebral artery (before Cerebrolysin administration): middle total energy level of EEG (<40–60 μV^2/Hz); moderate inter-hemispheric asymmetry (IHA) in α-rhythm in occipital lobes, D > S; θ-focus in the projection of the ischemic lesion; b) 40 min after the first administration of 2 ml of Cerebrolysin; c) 40 min after additional administration of 8 ml of Cerebrolysin; d) on day 6 of stroke (after the end of treatment with 10 ml/day Cerebrolysin) – normalization of EEG pattern: steady bilateral increase in power of α-range with complete regress of IHA in α-rhythm; decrease in power of focal δ- and θ-rhythms.

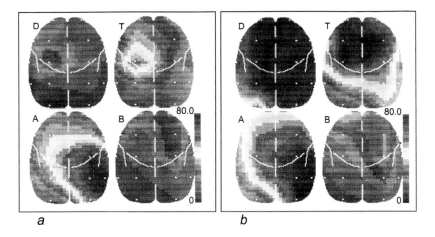

Plate 12. EEG maps of patient S., 63-year-old, on the background of Cerebrolysin treatment: a) the first 4 h after the stroke in the territory of the left middle cerebral artery (before Cerebrolysin administration): middle total energy level of EEG (40–60 μV^2/Hz); prominent IHA in α-1-range, D > S; early formation of focus of δ- and θ-activity in the projection of ischemic lesion; b) 2 h after the first administration of 30 ml of Cerebrolysin: prominent bilateral increase in power of all frequency ranges, especially of slow (δ-, θ-); "slowing" of δ-, θ-activity in the projection of ischemic lesion.

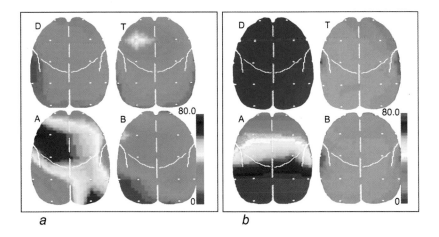

Plate 13. EEG maps of patient O., 56-year-old, on the background of Cronassial treatment: a) the first 5 h after the stroke in deep branches of the left middle cerebral artery (before treatment): middle total energy level of EEG; IHA in α-1-range in occipital (D > S) and central (S > D) areas; early formation of focus of α-θ-activity (6–10 Hz) in the projection of ischemic zone; b) on day 6 – complete normalization of EEG pattern: regress of focal EEG changes; normalization of zonal distribution of dominant α-rhythm.

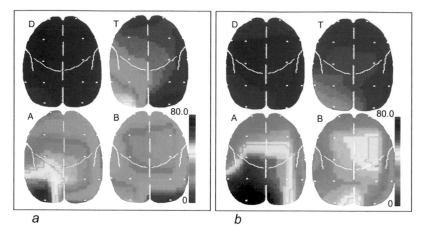

Plate 14. EEG maps of patient B., 48-year-old, on the background of piracetam treatment: a) first 10 h after the stroke in the territory of the right middle cerebral artery (before piracetam administration): middle total energy level of EEG (40–60 μV^2/Hz); prominent IHA in α-1-range, D < S; b) on day 30 after the stroke onset (after the end of piracetam treatment): bilateral increase in power and frequency ("acceleration") of α-range; improvement of zonal distribution of α-activity.

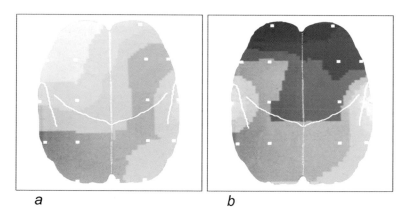

Plate 15. Zonal distribution of the amplitude of pre-Rolandic SSEP component N30 in patient P., 63-year-old, with ischemic stroke in the territory of the left internal carotid artery, treated with piracetam: a) first 12 h after the event (before piracetam administration); b) on day 30 after the stroke onset (after the end of piracetam treatment) – an increase in amplitude, a decrease in latency, and normalization of zonal distribution of N30.

Chapter 8

Microglial Activation, Cytokine Production, and Local Inflammation in Focal Brain Ischemia

Studies of Baron *et al.* [11], Chen *et al.* [29], and Fisher and Garcia [46] demonstrated that within the first hours after the onset of stroke there are neurons with structure and functions changed to different extent in the damaged brain area, and the extent of these changes decreases from the central zone of the ischemic core towards the peripheral parts of the penumbra. Regularities in the distribution of these structural and functional changes throughout the penumbra determine the final size of the brain infarction formed by the 3rd to 5th days after stroke onset. These changes must be studied to establish the basis for developing effective neuroprotective therapeutic approaches [7, 46, 80, 148, 149, 158].

In the penumbral area glial cells are damaged more prominently than neurons [136]. The "up-formation" of a brain infarction is to a considerable extent connected with the aggressive effects of glial cells excited by ischemia on viable neurons in the peri-infarction zone [53, 140].

Being the only immune competent compartment of the CNS, resident microglia has a very low threshold for activation and swiftly responds to neuronal damage, taking part in all metabolic ischemic cascades [47, 60, 185]. The ischemic process activates microglial cells, preparing them to proceed as phagocytes. In experimental models of ischemic stroke it was shown that microglia being activated causes a prominent neurotoxic effect contributing much to damage to the penumbra [58, 118]. According to Banati *et al.* [10], this neurotoxic effect of microglia occurs via three basic mechanisms: 1) production of direct neurotoxic factors; 2) production of microglial factors which trigger pathological biochemical cascades that lead to cell death, and 3) induction of local inflammatory response.

115

It has been demonstrated that microglia begins to produce a wide range of neurotoxic compounds under conditions of experimental cerebral ischemia. These agents are pro-inflammatory cytokines [3], ligands for glutamate NMDA receptor complex [55, 56, 114], proteases, cathepsin B, lysozymes, eicosanoids (including thromboxane B_2), superoxide anion, and NO$^\bullet$. Also, microglia initiates the cytotoxic effects of astrocytes [91, 163]. Analyzing the list of compounds being synthesized by microglial cells, one can see active and concordant with other cell pools effects in the participation of activated microglia in all basic events of the glutamate–calcium cascade. This participation maintains glutamate excitotoxicity, the activation of intracellular enzymes, free radical reactions, and lipid peroxidation (see Scheme 4.1). Along with this, microglia fulfils specific immune functions inducing and supporting local inflammatory reaction in the ischemized brain area that leads to delayed neuronal loss and to aggravation of changes of microcirculation and the blood–brain barrier [111, 112, 181].

An abrupt elevation of the level of pro-inflammatory agents was revealed in the experimental focal ischemic zone in animals' brains [13, 45, 66, 125], which formed the conditions for acute and delayed death of the cells encircling the initial necrosis. Also, it was shown that even a modest increase in pro-inflammatory cytokine concentrations mediates the progression of atherogenesis and the development of chronic ischemia of the brain leading to the formation of encephalopathy [61, 99, 110, 182]. Thus, inflammatory reaction as well as oxidant stress closes the vicious circle of the persistent extension of cerebrovascular disease.

The utmost significance of microglial reactions in the enlargement of brain infarction makes them a unique cellular therapeutic target in terms of stroke treatment [113, 154, 175, 176, 186]. This was confirmed by a decrease in neuronal loss in ischemized rat brain registered when microglia was pharmacologically inhibited [56, 98]. However, complete inhibition of microglia can by no means be applied in clinical practice as its functions required for normal life maintenance of the brain are so manifold. Glia provides a system of trophic support as well as being actively involved in the specific functioning of neural tissue [49]: in the normal state it inhibits neuronal hyperactivity [32] and mediates active consumption and re-uptake of mediators and other agents contributing to neuronal damage from the synaptic cleft [134]. Under conditions of ischemia, microglial cells induce not only the synthesis of neurotoxic compounds, but also of signal molecules, cell regulators, and trophic factors which promote neuronal survival and inhibit the formation of post-ischemic cicatrices [19, 100]. The substances which cause not only damage, but also the system of viability support in the ischemic lesion, are represented by the wide range of regulatory peptides including cytokines and neurotrophic and modulatory factors [96].

Cytokines are specific immune mediators of small soluble peptide structure. They are produced by both immunocompetent (T-cells, macrophages, microglial cells) and non-immunocompetent (neurons, astrocytes) cells [92]. To date over 40 cytokines have been identified. They can be subdivided onto several large families: interleukins (ILs), tumor necrosis factors (TNFs), interferons (IFNs), and trophic factors. It was conditionally decided to divide cytokines onto regulatory, or anti-inflammatory (IL-10, IL-4, trophic factors) and pro-inflammatory (IL-1, IL-6, IL-8, TNFs, etc.) [38, 63, 86] groups.

In brain, cytokines are arranged in neural and glial compartments. They are also widely distributed within discrete networks. They act in both an autocrine (e.g., stimulate further proliferation of monocytes) and a paracrine (e.g., influence on neurons and endothelial cells) mode. Binding to specific receptors, they activate systems of intracellular second messengers, including protein kinases and phosphatases [130].

It is noteworthy that IL-1, IL-6, TNF-α, and some other cytokines regulate the function of the hypothalamic–pituitary system, directly influencing the adenohypophysis and promoting conjunction between the neuro–immune and neuro–endocrine systems [4] (Fig. 8.1). It has been proven that one of the most powerful cytokines, IL-1, influences every level of the hypothalamic–pituitary–adrenal system increasing the secretion of corticotropin releasing hormone, adrenocorticotropic hormone, glucocorticoids [4–6], luteinizing hormone releasing factor [105], and possibly some other hormones: somatotropic, follicle stimulating, luteinic hormones, and prolactin [169]. Thus, it can be suggested that immune and neuro–endocrine cells use common signal systems. Tight interaction between cytokines and hormones at the central and the peripheral levels cause hormonal responses to the increase in expression of cytokines. This explains why stress-mediating endocrine systems are triggered in response to ischemia within the common reaction of the neuro–immune–endocrine system (see Chapter 11).

It is hard to simplify the concept of cytokine function because these agents are so pleiotropic. Cytokines have a multitude of different biological effects depending on the extent of cytokine release and the target cell population. Because of this inherent pleiotropy, only experimental manipulation of cytokine levels and the establishment of specific knockout mice have helped to identify their physiological and pathological roles. Such studies have revealed that cytokines control growth, differentiation, and functioning of the majority of cell types; they modulate neuro–endocrine and

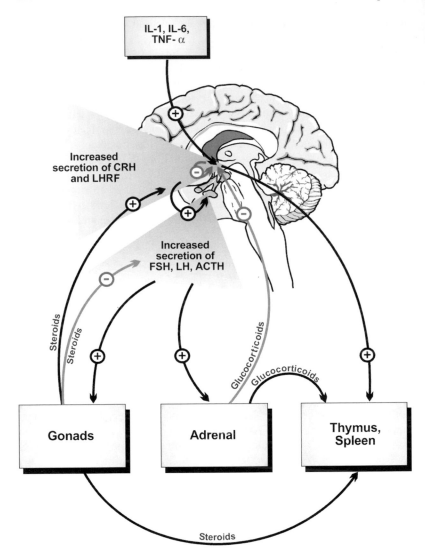

Figure 8.1. Influence of cytokines on the embodiment of hormonal response to acute brain ischemia. CRH – corticotropin releasing hormone; LHRF – luteinizing hormone releasing factor; FSH – follicle stimulating hormone; LH – luteinizing hormone; ACTH – adrenocorticotropic hormone. Black arrows indicate stimulating projections, grey – inhibitory projections.

thermal regulation, sleep, feeding, stable body weight maintenance, cognitive functions, behavior, and tolerability. Immune mediators play a crucial role in triggering inflammatory reactions and microcirculation disturbances as well as in changes in the blood–brain barrier permeability, and also in mechanisms of cell survival and death.

The family of IL-1 ligands comprises three isoforms: IL-1α, IL-1β, and IL-1ra (IL-1 receptor antagonist), which are encoded by three separate genes but bind to IL-1 receptor type I (IL-1RI) with comparable affinity [35]. IL-1α and IL-1β are biologically active and cause different reactions in target cells. IL-1β is synthesized as a large molecule that is biologically inactive; it is cleaved into the 17 kDa bioactive mature IL-1β by IL-1β converting enzyme (ICE). IL-1 is produced in microdoses in healthy individuals, mediates the induction of such neuronal antioxidant as superoxide dismutase, and promotes astroglial proliferation [92]. Under pathological conditions including ischemia, microglial IL-1 production is the main activating signal for the induction of other pro-inflammatory cytokines as well as for the stimulation of astrocytes to the production of potentially neurotoxic compounds such as NO$^{\bullet}$ and arachidonic acid metabolites [184]. As shown recently, marked IL-1 gene expression in the brain causes secondary marked expression of the gene encoding iNOS, which results in excessive synthesis of NO$^{\bullet}$ metabolites [184].

IL-1 can be either neurotoxic or neuroprotective [18] depending on the phase of development of the animal. In the adult animal, IL-1β increased under conditions of global and focal acute ischemia can be highly neurotoxic [25, 116]. Recombinant human IL-1β injected into cerebral ventricles of rat that underwent 60 min long occlusion of middle cerebral artery with subsequent reperfusion significantly elevated brain edema and the size of brain infarction as well as the number of neutrophils adherent to endothelium and infiltrated into the ischemized area [190]. The extent of these effects exhibited linear correlation with the injected dose of the cytokine.

However, in response to IL-1β expression, the IL-1ra isoform is secreted in the adenohypophysis. IL-1ra, in contrast to IL-1β, appears to be an endogenous neuroprotective agent. According to experimental data, the protective effect of exogenous IL-1ra is manifested in 40% of acute focal ischemia cases [184]. Whereas the biological activity of endogenous IL-1ra is neutralized during the course of acute neurodegeneration, there is considerable increase in the area of neuronal loss caused by middle cerebral artery occlusion [104]. It has been proven that IL-1ra mRNA always accompanies IL-1β mRNA. Neurotoxic effects of IL-1β, and inhibition by IL-1ra, may be due to actions of neurons or alternatively, those effects could be mediated at the level of non-neuronal cells such as glial and endothelial cells, leading to production of neurotoxic factors, and/or direct actions on neurons [183]. The understanding of molecular pathways of induction of IL-1β and IL-1ra expression in the brain within the first hours after stroke onset will allow new therapeutic strategies aimed at averting delayed cell death to be worked out.

Tumor necrosis factor (TNF) is structurally unrelated to IL-1 and binds to different receptors than those of IL-1, but both cytokines have similar biological effects on fever, sickness behavior, and neuroendocrine function [184].

Two forms of TNF, TNF-α and TNF-β, bind to the same receptor on target cells and have the same biological activities even though they have only 28% similarity in amino acid sequence [184]. The genes encoding both TNF-α and TNF-β are found within the Major Histocompatibility Complex (MHC) on chromosome 6. Inducers of TNF production include IL-1, IL-2, interferons, endotoxin, and mitogens. Suppressors of TNF production include IL-4, IL-6, transforming growth factor $β_1$ (TGF-$β_1$), and dexamethasone [184].

TNF-α, which is a pleiotropic polypeptide produced by several cell types in the immune system and in glial cells and astrocytes, as well as in endothelial cells and in some neurons [26], in many ways resembles the action of IL-1 [44, 102]. Due to wide distribution of receptors to TNF-α, it activates various pathways of signal transduction: it induces or inhibits the expression of transcription and growth factors, of cytokines and their receptors, as well as of acute phase proteins that can explain a complex immune modulatory function of this cytokine [106]. Jones and Selby [83] revealed mitogenicity of TNF-α for certain cell types.

Patients with ischemic stroke exhibit elevated serum contents of TNF-α versus healthy individuals [41]. Already 1 h after reperfusion in the ischemized cortex of rat after occlusion of the middle cerebral artery, significant and considerable increase in IL-1β and TNF-α mRNA has been found; increased levels persisted for 3-5 days following ischemic injury [44, 102]. Additional production of TNF-α in blood vessels and astrocytes is registered in the ischemized area. This production is accompanied by active inflammatory reaction [20], and TNF-α content in microglial cells varies in relation to acuity of inflammation. The immediate and delayed pattern of TNF-α expression following ischemia probably reflects different sources of TNF-α production. After neuronal injury TNF-α mRNA is rapidly (1–4 h) increased in affected neurons and reaches a peak 6–12 h after ischemia onset [21, 116, 161]; protracted expression (2–5 days) is detected in microglia and astrocytes surrounding the damaged tissue [87, 102]. Bruce-Keller and Mattson [22] noted that the precise mechanisms of TNF-α induction, in either neurons or microglia, are not well resolved, yet evidence suggests that TNF-α expression is responsive directly to signals from injured neurons. The promoter motifs of TNF-α contain elements responsive to the transcription factor NF-kappaB, which is known to be activated very rapidly following oxidative stress, such as that induced by ischemia [144, 162]. It was also shown that over-activation of NMDA receptors also triggers TNF-α expression in damaged neurons [22]. Intracellular signals, such as pro-inflammatory cytokines and growth factors released during ischemia, can also induce TNF-α expression. IL-1β induces TNF-α expression through a mechanism involving protein kinase C activation [22]. Moreover, once formed TNF-α induces its own expression via activation of NF-kappaB.

It was shown that TNF-α takes part in damage to myelin and oligodendrocytes [44, 102]. Several experiments conducted on models of acute focal ischemia demonstrated that such cytokines as IL-1α and IL-1β as well as TNF-α appear to be pivotal mediators of microglial neuro–immune functions [14, 16, 141] and are produced locally in response to brain ischemia [191]. However, the interrelation between the neuroprotective and the neurotoxic properties of these cytokines remains to be understood. In experiments carried out *in vitro* [184] it was shown that TNF-α did not only not cause any toxic effect on CNS neurons, but, moreover, mediated regeneration of damaged axons and protected cultured neurons. Bruce *et al.* [21] demonstrated in experiments that the size of brain infarction occurring after occlusion of the middle cerebral artery in mice was larger in animals genetically deficient in TNF-α receptors, and the injection of the cytokine 48 h before the occlusion appeared to be neuroprotective. The increase in the extent and the distribution of neuronal damage in the TNF-α-negative mice was due to escalation of oxidant stress and by decrease of antioxidant enzyme levels [21, 23]. This suggested that the neuroprotective properties of TNF-α was connected with stimulation of antioxidant pathways. Carlson at al. [27] also showed that TNF-α increases neuronal survival to a toxic influx of calcium mediated through NMDA glutamate-gated ion channels, which was a major contributor to neuronal death following ischemic stroke. Neuroprotection by this cytokine required both activation of the p55/TNF receptor type I and the release of TNF-α from neurons. They concluded that drugs directed to the signal pathways of TNF-α may be useful in stroke treatment.

Thus, at present it appears that the role of TNF-α in ischemic stroke is complex. Increase in TNF-α content in macrophages in the infarction tissue and in neurons of the ischemized brain area, on one hand, causes the expression of pro-adhesion molecules on endothelium that leads to accumulation and adhesion of leukocytes, to their migration out of capillaries into brain tissue; on the other hand, it mediates tissue remodeling and regulates gliosis and the formation of a cicatrix [44, 102]. This led Pantoni *et al.* [130] to conclude that in order to improve subsequent recovery of a neurological deficit it is necessary to maintain elevated levels of TNF-α within the closest time interval after ischemia onset.

The studies of IL-6, a pleiotropic protein produced by active monocytes, macrophages, and endothelial cells [79], in experimental cerebral ischemia proved its importance in the regulation of acute phase response and the formation of infarction together with the importance of IL-1 and TNF-α [1]. Wang *et al.* [172] revealed in the ischemized cerebral cortex of rats significant elevation of IL-6 mRNA 3 h after the constant occlusion of the middle cerebral artery which increased 10-fold 12 h later and remained stably elevated only in the ischemized area for not less than 24 h. It was shown that IL-1 and TNF-α are powerful inducers of IL-6-synthase in astrocytes [77]: a quick increase in microglial IL-1β content within the first 1-2 h after brain ischemia onset induces synthesis of IL-6, elevated concentration of which is detected on a delayed time interval, only 3–6 h later.

Increased concentrations of IL-8, a protein of C-X-C chemokines family, in patients with chronic ischemic brain disease are also of clear interest. Structurally, this protein represents a combination of cytokine and leukocyte chemoattractant [13, 43, 178]. IL-8 possesses marked pro-inflammatory properties [51, 145]. It leads to expression of molecules taking part in intercellular adhesion and enhances adherence of neutrophils to endothelial cells and sub-endothelial matrix proteins. These effects of IL-8 reflect its crucial role in mediation of inflammatory response and in brain edema development [15, 50, 95, 137, 138].

Thus, experimental studies revealed activation of pro-inflammatory cytokine (IL-1, TNF-α, IL-6, IL-8) systems under the conditions of acute focal brain ischemia which is accompanied by the development of local inflammation in the ischemized lesion. The levels of pro-inflammatory cytokines remain significantly elevated for several days after stroke onset, which reflects the intensity of inflammatory reactions and their role in brain tissue damage [171].

One of the most important stages of the arrangement of local inflammatory response and of secondary damage to brain tissue is the activation of endothelium via pro-inflammatory cytokines [87, 173]. As a result, the induction of intercellular adhesion molecule-1 (ICAM-1) and vascular cell adhesion molecule-1 (VCAM-1) ensue on the endothelial surface [122, 123]. TNF-α and interferon-γ (IFN- γ) secreted by the activated T-cells attract antigens of Classes I and II of the Major Histocompatibility Complex (MHC) onto the surface of neurons and astroglia [14]. The most powerful regulatory inducers of leukocyte–endothelium adhesion molecules (CD11a and CD18) are IL-1β and TNF-α. IL-8, which they activate, plays a suggestive pivotal role in triggering of leukocyte migration from the vascular lumen into the ischemized focus, which leukocytes then infiltrate [141, 190]. Laboratory studies of CSF in patients with ischemic stroke 2–3 days after onset of the illness confirmed that there is elevation of leukocyte level in CSF [155]. Studies conducted with a radioactive marker have revealed a large number of polymorphonuclear cells in the brain area with decreased perfusion already 6–12 h after stroke onset [2]. Neuropathological studies have revealed intense leukocyte infiltration of brain parenchyma on days 2–3 of stroke [30]. Leukocytes migrated from systemic blood flow (at first neutrophils and then monocytes) enhance the destruction of brain cells by their toxic products, phagocytosis, and immune reactions [70, 89, 91, 92, 191].

Astrocytes are also important target cells of cytokines as they sequentially distribute their effects. Cytokines increase the expression of colony-stimulating factors (CSF-1 and GM-CSF) by astrocytes that in turn activate microglial mitosis and change the surface of glial cells for antigen expression enhancing antibody-dependent cell-mediated cytotoxicity [17]. Thus, vicious excitatory circles ensue. In experimental brain damage demonstrating selective neuronal death, astrocytes abruptly increase the production of acute phase proteins which act as inhibitors of various proteins and growth factors [133]. Acute phase proteins of Class I, the production of which is stimulated by IL-1 and IL-6, comprise C-reactive protein (CRP) and complement factors [125, 126, 156]; acute phase proteins of Class II, which are synthesized after IL-6-mediated induction, include α_2-macroglobulin and α_1-antichymotrypsin [128]. The sequence of cascaded reactions triggered by the increased production of acute phase proteins is very complex. Reactions triggered by complement factors, which lead to cell lysis after membrane attack, are better studied. Various experimental works showed that CRP, a conventional marker of somatic inflammatory reactions, as well as pro-inflammatory cytokines, reflects the "activity" of ischemic events in brain and possesses certain value for the prognosis of acute focal ischemia outcome [85, 94, 131, 139, 150]. It is noteworthy that in therapeutic practice in patients with ischemic heart disease and chronic inflammation of connective and bone tissue a correlation has been found between the dynamics of CRP level within the first days of the disease and the severity of its clinical course and outcome [72, 103, 117, 125, 126].

Astroglia excited by cytokines enhances synthesis not only of acute phase proteins, but also of other regulatory molecules, such as NO, endothelial relaxation factor, cGMP [146, 179], and, thus, influences processes of oxidant stress and glutamate excitotoxicity.

Results from experimental studies suggest that the progression of damage in the penumbral zone on the background of acute focal ischemia may take place not only under the influence of absolute elevation of the contents of pro-inflammatory agents, but also under conditions when there exists a lack of anti-inflammatory and neurotropic factors [8, 19, 28, 54, 82, 100, 115, 124]. Chen and Manning [28] revealed that a preliminary injection of cytokines (TGF-β_1 and IL-10) significantly inhibits production of IL-1β, IL-6, IL-8, and TNF-α. Maximal inhibition was reached with relatively physiological doses (1–2 ng/ml) of TGF-β_1 and IL-10. A protective anti-ischemic effect of TGF-β_1 was demonstrated in experimental studies [8–10, 62, 87, 147, 170, 175] and was manifested by the inhibition of leukocyte adhesion to endothelium [154] and by the depression of production of injuring oxygen-derived and NO metabolites by macrophages [36, 75, 167]. It was shown that the neuroprotective properties of TGF-β_1 were also connected with the stabilization of neuronal calcium homeostasis [135] and with enhanced expression of *bcl-2* (see Chapter 10), which leads to significant inhibition of free radical damage and to suspension of necrosis and apoptosis [40, 52, 78, 84]. It was also demonstrated that TGF-β_1 plays an important role in the increase in vascular support of brain tissue after cerebral ischemia, especially in the penumbral area [97].

Studies devoted to IL-10 showed that this cytokine appears to be a powerful regulator of cell-mediated immune response of monocytes and macrophages. It decreases production of prostaglandin E_2 and inhibits expression of ICAM-1, both of which actively participate in the development of local inflammation [93]. It also depresses production of many pro-inflammatory cytokines including IL-1, TNF-α, IL-6, and IL-8 [90]. The inhibitory effect of IL-10 on the synthesis of both superoxide anion and the intermediate reactive products of oxygen metabolism [90] was also shown. This effect modulates the synthesis of NO in relation to the duration of exposure of activated macrophages to it [33]. IL-10 supports NO synthesis, which is one of its multiple activating effects on monocytes and macrophages; it also induces production of IL-1ra [180]. Thus, IL-10 is a factor modulating macrophagal function and possessing powerful anti-inflammatory activity.

Thus, microglia is well equipped to influence the type and extent of neuronal damage through the interaction with injured cells. Both CNS disease state and the stage of its pathogenesis are signaled to the periphery via the expression of cytokines and major histocompatibility proteins, the induction of brain–endothelial adhesion molecules, and the release of chemoattractants. However, the exact contribution of glial tissue to ischemic brain disease remains far from clear because at least some local microglia exchange actively with peripheral monocytes/macrophages [47].

8.1. Dynamics of pro-inflammatory and regulatory cytokines and C-reactive protein concentration in CSF of patients with acute ischemic stroke

Despite the voluminous experimental material gained, there is a paucity of clinical studies reviewed in literature concerning the phenomenon of local inflammation and its role in the development of acute focal cerebral ischemia in humans. And all such studies were of fragmentary character and were not performed longitudinally.

Together with the Institute of Molecular Genetics (N. F. Myasoyedov *et al.*) and the Department of Rheumatology of Moscow Medical Academy (E. L. Nasonov *et al.*) [66–68, 153, 193], we performed complex clinical and immunobiochemical study of cytokine status and C-reactive protein (CRP) in 50 patients (28 males and 22 females; average age 64 ± 1.8 years) with acute ischemic stroke in the carotid artery territory (left in 26 cases and right in 24 cases). Twenty patients were admitted to the Intensive Stroke Unit within the first 2–5 h and 30 patients within 6–12 h after stroke onset. We excluded patients with acute inflammatory, hereditary neurodegenerative, and autoimmune concomitant diseases. The main etiological factors of stroke were atherosclerosis with concomitant arterial hypertension (80.9%), atherosclerosis without evident arterial hypertension (11.8%), and arterial hypertension in young patients (7.3%). In 81.1% of patients we found signs of ischemic heart disease. In 43 (86%) patients stroke exhibited atherothrombotic origin, in 6 (12%) it followed cardiac embolism, and in 1 patient it followed the depression of systemic blood supply. There were no lacunar or microvascular strokes in this group.

From admission patients received multi-dimensional and at the same time maximally standardized treatment aimed to correct respiratory and cardiovascular dysfunction, to recover acid–base and water–electrolyte balance if indicated, to lessen brain edema if indicated, and to improve brain tissue perfusion. Hemodilution, anti-aggregants, angioprotectors and, where necessary, anticoagulants were used under laboratory control. Neuroprotective agents were not administered. To objectify the severity and the extent of neurological deficit, we used three complementary clinical scales: Scandinavian [143], Orgogozo [129] and original [64].

At the time of admission the neurological deficit of 29 patients was diagnosed as moderate (total clinical scores were 32.3 ± 1.8, 52.1 ± 2.7, and 39 ± 0.5, respectively). In these cases focal neurological deficit predominated in the clinical picture, and only in 15% of patients it was accompanied by mild consciousness disorders. In 21 patients severe stroke was observed on admission (total clinical scores were 22.3 ± 2.3, 35.6 ± 3.5, and 32.06 ± 2.7, respectively). In these cases focal neurological deficit was accompanied by clouding of consciousness: somnolence was diagnosed in 3 patients, stupor in 15, and coma I-II in 3 patients.

Quantitative study of the concentrations of pro-inflammatory (IL-1β, TNF-α, and IL-8) and anti-inflammatory (TGF-β₁ and IL-10) cytokines, as well as CRP level in CSF was performed by enzyme-linked immunosorbent essay (ELISA) repeatedly: on admission (before treatment was started) and on day 3 after stroke onset. Commercial immunoenzyme sets manufactured by Sigma (USA) were used for the detection of cytokines and CRP.

Monoclonal cytokine-specific antibodies were immobilized on 96-well polystyrene microplates for immunosorbent essay. The wells were saturated with standard solution and the samples. The studied cytokine present in the medium was bound to immobilized monoclonal antibodies. After the substances that failed to bind to each other were washed off, enzyme-linked polyclonal cytokine-specific antibodies were added to the wells. After the substances that failed to bind were washed off repeatedly, substrate solution was added. A color reaction developed. The color reaction depended on the quantity of cytokine bound in the previous stage. The reaction was terminated with 2 N solution of sulfuric acid. The reaction intensity was measured using an ELISA reader. Then a logarithmic curve was constructed for the detection of the contents of the studied substances in CSF in relation to optical density values acquired from the ELISA reader. The latter was performed using the computer program TITERSOFT, promoting optimal accuracy of calculations.

Parameters were compared with normal levels obtained from peridural anesthesia in a control group consisting of 20 patients (10 male, 10 female, mean age 56.7 years) with orthopedic pathology without neurological and inflammatory disorders, as well as without systemic diseases. Average values of cytokines and CRP in the control group comprised 0.7 ± 0.1 pg/ml for IL-1β, 11 ± 0.9 pg/ml for TNF-α, 101.8 ± 34.7 pg/ml for IL-8, 30.27 ± 6.3 pg/ml for IL-10, 91.16 ± 3.34 pg/ml for TGF-β₁, and 9 ± 0.01 μg/ml for CRP.

To perform clinical and immunobiochemical correlation, we assessed shifts of total clinical scores by the scales applied by day 5 (the end of the very acute period connected with brain infarction formation) and by day 21 (the end of acute period of stroke connected with remaining brain edema) and relative values of dynamics of cytokines and CRP concentrations by day 3 (K_3), calculating them according to the following formula:

$$K_3 = (n_3 - n_1) / n_1 \times 100\%,$$

where n_1 and n_3 are the concentration values on day 1 and day 3 after stroke onset.

Descriptive statistics and frequency distributions were generated for the study and point data. The Wilcoxon matched-pairs signed-rank test was used to analyze the statistical significance for the changes in measured parameters and the Mann–Whitney U test was used for pairwise and group comparisons. Mortality rates and disability levels on the Barthel Index among the treatment groups were compared with the use of Fisher's exact test.

Correlation analysis included shift values of total clinical scores by the scales applied and baseline values of shifts of cytokines and CRP concentrations. The differences between values were assumed to be significant at the 5% level in every comparison conducted.

All the patients on admission (within the first 2–12 h after stroke onset) exhibited imbalance between pro- and anti-inflammatory cytokines. By the time of admission all the patients had significantly elevated levels of triggering pro-inflammatory cytokines (4.4 ± 0.92 versus 0.7 ± 0.1 pg/ml for IL-1β, $p < 0.001$; and 22.8 ± 4.7 versus 11 ± 0.9 pg/ml for TNF-α, $p < 0.05$), and CRP (31.42 ± 9.21 versus 9 ± 0.01 µg/ml) versus controls while the levels of anti-inflammatory cytokines (IL-10 and TGF-β_1) tended to decrease (Fig. 8.2). The elevation of pro-inflammatory cytokine levels revealed in our study is consistent with the results of Tarkowski *et al.* [160], who found CSF IL-6 concentration elevated on day 1 after ischemic stroke onset. This initial increase of IL-6 was significantly correlated ($r = 0.65$; $p = 0.002$) with the volume of infarction measured by MRI 1–2 months later. Recent clinical research by Perini *et al.* [132] also showed increased IL-6 serum levels in patients with stroke between day 1 and day 14 ($p < 0.04$ versus controls) with a maximum elevation of its concentration on day 3, which correlated with clinical outcome.

In patients admitted to the Intensive Stroke Unit within the first 2–5 h after stroke onset, IL-1β and TNF-α levels were significantly lower than those in patients admitted within 6–12 h after stroke onset (2 ± 0.37 versus 6.73 ± 0.4 pg/ml for IL-1β, $p < 0.001$, and 18.5 ± 5.6 versus 38.7 ± 9.2 pg/ml for TNF-α, $p < 0.05$). A similar tendency was observed for IL-8 (134.8 ± 36.3 versus 187 ± 69.2 pg/ml), which proves the delayed character of local inflammatory reactions compared with quickly developed processes of the glutamate–calcium cascade, which occur within the period of the therapeutic window (see Fig. 8.2).

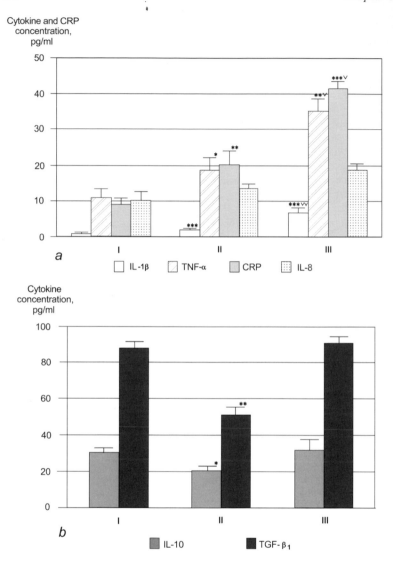

Figure 8.2. Concentration of cytokines and C-reactive protein (CRP) in CSF of patients with carotid ischemic stroke. CRP concentration in μg/ml; I – control; II – 2–5 h; III – 6–12 h after the onset of stroke. Significance of the differences versus controls: * $p < 0.05$, ** $p < 0.01$, *** $p < 0.001$. Significance of the differences versus values obtained within 2–5 h: $^v p < 0.01$, $^{vv} p < 0.001$.

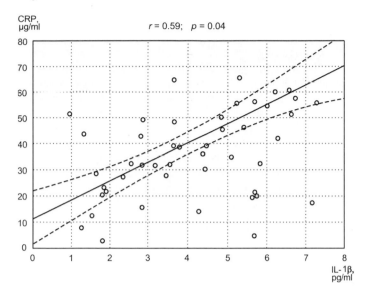

Figure 8.3. Correlation between IL-1β and CRP concentrations in CSF of patients within the first hours after carotid ischemic stroke onset.

A correlation between IL-1β and CRP concentrations on day 1 after stroke onset was shown ($r = 0.59$, $p = 0.04$) (Fig. 8.3). However, according to correlation analysis the extent of shift of IL-1β concentration by day 3 did not influence the dynamics of CRP and the stroke outcome, this indirectly reflecting only the triggering role of IL-1β in the initiation of inflammatory process (Fig. 8.4).

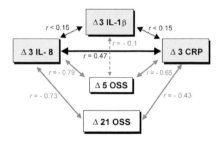

Figure 8.4. Correlation between the dynamics of IL-1β, IL-8, and CRP concentration in CSF by day 3 and the shifts of the total clinical score by the Orgogozo Stroke Scale by days 5 and 21 after carotid ischemic stroke onset. Black arrows mean positive correlations, grey – negative correlations.

Nevertheless, close correlation between the relative increase in IL-8 content by day 3 and the increase in total clinical score by day 5 ($r = -0.79$, $p = 0.031$) and day 21 ($r = -0.73$, $p = 0.035$) (see Fig. 8.4) was revealed, possibly reflecting the participation of chemoattractant IL-8 in the development of inflammatory process within the first days after stroke onset: in the attraction of potentially toxic leukocyte pool into the ischemic focus, as well as in the increase of blood–brain barrier permeability and in the formation of brain edema [50, 95, 137, 138].

CRP concentration in CSF continued to increase up to day 3 (by 283.6% versus control value, $p < 0.05$). The extent of its increase positively correlated with the increase in total clinical score by day 21 ($r = -0.43$, $p < 0.05$ by OSS, and $r = -0.54$, $p = 0.045$ by SSS), which emphasizes the unfavorable prognostic significance of elevated CRP level within the first days after stroke onset (see Fig. 8.4). According to the retrospective analysis of the dynamics of CRP level its maximal shift was registered in patients with severe disability by the end of the acute period of stroke (by 243.7%) and with minimal increase in total clinical score by day 5 (by 249.9%) and by day 21 (by 253.6%) after stroke onset. Along with this, minimal shift of CRP concentration was registered in patients with good neurological recovery by the end of the acute period (106.8%, $p < 0.05$ versus the group of severely disabled) and with maximal increase in total clinical score by day 5 (148.6%, $p < 0.05$) and by day 21 (161.1%, $p < 0.05$) after stroke onset (Fig. 8.5). Thus, the significant increase in CRP level in CSF by day 3 appears to be a valid criterion of unfavorable prognosis in relation to further neurological recovery, reflecting the significance of local inflammatory processes in the formation of brain infarction.

No correlation was found between the change in TNF-α level day 3 after stroke onset and the clinical outcome. However, the study of TNF-α dynamics allowed elucidation of the complicated role of this cytokine in the pathogenesis of ischemic stroke. A reverse correlation between the shifts of TNF-α and CRP levels ($r = -0.44$, $p = 0.05$) was found, i.e. in case of progressive inflammatory reaction present on day 3 after stroke onset, further elevation of CRP level was accompanied by a decrease in TNF-α level (Fig. 8.6). This indirect evidence of unfavorable predictive value of decrease of TNF-α level in CSF on day 3 agrees with the experimental data about immune modulatory action of this cytokine which takes part in the initiation of local inflammation right after the onset of ischemia, but then in its restriction and in reduction of brain infarction volume, as well as in recovery of tissue homeostasis [106, 130].

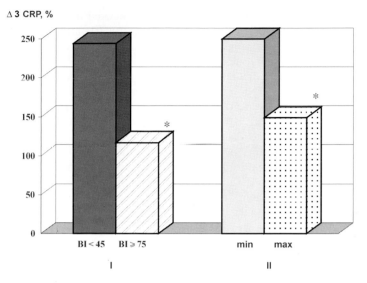

Δ 3 CRP, %

Figure 8.5. Connection between the degree of the increase in CRP concentration in CSF by day 3 and clinical dynamics by the Barthel Index (BI) and the Orgogozo Stroke Scale (OSS) in patients with carotid ischemic stroke. I – functional outcome on day 21 (assessed by BI); II – increase in the total clinical score (assessed by OSS) on day 5. Significance of the differences: * $p < 0.05$.

The level of the pleiotropic anti inflammatory cytokine IL-10 within the first hours after stroke onset (24.5 ± 5.3 pg/ml) did not significantly differ from the normal value (30.2 ± 6.3 pg/ml). By day 3 the relative increase in IL-10 concentration was 34%, which positively correlated with the increase in the neurotropic cytokine TGF-β_1 ($r = 0.86$, $p = 0.001$) and negatively with CRP shift ($r = -0.51$, $p = 0.04$) (Fig. 8.7). These findings suggest a synergy of IL-10 and TGF-β_1 action that is consistent with conclusions drawn from experimental work [28]. Increased production of IL-10 can positively influence the outcome of the illness.

TGF-β_1 concentration in CSF within the first hours after stroke onset also did not significantly differ from control value (79.3 ± 31.4 versus 91.1 ± 33.4 pg/ml, respectively; $p > 0.05$) and increased by day 3 by 63.2%. We found reverse correlations between shifts of TGF-β_1 and IL-1β concentrations ($r = -0.56$, $p = 0.045$), as well as between shifts of TGF-β_1 and CRP levels ($r = -0.51$, $p = 0.049$) and positive correlation between

TGF-β_1 dynamics and the extent of recovery, estimated by OSS, by day 5 ($r = 0.49$, $p < 0.05$) and by day 21 ($r = 0.91$, $p = 0.03$) after stroke onset. Simultaneously we found tight and highly significant linkage between the increase in IL-10 and TGF-β_1 concentrations with TNF-α ($r = 0.91$–0.96, $p = 0.001$–0.003) by day 3, that confirmed the reparative and immunomodulatory properties of TNF-α (see Fig. 8.7). It is noteworthy that changes in CRP level exhibited reverse correlation with the increase in TNF-α, the anti-inflammatory cytokine IL-10, and neurotropic cytokine TGF-β_1 concentrations that reflect the enhancement of inflammatory response on the background of the insufficiency of protective factors.

We found no significant differences between immunobiochemical patterns of patients with left and right hemispheric strokes.

Comparing the dynamics of immunobiochemical pattern by day 3 after stroke onset in relation to the severity of the illness, we found that in patients with moderately severe stroke versus severe stroke there was a tendency towards the decrease of IL-1β level, significantly lower increase in IL-8 (by 9.7 versus 36.8%, respectively, $p < 0.05$) and CRP (by 156.2 versus 460.5%, respectively, $p < 0.01$) (see Table 8.1 and Fig. 8.8), as well as of IL-10 (by 13.1 versus 55.1%, respectively, $p < 0.05$) and TGF-β_1 (by 34.7 versus 91.8%, respectively, $p < 0.01$) (Fig. 8.9). TNF-α concentration did not significantly change in patients with moderately severe stroke, but decreased (by 35.6%; $p < 0.01$ versus the result on day 1) in patients with severe stroke (see Fig. 8.8). This is consistent with the clinical observation by Efstratopoulos *et al.* [39] who found correlation between the increase of TNF-α serum level and good recovery and favorable prognosis of ischemic stroke. The increase in CSF concentration of pro-inflammatory and neurotropic cytokines in severe stroke might follow the prominent elevation of IL-8 level in cases of vast ischemic brain damage.

Thus, our clinical and immunobiochemical study revealed the participation of immune reactions and local inflammation in the pathogenesis of ischemic stroke and in the formation of brain infarction. It demonstrated the significance not only of excessive release of pro-inflammatory cytokines, but also of the deficit of anti-inflammatory and neurotropic factors in the development of inflammatory response, this being in agreement with experimental findings acquired with animal models of acute focal cerebral ischemia [12, 53, 54, 107, 130, 179].

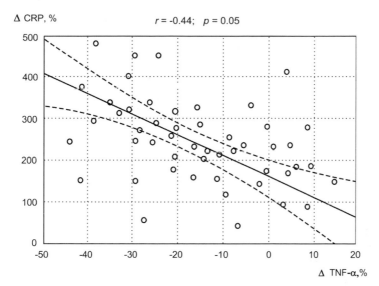

Figure 8.6. Correlation between the dynamics of TNF-α and CRP concentrations in CSF of patients with carotid ischemic stroke by day 3.

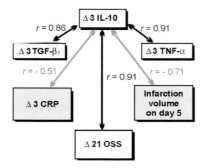

Figure 8.7. Correlation between the dynamics of cytokine concentrations in CSF by day 3, the shifts of the total clinical score by the Orgogozo Stroke scale (OSS) by day 21 and the infarction volume on day 5 after the onset of stroke. Black arrows mean positive correlations, grey – negative correlations.

Table 8.1. Increase (%) in cytokines and CRP concentrations in CSF of patients with carotid ischemic stroke by day 3 in relation to the stroke severity

Stroke severity	IL-1β	TNF-α	IL-8	IL-10	TGF-β₁	CRP
Moderately severe	−27.6±42.6	1.7± 0.34	9.7± 0.1	13.1±9.5	34.7±25.1	156.2±73.0
Severe	20.3±33.1	−35.6±10.5	36.1±10.5	55.1±17.6	91.7±13.0	460.5±99.4
p	>0.05	<0.01	<0.05	<0.05	<0.01	<0.01

We found close correlation between the extent of immunobiochemical change and both the severity and outcome of ischemic stroke. We found high predictive value of the dynamics of CRP concentration in CSF that is of doubtless of practical importance as it is connected with simplicity, speed, and economic feasibility of its measurement under laboratory conditions. The prognostic significance of CRP in ischemic stroke has recently been confirmed in a clinical study by Muir *et al.* [120] who showed that survival in stroke patients with serum CRP level more than 10.1 mg/liter was significantly worse than in those with CRP less than this level. Ischemic strokes accounted for 63% of deaths in patients with high CRP level.

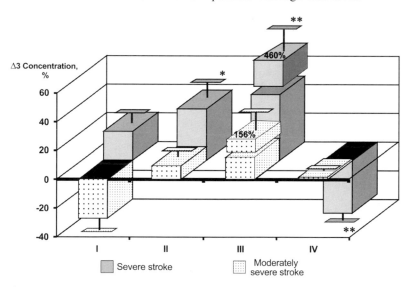

Figure 8.8. Dynamics of IL-1β (I), IL-8 (II), CRP (III) and TNF-α (IV) levels in CSF by day 3 in relation to ischemic stroke severity. Significance of the differences versus patients with moderately severe stroke: * $p < 0.05$, ** $p < 0.01$.

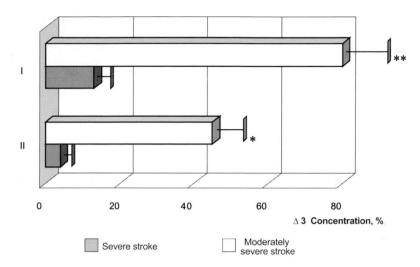

Figure 8.9. Dynamics of TGF-β_1 (I) and IL-10 (II) in CSF by day 3 in relation to ischemic stroke severity. Significance of the differences versus patients with severe stroke: * $p < 0.05$, ** $p < 0.01$.

From the experimental and clinical findings, we suggest that elaboration and application of new neuroprotective therapeutic techniques directed to the inhibition of local inflammatory reaction via quantitative decrease in leukocytes circulating in blood, blockage of pro-inflammatory cytokines and adhesion molecules, immune suppression, as well as via activation of protective anti-inflammatory systems, would allow significant improvement in treatment efficacy in patients with acute ischemic stroke, resulting in amelioration of clinical outcome of the disease and in restriction of the infarction area.

8.2. Microvascular–cellular cascade and secondary brain damage

Studies by Siren *et al.* [151] showed that in spontaneous hypertensive rats hypertension can cause activation of intercellular adhesion molecules-1 (ICAM-1) on endothelial cells and perivascular accumulation of leukocytes in the brain parenchyma. This reflects possible predisposition of patients with chronic brain ischemia due to arterial hypertension to the activation of polymorphonuclear leukocytes (PMNL) before ischemic stroke [163].

Cerebral artery occlusion leads to acute (in the first minutes) disturbances of microcirculation in the affected vessel area causing focal ischemia of brain tissue. The following events taking place in neurons and glia (the reactions of the glutamate–calcium cascade, especially connected with oxidative stress, destruction of phospholipids, excessive synthesis of eicosanoids and platelet activating factor; metabolic acidosis; leukocyte reaction caused by the activation of microglia) influence secondarily on microcirculation aggravating its change even under conditions of early reperfusion and leading to additional disturbances of the blood–brain barrier permeability. Experimental study [164] of microcirculation and cellular reactions in the ischemic focus allowed them to be presented in chronological sequence (microvascular–cellular cascade) (Table 8.2).

In the first minutes after cerebral artery occlusion (**Stage I of the microvascular–cellular cascade**) cerebral blood flow decreases and this disturbs tissue microcirculation. From the 2nd min after reperfusion reactive hyperemia of various degrees occurs. Reperfusion produces multimodal

Table 8.2. Microvascular–cellular cascade

Stage	Time interval after stroke onset	Basic events
I	First minutes	Reactive hyperemia
		Post-ischemic hypoperfusion
		Beginning of release of pro-inflammatory mediators and aggregants
II	3–6 h	Low perfusion hyperemia
		Venous dilatation and dystonic disturbances of small vessels
		Beginning of acute inflammatory reaction
		Beginning of PMNL migration into brain tissue
		Secondary enhancement of oxidant stress
		Reperfusion brain damage
III	6–72 h	Active formation of cytotoxic edema
		Maximal activation of microglia and PMNL migration
		Alteration of BBB permeability
		Aggravation of rheological disturbances
IV	Days 3–7	Irreversible hemorheological changes
		Luxuriant perfusion
		Macrophagal migration into the brain tissue
		Asrtoglial proliferation
		Infarction formation terminates
V	Weeks, months	Cyst formation and astrogliosis in brain tissue

individual time courses ranging from recovery after marked reactive hyperemia to severe post-ischemic hypoperfusion [163].

There are insufficient data available on when and how PMNLs are activated and when and how endothelial cells upregulate adhesion molecules during ischemia [163]. In the first minutes the release of inflammatory mediators together with an elevation of ratio between thromboxane A_2 (TxA_2) and prostaglandin I_2 (PGI_2) occurs that may increase platelet aggregation and trigger a cascade of further events in microvascular–cellular derangement [168]. Platelet activating (PAF), which is a pro-inflammatory mediator, could be one of the candidates for such mediators to stimulate both leukocytes and endothelial cells [101, 159].

During 3–6 h after ischemia onset (**Stage II of the microvascular– cellular cascade**) CBF volume slightly exceeds the pre-occlusive level (low perfusion hyperemia), which is mainly connected with venous dilatation [88]. The phenomenon of temporal–spatial dynamics of microcirculation occurs in the ischemized brain area and dystonic astroglia appear manifesting by alternation of dilated and constricted sections of the same vessel that leads to partial loss of vessel integrity, and from that the level of blood flow no longer corresponds to brain metabolic demand ("miserly" perfusion syndrome). At the same time, the synthesis of pro-inflammatory mediators increases. PAF enhances adhesion of PMNLs to the endothelium [23, 163]. As a result, damage to the endothelium (disruption, shrinkage, and coagulation necrosis) occurs [192]. After the activation by pro-inflammatory mediators, endothelial cells begin to express adhesion molecules (for instance, P-selectin, GMP140) which accumulate and cause additional release of IL-1, TNF-α, and PAF. PMNLs circulating in blood also become activated with upregulation of adhesion molecules (Sialyl-Lewis X, Lewis X, and CD11/18) on their cell surface. PMNLs influenced by IL-8 and other chemoattractants gather in the ischemized area, accumulate there around small vessels (predominantly venules), and roll on endothelial cells via P-selectin [163]. The adhesion of PMNLs to the endothelium is mediated by the binding adhesion molecules (CD11/18 and ICAM-1) and may trigger active generation of free radicals (for instance, the xanthine oxidase system) capable of destroying all the mentioned cellular elements [2, 157]. Continuously released from PMNLs, monocytes, and endothelial cells, inflammatory mediators (IL-1, TNF-α, and PAF) maintain destructive events. Under the influence of TNF-α, endothelial cells acquire hemostatic procoagulant properties [127].

Leukocytes play a very important role in brain reperfusion damage [71]. Being delivered from blood flow to tissues, they increase the level of toxic oxygen intermediates, granular constituents of leukocytes, and phospholipase products.

At **Stage III of the microvascular–cellular cascade** (6–72 h after the onset of ischemia) cytotoxic edema is actively formed. Glial cells accumulate intercellular liquid and begin to swell due to increase in the osmotic potential in the cells (more than 200 mm Hg) caused by cell membrane damage. They increase in volume, compressing both adjacent structures and microcirculation vessels. Neuronal swelling in this range of events is less pronounced [164].

Extravasation of PMNLs into the ischemized brain area occurs. Maximal migration of PMNLs is observed 24–72 h after ischemia onset. Leukocyte reaction decreases only by day 7.

Severe disturbances of the blood–brain barrier permeability develops belatedly, several hours after occlusion. PMNLs adhered to endothelium slowly permeate through the endothelial wall. Along with this, evidence has been shown that neutrophils can permeate through the blood–brain barrier via the transendothelial pathway without any disruption of the barrier [37]. PMNLs permeate through the basement membrane with the help of proteolytic enzymes.

The increase in permeability in the arterial portion and the depression of absorption in the venous portion of microcirculation territory enhance the retention of liquid in brain tissue. One can observe an increase in the distance between capillaries, compression of vessels, local plasmorrhagia, erythrocyte extravasates, and marked erythrocyte aggregation.

The abrupt increase in the production of many toxic compounds by PMNLs and endothelial cells is characteristic. Increased synthesis of leukotrienes, thromboxane A_2, prostacyclin, prostaglandin E_2, and vasoconstrictors is observed which induce further links of this chain reaction [42]. The increase in superoxide radical level leads to the activation of lipid peroxidation and to secondary damage and dysfunction of the endothelium. PAF is continuously released from the damaged brain tissue together with eicosanoids, such as thromboxane A_2, prostacyclin, and leukotrienes due to enhanced lipid peroxidation. Under the influence of leukotriene B_4, possessing pro-inflammatory properties, adhesion and chemotactic movement of PMNLs, stimulating their aggregation, and the additional production of superoxide in PMNLs ensue [142].

Dying neurons and their degenerating axons release IL-1 which stimulates microglial cells activating them [163]. In response the reactive microglia increases synthesis of other pro-inflammatory cytokines and regulators, including astrocyte growth factor, which stimulates glial proliferation [34]. Amoeboid microglial cells are monocytes which migrate from blood into the tissue through the disrupted blood–brain barrier [69].

During the following days (days 3–7) **Stage IV of the microvascular–cellular cascade** begins. The level of tissue ischemia increases, pointing to a decrease in cerebral blood supply [163]. Hemorrheological disturbances reach an irreversible level due to sludging of erythrocyte aggregates, platelet aggregation, plasmorrhagia, increase of blood viscosity, and transformation of endothelial cells surface into uneven coagulation surface with platelet and leukocyte adherence to vessel walls. Spotty obliteration of the terminal vascular bed favors the development of arteriolar–venular shunting under high pressure (red vein and luxurious perfusion syndromes). One can observe heterogeneous segmental stasis and the decrease of blood flow via the supplying capillaries. These changes progress until the central ischemic necrosis becomes well-demarcated [163].

At this stage macrophages (monocytes and microglial cells) migrate from the periphery of the ischemized area to the ischemic core. Several days after stroke activated monocytes reach their peak of secretion activity and release neurotoxins [58]. Excited microglia produces soluble pro-inflammatory factors that cause astrocytosis and neuronal degeneration.

Slowing down of microglial phagocytic activity is an important beneficial sign which predicts close termination of pathological events [58].

During the following weeks and months **(Stage V of the microvascular–cellular cascade)** all the mentioned abnormal changes terminate and a permanent morphological lesion forms. There is no longer blood supply in the damaged area. The tissue softens and liquefies; decay products are removed by macrophages with cyst formation. Proliferation of fibroblasts and following fibrosis are rarely seen in the infarction. Astrogliosis prevails in lesions with new capillaries [163].

Thus, systematization of microcirculation and cellular reactions developing on the background of acute focal ischemia again emphasize that the disturbances of microcirculation reflect all basic events taking place in the brain tissue. Apparently, these microcirculation and cellular events are components of a united tissue system with parts so closely interconnected with each other that it is impossible to consider them in terms of reasons and consequences.

Under normal conditions the state of microcirculation is determined by the demands of tissue metabolism. However, such coordination alters in ischemic stroke, and microcirculation changes begin to take part in the ischemic cascade reactions. The chronology of microcirculation disturbances takes days and weeks after stroke onset and is accompanied by elongated reconstitution of both metabolism and morphology of the ischemized brain area.

The discovery of common biological regulators modulating microcirculation as well as cellular events confirms the existence of the unique triggering molecular mechanism responsible for the universal response of brain tissue to a damaging stimulus.

REFERENCES

1. Akira, S., Togo, T., Kishimoto, T., 1993, *Adv Immunol.* **54**: 1–78.
2. Akopov, S. E., Simonian, N. A., Grigorian, G. S., 1996, *Stroke.* **27**: 1739–1743.
3. Arvin, B., Neville, L. F., Barone, F. C., Feuerstein, G. Z., 1996, *Neurosci Biobehav Rev.* **20** (3): 445–452.
4. Arzt, E., Stalla, G. K., 1996, *Neuroimmunomodulation.* **3**: 28–34.
5. Arzt, E., Buric, R., Stelzer, G., *et al.*, 1993, *Endocrinology.* **132**: 459–467.
6. Arzt, E., Sauer, J., Buric, R., *et al.*, 1995, *Endocrine.* **3**: 113–119.
7. Astrup, J., Siesjo, B., Symon, L., 1981, *Stroke.* **12**: 723–725.
8. Banati, R. B., Graeber, M. B., 1994, *Dev Neurosci.* **16**: 114–127.
9. Banati, R. B., Gehrmann, J., Kreutlberg, G. W., 1996, In *Cellular and Molecular Mechanisms of Ischemic Brain Damage* (Siesjo, K., Wieloch, T., eds.) pp. 329–337.
10. Banati, R. B., Gehrmann, S., Schubert, P., Kreutzberg, G. W., 1993, *Glia.* **7**: 111–118.
11. Baron, J.-C., von Kurnmer, R., del Zoppo, G. J., 1995, *Stroke.* **26**: 2219–2221.
12. Beamer, H. B., Coull, B. M., Clark, W. M., *et al.*, 1995, *Ann Neurol.* **37**: 800–805.
13. Beamer, N. B., Sexton, G., Wym, M., *et al.*, 1998, *Neurology.* **50** (6): 1722–1728.
14. Benveniste, E. N., 1992, *Am J Physiol.* **263**: 1–16.
15. Bishayi, B., Samanta, A. K., 1996, *Scand J Immunol.* **43** (5): 531–536.
16. Blasi, E., Badnai, R., Bocchini, V., *et al.*, 1990, *J Neuroimmunol.* **27**: 229–237.
17. Blinzinger, K., Kreutzberg, G., 1968, *Z Zellforsch Mikrosk Anat.* **85**: 145–157.
18. Brenneman, D. E., Hill, J. M., Glazner, G. W., *et al.*, 1995, *Int J Dev Neurosci.* **13**: 187–200.
19. Brenneman, D. E., Schultzberg, M., Bartfai, T., Oozes, I., 1992, *J Neurochem.* **58**: 454–460.
20. Brosnan, C. F., Cannella, B., Batistini, L., Raine, C. S., 1995, *Neurology.* **45** (6): 16–21.
21. Bruce, A. J., Baling, W., Kindy, M. S., *et al.*, 1996, *Nat Med.* **2**: 788–794.
22. Bruce-Keller, A. J., Mattson, M. P., 1999, Mechanisms of neuroprotective cytokines. In *Cerebral Ischemia* (Walz, W., ed.) Humana Press, Totowa, New Jersey, pp. 125–142.
23. Bruce-Keller, A. J., Geddes, J. W., Knapp, P. E., *et al.*, 1999, *J Neuroimmunol.* **93** (1–2): 53–71.
24. Bussolino, F., Alessi, D., Turello, E., *et al.*, 1991, In *New Aspects of Human Polymorphonuclear Leukocytes* (Horl, W. H., Schollmeyer, P. J., eds.) Plenum, New York, pp. 55–64.
25. Buttini, M., Sauler, A., Boddeke, H. W., 1994, *Mol Brain Res.* **23**: 126–134.
26. Canella, B., Raine, C. S., 1995, *Ann Neurol.* **37**: 424–435.
27. Carlson, N. G., Wieggel, W. A., Chen, J., *et al.*, 1999, *J Immunol.* **163** (7): 3963–3968.
28. Chen, C. C., Manning, A. M., 1996, *Cytokine.* **8** (1): 58–65.
29. Chen, S. C., Soares, H. D., Morgan, J. J., 1996, *Adv Neurol.* **71**: 433–450.
30. Chuaqui, R., Tapia, J., 1993, *J Neuropathol Exp Neurol.* **52**: 481–489.
31. Clark, R. S., Kochanek, P. M., Schwan, M. A., *et al.*, 1996, *Pediatr Res.* **39** (5): 784–790.
32. Cornell–Bell, A. H., Kinkbeiner, S. M., 1991, *Cell Calcium.* **12**: 185–204.
33. Corradin, S. B., *et al.*, 1993, *Eur J Immunol.* **23** (8): 2045–2048.
34. Davis, E. J., Foster, T. D., Thomas, W. E., 1994, *Brain Res Bull.* **34** (1): 73–78.
35. Dinarello, C. A., 1991, *Blood.* **77**: 1627–1652.
36. Ding, A., Nathan, C. F., Graycar, J., *et al.*, 1990, *J Immunol.* **145**: 940–944.
37. Dorovini-Zis, K., Bowman, P. D., 1992, *J Neuropathol Exp Neurol.* **51**: 194–205.
38. Dunn, A. J., 1994, *Abst Int Symp "Physiological and Biochemical Basis of Brain Activity"* (June 22–24). St. Petersburg, 48.

39. Eftsratopoulos, A., Tsiodra, P., Vojaki, S., *et al.*, 1996, *Cerebrovasc Dis.* **6**: 52.
40. Ellerby, L. M., Ellerby, H. M., Park, S. M., *et al.*, 1996, *J Neurochem.* **67**: 1259–1267.
41. Elneihoum, A. M., Falke, P., Axelsson, L., *et al.*, 1996, *Stroke.* **27**: 1734–1738.
42. Ernst, E., Matrai, A., Paulsen, F., 1987, *Stroke.* **18**: 59–62.
43. Fassbender, K., Rossol, S., Kammer, T., *et al.*, 1994, *J Neurol Sci.* **122**: 135–139.
44. Feuerstein, G. Z., Liu, T., Barone, F. C., 1994, *Cerebrovasc Brain Metab Rev.* **6**: 341–360.
45. Fisher, M., 1997, *Stroke.* **8**: 866–872.
46. Fisher, M., Takano, K., 1995, *Ballierie's Clinical Neurology, Cerebrovascular Disease* (Hachinski, V., ed.) London, pp. 279–296.
47. Fisher, M., Garcia, J. H., 1996, *Neurology.* **47**: 884–888.
48. Flugel, A., Priller, J., Graeber, M. B., 2000, *Eur J Neurol.* **7** (3): 158 (FW14-5).
49. Flyorov, M. A., 1996, *Neurochemistry* (Ashmarin, I. P., Stukalov, P.V., eds.) Moscow, pp. 193-200 (in Russian).
50. Foster, S. J., Aked, D. M., Schroder, J.-M., Christophers, E., 1989, *Immunology.* **67**: 181–183.
51. Furie, M. B., Randolph, G. J., 1995, *Am J Pathol.* **146**: 1287–1301.
52. Garcia, I., Martinou, I., Tsujimoto, Y., Martinou, J. C., 1992, *Science.* **258**: 302–304.
53. Garcia, J. H., Liu, K. F., Fisher, M., Tatlisumak, T., 1996, *Cerebrovasc Disease.* **6**: 180.
54. Garcia, J. H., Liu, K. F., Yoshida, Y., *et al.*, 1994, *Am J Pathol.* **195**: 721–740.
55. Giulian, D., 1993, *Glia.* **7** (1): 102–110.
56. Giulian, D., Robertson, C., 1990, *Ann Neurol.* **27**: 33–42.
57. Giulian, D., Corpus, M., 1993, *Adv Neurol.* **59**: 315–320.
58. Giulian, D., Vaca, K., 1993, *Stroke.* **24**: 184–190.
59. Giulian, D., Li, J., Leara, B., Keenen, C., 1994, *Neurochem Int.* **25**: 227–233.
60. Graeber, M. D., Streit, W. J., 1990, *Brain Pathol.* **1**: 2–5.
61. Grau, A. J., Hacke, W., Werle, E., *et al.*, 1996, *Thromb Res.* **82** (3): 245–255.
62. Gross, C. E., Bedhar, M. M., Howard, D. B., Spom, M. B., 1993, *Stroke.* **24**: 558–562.
63. Gurvits, B. Y., Galoyan, A. A., 1994, *Abst Int Symp "Physiological and Biochemical Basis of Brain Activity"*. St. Petersburg, 53.
64. Gusev, E. I., Skvortsova, V. I., 1990, *Case Record Form for Examination and Treatment of Patients with Ischemic Stroke*. Moscow, pp. 1–44 (in Russian).
65. Gusev, E. I., Skvortsova, V. I., Komissarova, I. A., *et al.*, 1999, *Korsakoff J Neurol Psychiatr.* **2**: 18–26 (in Russian).
66. Gusev, E. I., Skvortsova, V. I., Kovalenko, A. V., Sokolov, M. A., 1999, *Korsakoff J Neurol Psychiatr.* **2**: 71–76 (in Russian).
67. Gusev, E. I., Skvortsova, V. I., Nasonov, E. L., *et al.*, 1998, *Cerebrovascular Disorders* (Odinak, M. M., Kuznetsova, A. N., eds.) St. Petersburg, pp. 187–189 (in Russian).
68. Gusev, E., Skvortsova, V., Raevsky, K., *et al.*, 1998, *Actuelle Neurologie.* **3**: 25.
69. Haapaniemi, H., Tomita, M., Tanahashi, N., *et al.*, 1995, *Neurosci Lett.* **193**: 121–124.
70. Hallenbeck, J. M., 1996, In *Mechanisms of Secondary Brain Damage in Cerebral Ischemia and Trauma*. New York, pp. 27–31.
71. Hallenbeck, J. M., Dutka, A. J., 1990, *Arch Neurol.* **47**: 1245–1254.
72. Haverkate, F., Thompson, S. G., Pyke, S. D. M., *et al.*, 1997, *Lancet.* **73**: 374–378.
73. Heiss, W. D., Graf, R., 1994, *Curr Opin Neurobiol.* **1**: 11–19.
74. Heiss, W. D., Fink, G. R., Pietnyk, V., *et al.*, 1992, *Pharmacology of Cerebral Ischemia* (Kriegistein, J., Oberpich-ler-Schwenk, H., eds.) Stuttgart, pp. 529–536.
75. Henric-Noack, P., Prehn, J. H. M., Kreiglestein, J., 1996, *Stroke.* **27**: 1609–1615.
76. Hetier, E., Ayala, J., Denefle, P., *et al.*, 1988, *J Neurosci Res.* **21**: 391–397.

77. Hirano, T., 1992, *Chem Immunol.* **51**: 153–180.
78. Hockenbery, D. M., Oltvai, Z. N., Xiao-Ming, Y., *et al.*, 1993, *Cell.* **75**: 241–251.
79. Howard, M. C., Miyajima, A., Coffmann, R., 1993, *Fundamental Immunology* (We, P., ed.) New York, pp. 763–800.
80. Huber, M., Heiss, W. D., 1996, *Semin Thromb Hemost.* **22** (1): 53–60.
81. Isayama, K., Pilts, L. H., Nishimura, M. C., 1991, *Stroke.* **22**: 1394–1398.
82. Issawdeh, S., Lorentien, J. C., Mustafa, M. I., *et al.*, 1996, *J Neuroimmunol.* **69** (1–2): 103–115.
83. Jones, A., Selby, P., 1989, *Progr Growth Factor Res.* **1** (2): 107–122.
84. Kane, D. J., Ord, T., Anton, R., *et al.*, 1995, *J Neurosci.* **40**: 269–275.
85. Kanoh, Y., Ohtani, H., 1997, *Biochem Mol Biol Int.* **43** (2): 269–278.
86. Kees, L., Reinhard, H., 1995, *Neurology.* **45** (6): 4–5.
87. Kim, J. S., 1996, *J Neurol Sci.* **137** (2): 69–78.
88. Kobari, M., Gotoh, F., Tomita, M., *et al.*, 1983, In *The Cerebral Veins* (Auer, L. M., Loew, F., eds.) Springer, Vienna-New York, pp. 287–291.
89. Kochanek, P. M., Hallenbeck, J. M., 1992, *Stroke.* **23**: 1367–1379.
90. Koedel, U., Bernatowici, A., Frei, K., *et al.*, 1996, *J Immunol.* **157** (II): 5185–5191.
91. Kogure, K., Arai, H., Abe, K., Nakano, M., 1985, *Progr Brain Res.* **63**: 237–259.
92. Kogure, K., Yamasaki, Y., Malsuo, Y., *et al.*, 1996, *Acta Neurochir.* **66**: 40–43.
93. Kornelisse, R. F., Savelkoul, H. F., Mulder, P. H., *et al.*, 1996, *J Infect Dis.* **173** (6): 1498–1502.
94. Kossmann, T., Morganti–Kossmann, C., Trent, O., *et al.*, 1995, *Shock.* **4** (5): 311–317.
95. Kossmann, T., Stahel, P. F., *et al.*, 1997, *J Cerebr Blood Flow Metab.* **17** (3): 280–289.
96. Kreutzberg, G. W., 1996, *Trends Neurosci.* **19** (8): 312–318.
97. Krupinski, J., Kumar, P., Kumar, S., *et al.*, 1996, *Stroke.* **27**: 852–857.
98. Lee, M. R., Sakatani, K., Young, W., 1993, *Exp Neurol.* **119**: 140–145.
99. Libby, P., Sukhova, G., Lee, R. T., Galis, Z. S., 1995, *J Cardiovasc Pharmacol.* **25** (2): 9–12.
100. Lindholm, D., Castren, E., Kiefer, R., *et al.*, 1992, *J Cell Biol.* **1** (17): 395–400.
101. Lindsberg, P. J., Hallenbeck, J. M., Feuerstein, G., 1991, *Ann Neurol.* **30**: 117–129.
102. Liu, T., Clerk, R. K., McDomell, P. C., *et al.*, 1994, *Stroke.* **25**: 1481–1488.
103. Liuzzo, G., Biasucci, L. M., Gallimore, J., *et al.*, 1994, *N Engl J Med.* **331**: 417–424.
104. Loddick, S. A., Wong, M. L., Bongiorno, P. B., *et al.*, 1997, *Biochem Biophys Res Commun.* **234**: 211–215.
105. Marchetti, B., Gallo, F., Romeo, C., *et al.*, 1996, *Ann NY Acad Sci.* **784**: 209–236.
106. Marino, M. W., Dunn, A., *et al.*, 1997, *Proc Natl Acad Sci USA.* **94**: 8093–8098.
107. Matsumoto, K., Graf, R., *et al.*, 1993, *J Cerebr Blood Flow Metab.* **13**: 586–594.
108. Matsumoto, M., Yamamoto, K., Hamburger, H. A., *et al.*, 1987, *Mayo Clinic Proc.* **62**: 460–472.
109. Matsumoto, K., Yamada, K., Kohmura, E., *et al.*, 1994, *Neurol Res.* **16** (6): 460–464.
110. Mattila, K. J., Asikainen, S., Nieminen, M. S., Valtonen, V. V., 1998, *Clin Infect Dis.* **26** (93): 719–734.
111. McGeer, P. L., Itagaki, S., Togo, H., McGeer, E. G., 1987, *Neurosci Lett.* **9**: 195–200.
112. McGeer, P. L., Kawamala, T., Walker, D. G., *et al.*, 1993, *Glia.* **7**: 84–92.
113. McIntosh, T. K., Smith, D. H., Garde, E., 1996, *Eur J Anaesthesiol.* **13** (3): 291–309.
114. Michaels, R. L., Rothman, S. M., 1990, *J Neurosci.* **10**: 283 292.
115. Mihajlovic, R., Jovanovic, M., Djordjevic, D., Jovicic, A., 1996, *Cerebrovasc Dis.* **6** (2): 13.

116. Minami, M., Kuraishi, Y., Yabuuchi, K., *et al.*, 1992, *Acta Neuropathol* (Berl). **58**: 390–392.

117. Mishra, O. P., Das, B. K., Chandra, L., *et al.*, 1995, *Indian Pediatr.* **32** (8): 886–889.

118. Morioka, T., Kalehua, A. N., Streit, W. I., 1991, *J Cerebr Blood Flow Metab.* **10**: 850–859.

119. Morioka, T., Kalehua, A. N., Streit, W. J., 1993, *J Comp Neurol.* **327**: 123–132.

120. Muir, K. W., Weir, C. J., Alwan, W., *et al.*, 1999, *Stroke.* **30** (5): 981–985.

121. Murota, S., Fujita, H., Morita, I., 1993, Involvement of adhesion molecules in vascular endothelial cell injury by oxygen radicals released from activated leukocytes. In *Intractable Vasculitis Syndromes* (Tanabe, T., ed.) Hokkaido University Press, pp. 115–122.

122. Nagy, Z., 2000, *Eur J Neurol.* **7** (3): 158 (FW14-4).

123. Nagy, Z., Vastag, M., Skopal, J., *et al.*, 1996, *Keio J Med.* **45**: 200.

124. Nakajima, Y., Mori, A., Maeda, T., Fujimiya, M., 1997, *Brain Res.* **765**: 113–121.

125. Nasonov, E. L., 1994, *Clinical and Immunopathological Picture of Rheumatic Disorders.* Moscow, pp. 193–200 (in Russian).

126. Nasonov, E. L., Tsvetkova, E. S., *et al.*, 1998, *Ther Arch.* **5**: 8-14 (in Russian).

127. Nawroth, P. P., Stern, D. M., 1986, *Exp Med J.* **163**: 740–745.

128. Nordan, R., Potter, M., 1986, *Science.* **233** (4763): 566–569.

129. Orgogozo, J. M., Dartigues, J. F., 1986, *Acute Brain Ischemia. Medical and Surgical Therapy.* Raven Press, New York, pp. 282–289.

130. Pantoni, L., Sarti, K., Lnytari, D., 1998, *Arterioscler Thromb Vase Biol.* **18**: 503–513.

131. Paradowski, M., Kubasiewicz Ujma, B., *et al.*, 1995, *Clin Biochem.* **28** (4): 459–466.

132. Perini, F., Morra, M., Alecci, M., *et al.*, 2001, *Neurol Sci.* **22** (4): 289–296.

133. Petito, C. K., Morgello, S., Felix, J., Lesser, M. L., 1990, *J Cerebr Blood Flow Metab.* **10**: 850–859.

134. Piani, D., Frei, K., Pfister, H. W., Fontana, A., 1993, *J Neuroimmunol.* **48**: 99–104.

135. Prehn, J. H. M., Bindokas, V. P., Marcuccilli, C. J., *et al.*, 1994, *Proc Natl Acad Sci. USA*, **91**: 12599 12603.

136. Pulsinelli, W. A., 1995, *Am Sci Med.* **2**: 16–25.

137. Quan, J. M., Martin, T. R., *et al.*, 1996, *Biochem Biophys Res Commun.* **219** (2): 405–411.

138. Ransohoff, P. M., Glabinski, A., Tani, M., 1996, *Cytokine Growth Factor Rev.* **7** (1): 35–46.

139. Ridker, P. M., Hennekens, C. H., Tracy, P. P., *et al.*, 1998, *Circulation.* **97** (5): 425–428.

140 Rinner, W. A., Bauer, J., Schmidts, M., *et al.*, 1995, *Glia.* **14** (4): 257–266.

141. Saito, K., Suyama, K., Nishida, K., *et al.*, 1996, *Neurosci Lett.* **206** (2–3): 149–152.

142. Samuelsson, B., 1983, *Science.* **220**: 568–575.

143. Scandinavian Stroke Study Group, 1985, *Stroke*, **16**: 885–890.

144. Schreck, R., Albermann, K., Baeuerle, P., 1992, *Free Rad Res Commun.* **17**: 221–237.

145. Schroder, J. M., Sticherling, M., Henneicke, H. H., *et al.*, 1990, *J Immunol.* **144**: 2223–2232.

146. Schwanzel-Fukuda, M., Abraham, S., Crossin, K. L., *et al.*, 1992, *J Comp Neurol.* **321**: 1–7.

147. Shull, M. M., Ormsby, Y., Kier, A., 1992, *Nature.* **359**: 693–699.

148. Siesjo, B. K., 1981, *J Cerebr Blood Flow Metab.* **1**: 155–185.

149. Siesjo, B. K., Bengtsson, F., 1989, *J Cerebr Blood Flow Metab.* **9**: 1579–1590.

150. Singh, N., Kahlon, P. S., Arora, S., 1995, *Indian Pediatr.* **32** (6): 687–688.

151. Siren, A.-L., McCarron, R. M., Liu, Y., *et al.*, 1993, In *Microcirculatory Stasis in the Brain* (Tomita, M., Mchedlishvili, G., Rosenblum, W. L., Heiss, W.-D., Fukuuchi, Y., eds.) Excerpta Medica, Amsterdam, pp. 169–175.

152. Skvortsova, V. I., Rayevsky, K. S., Kovalenko, A.V., *et al.*, 1999, *Korsakoff J Neurol Psychiatr.* **2**: 34–39 (in Russian).

153. Skvortsova, V. I., Nasonov, E. L., Zhuravlyova, E. Yu., *et al.*, 1999, *Korsakoff J Neurol Psychiatr.* **5**: 27–32 (in Russian).

154. Smith, W. B., Noack, L., Khew-Goodall, Y., *et al.*, 1996, *J Immunol.* **157** (1): 360–368.

155. Sornas, R., Ostlund, H., Muller, R., 1972, *Arch Neurol.* **26**: 489–501.

156. Steel, D. M., Alexander, A. S., 1994, *Immunol Today.* **74**: 80–87.

157. Suematsu, M., Schmid-Schonbein, G. W., Chavez–Chavez, R. H., *et al.*, 1993, *Am J Physiol.* **264**: 881–891.

158. Symon, L., Branston, N. M., Strong, A. J., *et al.*, 1977, *J Clin Pathol.* **30**: 149–154.

159. Tanahashi, N., Fukuuchi, Y., Tomita, M., *et al.*, 1993, In *Microcirculatory Stasis in the Brain* (Tomita, M., Mchedlishvili, G., Rosenblum, W. L., Heiss, W.-D., Fukuuchi, Y., eds.) Excepta Medica, ICS 1031, Amsterdam, pp. 203–210.

160. Tarkowski, E., Rosengren, L., Blomstrand, C., *et al.*, 1995, Materials of European Stroke Conference (June 1–3), *Cerebrovasc Dis.* **4–5**: 240–293.

161. Taupin, V., Toulmond, S., Serrano, A., *et al.*, 1993, *J Neuroimmunol.* **42**: 177–186.

162. Terai, K., Matsuo, A., McGeer, E. G., *et al.*, 1996, *Brain Res.* **739**: 343–349.

163. Tomita, M., Fukuuchi, Y., 1996, *Acta Neurochir.* **66**: 32–39.

164. Tomita, M., Fukuuchi, Y., Terakawa , S., 1994, *Acta Neurochir (Wien).* **1** (60): 31–33.

165. Tomita, F., Fukuuchi, Y., Tanahashi, N., *et al.*, 1995, *J Cardiovasc Pharmacol.* **25** (2): 34–39.

166. Touzani, O., Young, A. R., Derlon, J. M., *et al.*, 1995, *Stroke.* **26**: 2112–2119.

167. Tsunawasi, S., Sporn, M., Ding, A., Nathan, C., 1988, *Nature.* **334**: 260–262.

168. Turcani, P., Gotoh, F., Tomita, M., *et al.*, 1987, Role of platelets and leukocytes in the development of low perfusion hyperemia in the cerebral ischemic area of cats. In *Cerebral Vascular Disease* (Meyer, J. S., Lechner, H., Reivich, M., Ott, E. O., eds.) Excepta Medica, Amsterdam, pp. 285–289.

169. Uehara, A., Gillis, S., Arimura, A., 1987, *Neuroendocrinology.* **45**: 343–347.

170. Wahl, S. M., 1992, *J Clin Immunol.* **12**: 61–74.

171. Wang, X., Yue, T. L., Barone, F. C., *et al.*, 1994, *Mol Chem Neuropathol.* **23** (2–3): 14–103.

172. Wang, X., Yue, T. L., Barone, F. C., Feuerstein, G. Z., 1995, *Stroke.* **15**: 661–666.

173. Wang, X., Yue, T. L., Ohlstein, E. H., *et al.*, 1996, *J Biol Chem.* **271**: 24286–24293.

174. Warach, S., Gaa, J., Siewert, B., *et al.*, 1995, *Ann Neurol.* **37**: 231–241.

175. Weissner, C., Gehrmann, J., Lindholm, D., *et al.*, 1993, *Acta Neuropathol Beri.* **86**: 439–446.

176. Wiessner, C., Vogel, P., Neumann-Haefelin, T., Hossmann, K. A., 1996, In *Mechanisms of Secondary Brain Damage in Cerebral Ischemia and Trauma.* New York, pp. 1–7.

177. Welch, K. M. A., Windham, J., Knight, R. A., *et al.*, 1995, *Stroke.* **26**: 1983–1989.

178. Wells, T. H., Power, C. A., Lusti Narasimhan, M., *et al.*, 1996, *J Leukoc Biol.* **59** (1): 53–60.

179. Wetsel, W. C., Eraly, S. A., Whyte, D. B., *et al.*, 1993, *Endocrinology.* **132**: 2360–2370.

180. Williams, K., Dooley, N., Ulvestad, E., *et al.*, 1996, *Neurochem Int.* **29** (I): 55–64.

181. Winfree, A., 1993, *SFI Studies in the Sciences of Complexity.* Reading, MA: Addison-Wesley, pp. 207–298.

182. Wisniewski, H. M., Wegiel, J., Wang, K. C., *et al.*, 1989, *Can J Neurol Sci.* **16**: 535–542.
183. Wong, M. L., 1997, In *Neuroscience, Neurology and Health*. WHO, Geneva, pp. 81–104.
184. Wong, M. L., Bongiorno, P. B., Rettori, V., 1997, *Exp Med J.* **163**: 740–745.
185. Wood, P., 1994, *Life Sci.* **55** (9): 666–668
186. Wood, P., Kreutdierg, G. W., 1996, *Trends Neurosci.* **19** (8): 312–318.
187. Wood, P., Choksica, S., Bocchini, V., 1994, *Neurol Report.* **5** (8): 977–980.
188. Yamasaki, Y., Matsuo, Y., Matsuura, N., *et al.*, 1995, *Stroke.* **26**: 318–323.
189. Yamasaki, Y., Suzuki, T., Yamaya, H., *et al.*, 1992, *Neurosci Lett.* **142**: 45–47.
190. Yamasaki, Y., Yamaya, H., Watanabe, M., *et al.*, 1991, *J Cerebr Blood Flow Metab.* **13** (1): 113.
191. Yanagimoto, S., Kinugawa, H., Oe, H., *et al.*, 1996, *Cerebrovasc Dis.* **6**: 121.
192. Yokoyama, M., Fukuuchi, Y., Tomita, M., *et al.*, 1994, In *Microcirculation Annual 1994* (Tsuchiya, M., Asano, M., Ohhashi, T., eds.) Nihon–Lgakukan, Tokyo, pp. 165–166.
193. Zhuravlyova, E. Yu., 1998, *The Role of Peptidergic Neurotransmitter Systems in Both Pathogenesis and Neuroprotective Therapy of Acute Ischemic Stroke (Clinical and Biochemical Study)*. Candidate's dissertation. Moscow, p. 190 (in Russian).

Chapter 9

Autoimmune Mechanisms of Trophic Dysfunction and Ischemic Brain Damage

The nervous, immune, and endocrine systems function together to support the dynamic homeostasis of the human body. As components of a unified system, they interact via mutual regulation involving neuromediators, neuropeptides, cytokines, trophic factors, and hormones and the corresponding receptor systems [95]. Mutual regulation of the nervous, endocrine, and immune systems determines the reliability of their common functioning. However, this entails the risk of systemic dysfunction when any one component of the whole system is affected. Such disorders can be defined as under-regulation pathology, the pathogenesis of which may be connected with primarily nervous or endocrine and/or immune mechanisms [83].

If events caused by activated phagocytosing microglia happen in peripheral tissues, the area of local inflammation would have been replaced by live tissue due to proliferation of surviving cells. Inflammation in peripheral tissues is essentially a healing process. However, the result of inflammation in brain tissue is the formation of a cyst or gliosis, since neurons lack the property of cell fission. It is known that the human body does not have immune tolerability to brain tissue, which is defended by the blood–brain barrier (BBB) to prevent immune conflict. Immune "alienation" and specific reactions of brain tissue to damage, to compare them with the well-developed mechanisms of immune defense in other tissues allow, according to Yamasaki and Kogure et al. [57, 117–119], regarding the brain as a "newcomer" in the evolutionary biological system. It would be recognized by the system of defense as an alien, were the BBB damaged due to either to destruction of endothelium or dysfunction of astrocytes.

Blood–brain barrier changes occur from the first minutes of acute focal ischemia. However, they become much more pronounced several hours after

147

stroke onset due to the complex cascade of microcirculation cellular reactions. Inadequate function of the BBB allows penetration of neurospecific antigens into the blood flow. Immune intolerance to brain protein components causes autoimmune reaction. According to Stark [98], the terms "neurospecific proteins" and "neurospecific antigens" are identical. Autoantibodies may penetrate through the disrupted BBB and additionally perturb the vital functions of normal cells.

According to the "immune net" theory proposed by Jerne in 1973 and confirmed later by several studies conducted over recent decades, the immune system produces antibodies to alien as well as to intrinsic human antigens. A wide range of variously directed natural autoantibodies have been described. It was shown that there is a large population of lymphocytes in healthy individuals (10–30% of all B-cells) that carry on their surface CD_5 antigen and are specialized in production of autoantibodies. Thus, the presence of natural autoantibodies in blood appears to be a normal physiological mechanism. However, in case of their under-regulation their quantity may abruptly increase, causing the development of a pathological autoimmune process.

Freely circulating antibodies to neurospecific antigens are rarely detected in the blood of healthy young people, because the brain is located beyond the BBB, which prevents the possibility for contacts between neurospecific antigens and immune competent cells that produce antibodies. The frequency of contact of freely circulating antibodies to brain antigens increases with age, suggesting that a "background" of destructive changes in the BBB and brain tissue appears [60, 66, 95]. The continued presence of antibodies to neuroantigens in blood indicates progressive development of destructive changes [13, 108]. Antibodies permeating into brain tissue are distributed throughout the intercellular space, being then transported by axonal flow into various cell structures inactivating corresponding neuroantigens, which finally leads to the enhancement of damage to neural tissue. Autoantibodies influence the metabolism of neural cells as well as overall specific functions of the CNS [64, 65, 86].

Thus, the development of autoimmune aggression occurs, and a chronic and progressive course of CNS pathology is connected with this. Autoimmune reactions present an important pathogenic link of encephalopathy of various origins—schizophrenia, multiple sclerosis, cerebellar degenerations, and other nervous diseases [23, 27, 42]. At the same time, the autoantibodies to neurospecific proteins as well as neurospecific antigens themselves may serve as unique markers of pathological states of the CNS [85]. Along with this, we ought to emphasize the complexity of adequate data acquisition when auto-sensitization to brain antigens is determined, as the process of auto-sensitization depends not only on the release of antigens into the CSF, but also on immunogenicity and tolerability of a neurospecific protein, as well as on the state of the immune system taken as a whole, which determines the immune status of the human body [7].

We monitored autoantibodies to phencyclidine binding protein of NMDA receptors in patients with acute ischemic stroke, which revealed steady and high elevation of its content in serum already 3 h from the stroke onset (see Section 4.1.3). Such fast generation of antibodies suggests prior sensitization of brain tissue to fragments of NMDA receptors, i.e. it points out the existence of "background" pre-stroke damage to both neuronal membranes and the BBB. Based on these findings, we concluded that there is a "preparedness" of brain tissue for the formation of infarction due to its preceding chronic ischemic state leading to the development of destructive events of encephalopathy.

Our contention was confirmed during clinical and immunological study of 40 patients with chronic ischemic brain disease having no history of stroke including "silent" stroke according to brain MR imaging. Detection of autoantibodies to phencyclidine binding protein of NMDA receptors in their serum showed that although the average level was close to the normal value (1.8 ± 0.25 ng/ml), more than 40% of patients exhibited its elevation up to 2.5–4.0 ng/ml. After detailed clinical analysis, we found that all patients with elevated autoantibody level had been chronically suffering from arterial hypertension (usual systolic blood pressure above 180 mm Hg) with repeated hypertensive crisis. Considering the influence of arterial pressure on BBB permeability, such connection between the extent of auto-sensitization to protein components of neuronal membranes and the damaging action of the progressive chronic ischemia appears rather certain.

Our clinical results are consistent with experimental data obtained from animals with global brain ischemia [29]. These experiments demonstrated how baseline BBB changes and the sensitization of the human body to structural components of brain tissue influence the severity and course of cerebral ischemia. Gannushkina found that in animals sensitized by extract of brain tissue the frequency of stroke development was elevated 3-fold versus intact animals after identical influences on blood flow: cerebral blood supply repeatedly reduced, baseline EEG changes caused by global ischemia recovered poorly; more prominent clinical deficit was registered. Antigen–antibody reaction was found within the affected brain area under the conditions of both altered blood supply and "anti-cerebral" antibody circulation. It is noteworthy that the appearance of this reaction took a rather short time (4–5 min).

9.1.   Nerve growth factor (NGF) and its autoantibody concentration in CSF and serum of patients with ischemic stroke

The level of brain trophic support plays a very important role in the development of ischemic damage. Trophic support to a considerable extent determines the choice of genetic programs of apoptosis and antiapoptotic defense as well as influences both necrotic death pathways and reparative events [39, 58, 60, 70, 104, 114] (see Chapters 10 and 12).

Neurotrophic properties are inherent in many proteins including structural proteins (for instance S100β). However, these properties are mainly found for growth factors, which represent a heterogeneous group of trophic factors comprising at least 7 families [67, 110, 111]: neurotrophins, cytokines, fibroblast growth factors, insulin-like growth factors, transforming growth factor β_1 family (TGF-β_1), epidermal growth factors, and others, including growth-associated protein 6 (GAP-6), platelet-derived growth factor, heparin-binding neurotrophic factor, erythropoietin, macrophagal colony stimulating factor, etc. (see Table 9.1).

The strongest trophic influence on every basic neuronal vital function is produced by neurotrophins, regulatory proteins of neural tissue that are synthesized in its cells (neurons and glia). They act locally where released and intensely induce dendritic growth (arborization) as well as axonal growth (sprouting) directed onto target cells. Synaptic sprouting promotes the re-enhancement of pre-existing neuronal flows [53] and the formation of new synaptic contacts [14, 49], providing neuronal tissue plasticity and arranging pathways involved in neurological recovery.

To date three neurotrophins have been well studied. They are similar to each other in structure: nerve growth factor (NGF), brain-derived neurotrophic factor (BDNF) and neurotrophin-3 (NT-3) [14]. In the developing human body, they are synthesized by a target cell (e.g., muscle fuse), are then distributed via diffusion towards a neuron, bind to receptor molecules on its surface, which leads to active growth of a neuronal axon. As a result, the axon reaches a target cell and establishes a synaptic contact with it. Growth factors improve the survival of neurons, which cannot exist without such trophic support.

It is interesting that even neurotrophins can manifest ambiguous nature, fulfilling modulatory apoptotic/antiapoptotic functions. NGF exerts its action via two receptors, P75NTR and TrkA, the expression of which varies in a cycle phase-specific manner [11]. NGF exhibits pro-apoptotic properties

Table 9.1. Contemporary classification of neurotrophic factors (modified from M. H. Tuszynki [110])

Trophic factor family	Representative	Basic targets
Neurotrophins	Nerve Growth Factor (NGF)	Basal forebrain cholinergic neurons, striatum, nociceptive sensory neurons, sympathetic neurons
	Brain-Derived Neurotrophic Factor (BDNF)	Basal forebrain cholinergic neurons, hippocampus, cortical, mechanoreceptors sensory, alpha-motor, vestibular, auditory, retinal ganglion neurons
	Neurotrophin-3 (NT-3)	Basal forebrain cholinergic neurons, hippocampus, cortical neurons, proprioceptor sensory, muscle spindle, auditory neurons, oligodendrocytes
	Neurotrophins-4/5 (NT-4/5)	Motor (alpha), retinal ganglion, sensory neurons
	Neurotrophin-6 (NT-6)	Not elucidated
	Neurotrophin-7 (NT-7)	Not elucidated
Cytokines	Ciliary Neurotrophic Factor (CNTF)	Central (cortical) and peripheral (alpha) motor neurons, striatum, parasympathetic, sensory neurons, hippocampus, basal forebrain cholinergic neurons, oligodendrocyte/astrocyte progenitor cells
	Leukemia Inhibitory Factor (LIF) or Cholinergic Differentiation Factor (CDF)	Central (cortical) and peripheral (alpha) motor neurons, glial cells, cortical neurons
	Cardiotrophin-1	Alpha-motor neurons
	Interleukins 1, 2, 3, 6, (IL-1,2,3,6)	Basal forebrain cholinergic neurons
Fibroblast Growth Factors (FGFs)	Fibroblast Growth Factor-1 (FGF-1, acidic FGF)	Cortical neurons, spinal cord and brain stem cells

(*continued*)

Table 9.1 (*continued*)

Trophic factor family	Representative	Basic targets
	Fibroblast Growth Factor-2 (FGF-2, basic FGF)	Basal forebrain cholinergic neurons, dopaminergic neurons, cortical neurons, retinal ganglion, brain stem cells
	Fibroblast Growth Factors 3-15 (FGFs 3-15)	Not elucidated
Insulin-like Growth Factors (IGFs)	Insulin-like Growth Factor I (IGF-I)	Alpha-motor, sensory, multiple brain neurons, oligodendrocytes
	Insulin-like Growth Factor II (IGF-II)	Sensory, sympathetic neurons
Transforming Growth Factors β (TGF β)	Transforming Growth Factors β$_1$, β$_2$, β$_3$ (TGF-β$_{1-3}$)	Alpha-motor, sensory neurons
	Glial Cell-Line Derived Neurotrophic Factor (GDNF)	Dopaminergic, central (cortical) and peripheral (alpha) motor, sensory neurons
	Neurturin	Dopaminergic, central (cortical) and peripheral (alpha) motor, sensory neurons
	Persephin	Dopaminergic, central (cortical) and peripheral (alpha) motor neurons
	Activin	Alpha-motor neurons
	Bone Morphogenetic Proteins (BMPs)	CNS and PNS formation during development
Epidermal Growth Factors (EGFs)	Epidermal Growth Factor-α (EGF)	Dopaminergic neurons, hippocampus, cortical neurons, brain stem cells
	Transforming Growth Factor-α (TGF-α)	Dopaminergic neurons, hippocampus, cortical neurons (*in vitro*)
	Neuregulins	Maturation of glial cells and synapses (*continued*)

Table 9.1 (*continued*)

Trophic factor family	Representative	Basic targets
	Neutral- and Thymus-derived Activator for ErbB Kinases (NTAK)	Not elucidated
Others	Growth-Associated Protein 6 (GAP-6)	Not elucidated
	Platelet Derived Growth Factor (PDGF)	Not elucidated
	Macrophagal Colony Stimulating Growth Factor (MCSF)	Not elucidated
	Erythropoietin	Basal forebrain cholinergic neurons
	Heparin-binding Neurotrophic Factor	Not elucidated
	Hepatocyte Growth Factor (HGF)	Hippocampus, alpha-motor, sensory neurons, Schwann cells

on growing cells expressing preferentially P75NTR and exhibits anti-apoptotic effect on quiescent cells, when TrkA is prevalent at the cell surface. NGF can have a dual action on cells depending on the relative cell surface expression of TrkA and P75NTR. The pro-apoptotic activity of NGF but not its anti-apoptotic activity is abrogated by an antibody against the extracellular domain of P75NTR and in cells isolated from P75NTR knock-out mice indicating that NGF exhibits a proapoptotic activity via P75NTR exclusively. On the other hand, the anti-apoptotic activity of NGF is specifically mediated by an interaction with TrkA, with no contribution of P75NTR. So, depending on the growth state of the cells, NGF exhibits dual pro- or anti-apoptotic properties via P75NTR and TrkA, respectively.

Trophic under-regulation is one of the universal components in the pathogenesis of nervous system damage [20, 28]. Lacking trophic support, mature cells exhibit such biochemical and functional de-differentiation that properties of the innervated tissues change [16, 28, 49, 53]. Trophic under-regulation worsens the state of macromolecules involved in membrane-associated electrogenesis, active ion transport, synaptic transduction (enzymes of mediator synthesis, post-synaptic receptors), and effector function (muscle myosin). The assemblies of de-differentiated central neurons generate focuses of pathological excitation, triggering biochemical cascades leading to neuronal necrosis or apoptosis [43, 50]. Oppositely, when the level of trophic support is sufficient the improvement of a neurological deficit after ischemic brain damage is often observed even when a morphological lesion that caused the deficit remains [111], which shows the high adaptive capacity of brain function.

Considering that the insufficiency of neurotrophins promotes cell death in the penumbral area, interest should be attracted to study of mechanisms causing trophic under-regulation. It was shown that in the development of trophic under-regulation such mechanism as changes in potassium, calcium, and zinc homeostasis is involved. The metal modulators alter the conformation of neurotrophins, rendering them unable to bind to their receptors or to activate signal transduction pathways and biological outcomes normally induced by these proteins [87]. The excessive synthesis of NO, which blocks tyrosine kinase incorporated in the active center of trophic factors, and the imbalance of cytokines also participate in the development of neurotrophin insufficiency. One of the mechanisms proposed is autoimmune aggression against intrinsic neurotrophins and structural neurospecific proteins possessing neurotrophic properties. Such aggression becomes possible when the disruption of the BBB protective function ensues after ischemic damage [31].

To study the role of autoimmune mechanisms in the pathogenesis of trophic under-regulation, a collaborative study was performed by our neurological clinic together with the Institute of Molecular Genetics (N. F. Myasoyedov *et al.*) and the Mental Health Research Center of the Russian Academy of Medical Sciences (T. P. Klushnik *et al.*). We studied NGF and its autoantibody contents in CSF and blood of patients with acute ischemic stroke [100–102].

We enrolled 25 patients (15 males and 10 females; average age 66.2 ± 4.1 years) with acute ischemic stroke in the carotid artery territory (left in 16 cases and right in 9 cases). Patients were admitted to the neurological resuscitation department within the first 12 h of stroke. We excluded patients with acute inflammatory, hereditary neurodegenerative, and autoimmune concomitant diseases. The main etiological factors of stroke were atherosclerosis with concomitant arterial hypertension (21 patients or 84%), atherosclerosis without evident arterial hypertension (3 patients or 12 %) and arterial hypertension in young patients (1 patient or 4%). In 96% of the patients, we found signs of ischemic heart disease. In 17 (68%) of patients stroke exhibited atherothrombotic origin, in 7 (28%) it followed cardiac embolism (atrial fibrillation) and in 1 patient (4%) it followed the depression of systemic blood supply after myocardial infarction. There were no lacunar strokes in this group.

From admission patients received multi-dimensional and at the same time maximally standardized treatment without neuroprotectors.

To objectify the severity and the extent of neurological deficit and to standardize the processing of clinical data, we used three complimentary scales: Scandinavian [89], Orgogozo [77] and original [32]. At the time of admission, the neurological deficit of 12 patients was diagnosed as moderate (total clinical scores were 32.3 ± 1.82; 52.1 ± 2.7; 39 ± 0.48, respectively). Focal neurological deficit predominated in the clinical picture, and in 41.7% of cases it was accompanied by mild general cerebral symptomatics. In 13 patients severe neurological deficit was observed at the time of admission (total clinical scores were 22.3 ± 2.33; 35.6 ± 3.5; 32.06 ± 2.7, respectively) along with clouding of consciousness to various extent (obnubilation was diagnosed in 7 patients, stupor in 4, and initial coma in 1 patient).

Quantitative detection of NGF content and its autoantibodies in CSF and serum was performed by enzyme-linked immunosorbent assay (ELISA) repeatedly: on admission (before treatment was started) and on the 3rd day of stroke. Commercial immunoenzyme sets manufactured by Promega (USA) were used for the detection of NGF.

To detect the level of autoantibodies to NGF, we activated polystyrene plates using NGF in 5 µg/ml concentration diluted in 0.01 M phosphate buffer with pH 7.2 containing 0.15 M Cl⁻ and 0.05% solution of Tween-20. In each plate 100 µl of this solution was added and then the mixture was incubated for 16–24 h at 4°C. After that wells of the plate were washed several times with phosphate buffer (PBS) and 100 µl of the serum (or CSF) tested was placed into each well diluted 1/20 with PBS containing 0.15 M Cl⁻, 0.5% gelatin solution, and 0.1% Tween-20; the mixture was incubated for 2 h in 37°C. Then the plates were washed and 100 µl of antibodies against human immunoglobulins conjugated with horseradish peroxidase (Sigma, USA) was placed in each well. After incubation for 2 h and washing of the plate, we determined bound antibodies by injecting into each well 100 µl of substrate mixture that contained 1.1% solution of *o*-phenylenediamine and 0.006% solution of H_2O_2 in 0.1% citrate buffer with pH 5.2. The reaction was terminated by the injection of 25 µl 0.05 M sulfuric acid into each cell. Optical density was measured using a multi-channel spectrophotometer.

Normal values were obtained from CSF samples taken from 20 orthopedic patients having neither neurological nor inflammatory disorders (control group); serum values were obtained from 20 healthy individuals of comparable age.

Statistical analysis was performed by nonparametric tests (Mann–Whitney U-test, Spearman's *r*-coefficient of rank correlation). Fatality was measured by Fisher's test. The differences between values were assumed to be significant at the 5% level in every comparison conducted.

The study of NGF content within the first day after stroke revealed its significant increase in CSF in all patients with moderately severe stroke, up to 189 ± 81 ng/ml, which exceeded the value in patients with severe stroke by 112% ($p < 0.04$) and by 188.8% in controls ($p < 0.001$). By the 3rd day the level of NGF was moderately elevated (by 21% versus the 1st day, $p = 0.046$) and comprised 230 ± 74 ng/ml. In severely affected patients with lower level of NGF in CSF (89 ± 39 ng/ml) on baseline (the 1st day), the neurotrophin content increased by the 3rd day by 124% (199.4 ± 61 ng/ml, $p = 0.054$). Thus, in the 1st day of stroke the level of NGF in CSF correlated with the extent of neurological deficit: the lower was NGF content, the lower was total clinical score and more severe was the deficit ($r = 0.42$; $p = 0.05$).

When compared to the volume of ischemic lesion (according to MR images) in the 1st day, NGF content exhibited no correlation. However, significant reverse correlation was found by the 5th–7th day of stroke between the infarction volume and the 1st day NGF content ($r = -0.55$; $p = 0.04$). The lower the NGF content, the larger was the infarction volume on the 5th–7th day of stroke in cases with both comparable location of vascular lesion and sizes of ischemic lesion on the 1st day (Fig. 9.1).

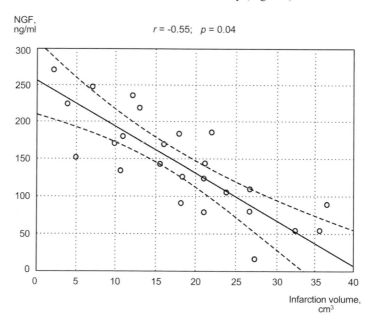

Figure 9.1. Correlation between NGF concentration in CSF on day 1 and the infarction volume on days 5–7 in patients with carotid ischemic stroke.

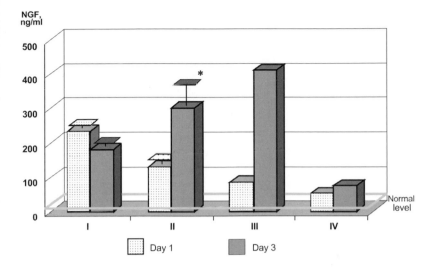

Figure 9.2. Dynamics of NGF concentration in CSF of patients with carotid ischemic stroke in relation to stroke outcome on day 21. I – good functional recovery (BI ≥ 75); II – severe disability (BI < 50); III and IV – death. Significance of the differences versus the first examination (on day 1): * $p < 0.05$.

The retrospective study of NGF in CSF in relation to both clinical outcome of stroke and the extent of neurological recovery by the 21st day (the end of the acute period) allowed the patients to be subdivided into 3 groups. In all patients with good recovery from neurological deficit (BI 75–100), NGF content within the first hours of ischemia exceeded 200 ng/ml and did not significantly change by the 3rd day of stroke ($p > 0.06$). In cases with severe disability (BI 0–45) baseline (1st day) NGF level was <150 ng/ml and increased ($p = 0.05$) by the 3rd day (Fig. 9.2). The third group contained 2 patients with subtotal hemispheric ischemia, progressive course of brain edema, and dislocation syndrome, who died within the acute period of stroke. One of them (died on the 16th day) exhibited low value of NGF content (84 ng/ml) on admission, but it abruptly elevated by the 3rd day (by 397.6%) to the maximal level of 418 ng/ml in the group. Such prominent elevation the 10th day before death may have been the last desperate attempt to compensate the triggered cell death pathways undertaken by the system of trophic support. The other patient (died on the 5th day) NGF levels registered on admission as well as on the 3rd day were the lowest in the group (45 and 72 ng/ml, respectively) (see Fig. 9.2). Our results were in accordance with the data of Stanzani *et al.* [105] showing that

NGF serum levels in stroke patients were significantly associated with clinical and neuroradiological parameters of brain injury.

The study of the level of autoantibodies to NGF in serum of stroke patients revealed values close to normal (0.7 ± 0.15 optical density (OD) units): within the first 2–12 h in patients with moderately severe stroke the average level was 0.77 ± 0.06 OD units and 0.8 ± 0.08 OD units in patients with severe stroke; no dynamics of these levels was observed by the 3rd day. At the same time, in CSF of all patients we found elevation of the level of autoantibodies to NGF already on admission versus control level (0.1 ± 0.06 OD units): by 190% in patients with moderately severe stroke (0.29 ± 0.07 OD units; p < 0.0001) and by 180% in severe stroke (0.28 ± 0.08 OD units; p < 0.001) (see Fig. 9.3). On the 1st day the level of autoantibodies to NGF neither correlated with the extent of neurological deficit (r = 0.06) nor with the ischemized volume according to MR images (r = 0.08). By the 3rd day of stroke in patients with moderately severe stroke the level of autoantibodies to NGF in CSF was practically unchanged. All those patients exhibited good recovery by the end of the acute period. In patients with minor strokes with complete neurological recovery within the acute period

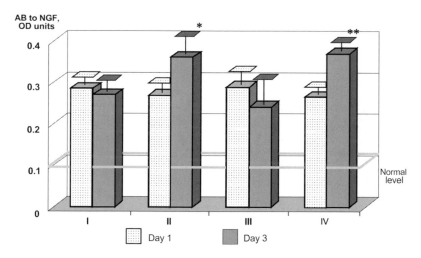

Figure 9.3. Levels of autoantibodies (AB) to NGF in CSF of patients with carotid ischemic stroke in relation to neurological deficit and functional recovery on day 21. I – moderately severe stroke (OSS > 40); II   severe stroke (OSS < 39); III – good functional recovery (BI ≥ 75); IV – severe disability (BI < 50). Significance of the differences versus the first examination (on day 1): * p < 0.05, ** p < 0.01.

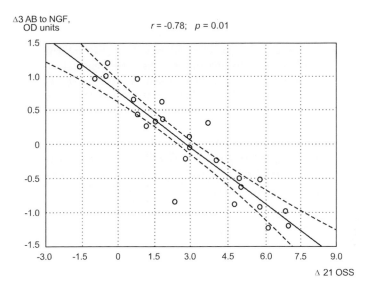

Figure 9.4. Correlation between the dynamics of CSF level of autoantibodies (AB) to NGF by day 3 and the increase in total clinical score (assessed by the Orgogozo Stroke Scale – OSS) by day 21 in patients with carotid ischemic stroke.

of stroke, there was a tendency towards a decrease in autoantibody level by 24% on average. Along with this, in the group with severe stroke we registered significant elevation of the level by the 3rd day (by 32.1% on average versus baseline values; $p < 0.045$) (see Fig. 9.3). It is noteworthy that the increase in autoantibody level was most prominent and significant in patients with the worst clinical outcome: in patients who later died by 120 and 165% and in later severely disabled by 71.2% ($p < 0.01$). Changes in the level of autoantibodies by the 3rd day of stroke exhibited reverse correlation with total clinical score by the 21st day ($r = -0.78$; $p = 0.01$) (Fig. 9.4).

Thus, from the first hours of acute focal ischemia the level of autoantibodies to NGF was elevated in CSF, whereas in blood serum no elevation was detected. We suppose that autoimmune reactions to neurotrophins have a local character and are connected with intracranial synthesis of antibodies within the first days of stroke. The absence of generalized autoimmune reaction without penetration of antibodies to NGF in blood can possibly be explained by relative preservation of the neurotrophin system in the pre-stroke period, when patients do not exhibit significant functional disturbances. Taking into consideration the increased BBB permeability that accompanies acute cerebral ischemia, we cannot rule out further generalization of autoimmune reaction to neurotrophins in the more distant period.

We found a close correlation between the dynamics of autoantibody level to NGF and functional outcome as well as neurological recovery. This emphasizes the role of auto-aggression against neurotrophins during the development of trophic under-regulation and the progression of ischemic events.

9.2. Neurospecific protein S100β and its autoantibody concentration in CSF and serum of patients with ischemic stroke

The data acquired have demonstrated the participation of autoimmune mechanisms in the formation of chronic destructive processes in brain tissue and its predisposition to the development of focal infarction [33–38]. So, it was of great interest for us to compare the revealed changes in NGF and its antibodies content with the reaction of S100, a membrane-associated neurospecific protein possessing trophic properties, onto acute ischemia.

The S100 protein was first isolated by Moore in 1965. The name "S100" is connected with the capacity of this protein to be dissolved in 100% saturated solution of ammonium sulfate at pH 7.2. It is now known that S100 is a group of acidic calcium-binding proteins unique for neural tissue; these proteins differ from each other in charge and molecular weight but have similar immune properties. Their concentration in brain 100,000-fold exceeds their content in other tissues and comprises up to 90% of the soluble protein fraction of neural cells [25, 30]. All fractions of S100 specifically interact with calcium, but they differ from each other in the number of calcium-binding centers (from 2 to 8) [25, 30, 44].

Most of the S100 proteins (up to 85–90% of their total content in neural tissue) are concentrated in astrocytes; 10–15% is located in neurons, and a negligible quantity is detected in oligodendrocytes. To date it has been shown that S100 proteins are synthesized by glial cells and then transported to neurons [74, 88]. In the cell they are predominantly located in cytosol as well as in synaptic protein and in chromatin [71, 83].

Studies performed in recent years indicate that S100 proteins are one of the pivotal molecular components of complex intracellular systems promoting functional homeostasis of brain cells via integration of variously directed metabolic pathways [83]. Four basic molecular processes underlying the biological activity of the S100 protein group have been found.

1. Binding of calcium ions and, hence, calcium-dependent specific intermolecular interaction of S100 proteins with other proteins involving conformational changes of the protein molecules. This interaction occurs as follows: S100 + Ca^{2+} → conformer + acceptor molecule → allosteric modulation of the acceptor molecule functional activity. About 20 proteins and over 25 endogenous neuropeptides have been described that specifically interact with S100 proteins. Among the protein ligands of S100, the following compounds have been identified: alkaline phosphatase, monoamine oxidase, myelin basic protein [72, 81], neurospecific enolase [121], various DNA-binding proteins, and glycoproteins [83].
2. Modulation of protein kinase and phosphoprotein kinase activities [30, 83]. The interaction occurs in a similar way as follows: S100 + Ca^{2+} > conformer + protein kinase → phosphoprotein kinase → influence on phosphorylation/dephosphorylation of protein molecules (post-translational modification).
3. Influence on microtubules of neural cells. As a result of association and dissociation of S100 molecules and calcium ions, the concentration of the latter changes, and this causes the dissociation and reconstitution of microtubules. Thus, S100 indirectly influences intracellular transport [83, 84].
4. Influence on metabolic pathways and on specific reception of neuromediators. S100 changes the level of specific binding with receptors for such ligands as acetylcholine, GABA, dopamine, serotonin, and noradrenaline.

Thus, the various isoforms and conformers of S100 proteins are the most universal macromolecules known; they are involved in the regulation of practically all basic membrane-associated, cytosolic, and nuclear metabolic processes connected with the support of perception and integration of data delivered to the nervous system [31, 95], contribute to early response gene reactions, and to embodiment of genetic programs of apoptosis and anti-apoptosis [90] (Fig. 7.1). The regulatory potential of S100 proteins is embodied via systems of second messengers, especially intracellular calcium ions. Calcium-dependent reconstitution of S100 protein spatial structure allow them in the form of a particular conformer to specifically bind with certain molecules of neural tissue, allosterically regulating the activity of the latter or forming supramolecular complexes with them possessing altered functional properties. Thus, S100 proteins, though they do not replace any key metabolic links, are involved in their systemic integration, constituting the molecular basis for the arrangement of specific physiological functions of the nervous system.

It has been proven experimentally that S100 proteins participate in regulation of directed growth of neuronal processes, in the completion of neural ontogenesis in a morphological as well as in a functional aspect, in the establishment of basic forms of inherited behavior, and in mechanisms of learning and memory.

S100 proteins are not vitally important components (such as glycolytic enzymes or enzymes of oxidative phosphorylation), necessary for the support of general homeostasis of living cells. Characteristically, experimental perturbation of S100 proteins are not usually accompanied by prominent worsening of somatic status of animals, but simultaneously they lead to abrupt and various disorders of integrative brain function, of informational homeostasis, the support and optimization of which are their general biological function [86].

S100β is a representative of the S100 protein family. It is the most brain tissue-specific protein. It is known that when brain tissue is destroyed S100β along with other proteins of this group can be detected in blood and CSF of patients. S100β protein is of special interest as it has been recently shown to possess properties inherent in neurotrophic and growth factors. It was demonstrated that the injection of low doses of S100β into neuronal culture promotes the generation and growth of neurites, whereas in control cultures cells never survived. S100β may be involved in regeneration of brain tissue after ischemic events.

The first application of S100 proteins (the main fraction) in clinical practice dates from 1973, when Haglid *et al.* [40] found their elevated concentrations in the brain tumor tissue. In the 1980s S100 proteins became used as markers of brain tissue damage when Michetti *et al.* [73] found S100 in CSF of only those patients whose disease was connected with brain damage. Using monospecific rabbit antiserum against S100 proteins, their threshold concentration accessible under normal conditions in CSF was limited, 6 ng/ml. In recent years, much attention has been given to S100 proteins as diagnostic markers. Martens *et al.* [68] detected the main fraction of S100 proteins in patients with acute stroke and revealed that they appeared in CSF in 1/3 of patients with large volume stroke, and when the stroke was ischemic their level reached a maximum by the 3rd day, and in case of hemorrhage by the end of the 1st day. Johnson and Liebiry [51], Blomquist *et al.* [10], Anderson *et al.* [2], Wong [115], and Pisa [82] proposed the use of S100 as a biochemical marker of central nervous system damage after cardiovascular surgery with artificial blood supply. High contents of S100 proteins or antibodies to them were found in craniocerebral injury [61], epilepsy [103], schizophrenia [46, 54], and demyelinating disorders [72]. Detection of S100 is applied in the histological diagnostics of brain tumors [109].

Studies conducted in recent years have proved that S100 proteins, including S100β, can be used as markers and a prognostic criterion of brain damage in ischemic stroke [4, 12, 82, 115, 116]. The total concentration of proteins of this group in large ischemic strokes comprises 0.4 ng/ml in serum (the normal value is about 0.2 ng/ml) and about 10 ng/ml in CSF (normally about 6 ng/ml) [1, 51]. A correlation was demonstrated between the content of total fraction of S100 and S100β protein and the infarction size as well as the clinical outcome of stroke [12, 26, 52, 68, 116].

Together with colleagues from Anokhin Institute of Normal Physiology of the Russian Academy of Medical Sciences (V. V. Sherstnyov, M. A. Gruden, *et al.*), we studied S100β and its autoantibody content in CSF and serum of 25 patients with carotid ischemic stroke (the clinical description of the group is given in the previous section) [24, 100, 101].

Quantitative detection of S100β protein and of the level of autoantibodies to it was performed by enzyme-linked immunosorbent assay (ELISA). Protein and autoantibody contents in CSF were assessed repeatedly: on admission (before treatment was started) and on the 3rd day of stroke. The monitoring of the level of primary and secondary (anti-idiotypic) antibodies in serum was performed on the 1st, 3rd, 7th, 14th, and 21st day.

To detect **S100β protein**, doubled CSF samples (in two replicas) of 50 µl were injected into each well except control of a 96-well polystyrene plate for immunoenzyme essay (VNII Polymerbyt, Russia). The samples were previously diluted in ratio 1:1 in antigen sorption buffer. Samples were incubated for 24 h at 4°C. After incubation the plate was washed thrice with PBS-Tween buffer with pH 7.2 and then distilled water. After that, to avoid nonspecific binding of antigen to polystyrene cover, in each well except control 100 µl of 1% albumin solution was added. The plate was incubated for an hour in 4°C and then washed as described. In the following stage of immunoenzyme assay we added the antibody to S100β solution (monospecific polyclonal affinity purified rabbit IgG) at 20 mg/ml concentration. It was incubated for 24 h in 4°C. The plate was then washed 6 times with Tris-HCl-buffer with pH 7.5, Tween-20 solution, and distilled water. Then in each well except control we injected 50 µl of IgG goat conjugate against rabbit IgG with horseradish peroxidase with standard 1:1000 titer. It was incubated for 2 h at 37°C. Then the plate was washed 6 times with Tris-HCl-buffer with pH 7.5, Tween-20 solution, and distilled water. Then into each well including control we injected 50 µl of the substrate: *o*-phenylenediamine in substrate buffer at 1 mg/ml content. Before placing onto the plate, H_2O_2 was added in proportion of 3.5 µl for 10 ml of the substrate. The reaction proceeded in darkness. Thirty minutes after the start of the reaction, we analyzed the results using an automatic immunoenzyme reader according to the optical density of the colored substrate. The protein content in CSF was determined according to a calibration curve that was constructed based on known standard solutions.

To detect **the level of autoantibodies to S100β protein** we injected 50 µl of S100β protein solution taken at the concentration of 25 µg/ml into each well of a 96-well polystyrene plate except control. Samples were incubated for 24 h at 4°C. After incubation the plate was washed thrice with PBS-Tween buffer at pH 7.2 and distilled water. After that, to avoid a phenomenon of nonspecific binding of antigen to polystyrene, in each well except control 100 µl of 1% albumin solution were added. The plate was incubated for an hour at 4°C and then washed as described. In the following stage of immunoenzyme assay we titrated serum samples from patients. In all rows except the first, we injected 50 µl of Tris-HCl-buffer with pH of 7.5. Then in the wells of the first and the second row, except control, we added 50 µl of baseline serum sample. Then using a multi-channel pipette of 50 µl volume the diluted sample was transposed from the second into the third row etc. Then mixtures were incubated for 24 h at 4°C. The plate with titrated samples was then washed 6 times with Tris-HCl-buffer with pH 7.5, Tween-20 solution, and distilled water. Then in each well except control we injected 50 µl of IgG goat conjugate against human IgG with horseradish peroxidase with standard 1:1000 titer. It was incubated for 2 h at 37°C. Then the plate was washed 6 times with Tris-HCl-buffer with pH 7.5, Tween-20 solution, and distilled water. Then in each well including control we injected 50 µl of the substrate: *o*-phenylenediamine in substrate buffer at 1 mg/ml content. Before placed onto the plate H_2O_2 was added in proportion of 3.5 µl for 10 ml of the substrate. The reaction was performed in darkness. Thirty minutes after the start of the reaction, we analyzed the results using the automatic immunoenzyme reader according to the optical density of the colored substrate.

To detect **the level of anti-idiotypic antibodies to S100β protein**, we injected 50 µl of antibodies to S100β protein solution taken in the concentration of 20 µg/ml into each well of a 96-well polystyrene plate except control. Samples were incubated for 24 h at 4°C. After incubation the plate was washed thrice with PBS-Tween buffer with pH 7.2 and distilled water. After that, to avoid nonspecific binding of antigen to the polystyrene, in each well except control 100 µl of 1% albumin solution were added. The plate was incubated for 1 h at 4°C and then washed as described. In the following stage of immunoenzyme assay, we titrated serum samples of patients. In all rows except the first, we injected 50 µl of Tris-HCl-buffer with pH 7.5. Then in the wells of the first and the second row, except control, we added 50 µl of baseline serum sample. Then using a multi-channel pipette of 50 µl volume the diluted sample was transposed from the second into the third row, etc., intensely shaking them. Then the mixtures were incubated for 24 h at 4°C. The plate with titrated samples was then washed 6 times with Tris-HCl-buffer with pH 7.5, Tween-20 solution, and distilled water. Then in each well except control we injected 50 µl of IgG goat conjugate against human IgG with horseradish peroxidase with standard 1:1000 titer. This was incubated for 2 h at 37°C. Then the plate was washed 6 times with Tris-HCl-buffer with pH 7.5, Tween-20 solution, and distilled water. Then in each well including control we injected 50 µl of the substrate: *o*-phenylenediamine in substrate buffer at 1 mg/ml content. Before placed onto the plate, H_2O_2 was added in proportion of 3.5 µl for 10 ml of the substrate. The reaction was performed in darkness. Thirty minutes after the start of reaction, we analyzed the results using the automatic immunoenzyme reader according to the optical density of the colored substrate.

Protein content in the solution was measured by spectrophotometry following Warburg [19] according to the absorption in the ultraviolet spectrum using a Beckman DU-70 spectrophotometer (USA). We measured optical density of protein solution at 260 (D^{260}) and 280 (D^{280}) nm, and then the data was analyzed using the formula: protein content (mg/ml) = $1.55D^{280} - 0.76D^{260}$.

The study of CSF on the 1st day of stroke revealed elevation of S100β content by 35.7% versus normal values (7.6 ± 0.6 ng/ml versus normal value of 5.6 ± 0.9 ng/ml) as well as normal levels of autoantibodies to S100β (1:2, 1:4). These values did not change significantly by the 3rd day of stroke ($p >$ 0.05).

The detection of S100β in CSF is a consequence of protein release into the intercellular space due to destruction of neuronal membranes, induced by both the intracellular accumulation of calcium ions and the reactions of the glutamate–calcium cascade. Possibly, several hours delayed elevation of S100β content in CSF after stroke onset may explain normal titer of its autoantibodies within the first 3 days of stroke. Clinical limitations hampered the observation of further dynamics of the level of autoantibodies in CSF, although it may be hypothesized that its content would have been elevated at longer times.

It is interesting that, in contrast to NGF, in all patients already within the first hours of stroke we found the elevation of autoantibodies to S100β in serum. In 15 (60%) of the patients the level was very high (1:128, 1:256), and 10 (40%) patients exhibited moderately increased levels (1:16, 1:32, 1:64). Repeated study on the 3rd day failed to reveal any significant dynamics of the level of autoantibodies. To elucidate the remoteness of primary autoantibodies generation the detection of the level of secondary (anti-idiotypic) autoantibodies to S100β protein was included in the study. The study of these autoantibodies on the 1st day of stroke found their low level in all patients (1:2, 1:4), confirming that the generation of primary autoantibodies was sudden.

Clinical and immunological analysis found significant changes in the serum level of autoantibodies to S100β in relation to pathogenic variant of stroke. It atherothrombotic stroke the level of primary autoantibodies was significantly higher than those in strokes following cardiac embolism or the depression of systemic blood supply ($p < 0.04$). Such differences are possibly not accidental. In durable and progressive development of atherothrombosis, the background vascular–cellular processes prepare the BBB and brain tissue for destructive changes. When the stroke develops with relative suddenness due to either embolism or abrupt depression of systemic blood supply, such background vascular–cellular processes can be expressed less.

The analysis of dynamics of autoantibody titers in patients with atherothrombotic stroke established distinct chronological conformities of primary and secondary generation of autoantibodies (Fig. 9.5). The level of primary autoantibodies remained stable until the 7th day and then exhibited a tendency towards decrease with average rate of 1 titer per 7 days. By the 21st day the titer has not been completely normalized. At the same time, from the 7th day we registered sequential elevation of secondary (anti-idiotypic) antibodies on average rate of 1 titer per 7 days with the maximal titer level reached by the 21st day of stroke. These results are consistent with experimental data [45, 80] that showed that the average time interval required for the initiation of antibody generation was 5–7 days.

Rapid elevation of the level of primary autoantibodies to S100β in serum may be the consequence of the previous sensitization of brain tissue to its structural component, as confirmed by the maximal representation of autoimmune changes in atherothrombotic variant of stroke. Along with that, very high level of antibodies (1:256, 1:512) in serum of several patients within the first hours of ischemia suggests that the generation of antibodies forerun the clinical manifestation of stroke. Thus, the participation of

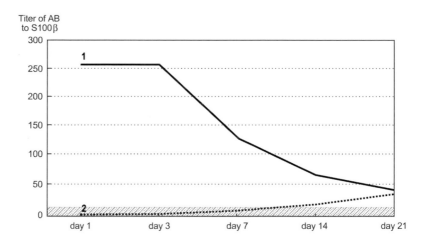

Figure 9.5. Dynamics of serum levels of primary and secondary antibodies (AB) to S100β in patients with atherothrombotic ischemic stroke. 1 — primary AB; 2 — secondary (anti-idiotypic) AB.

autoimmune mechanisms in preparation and embodiment of brain tissue response to CBF decrease and focal ischemia has become obvious.

9.3.   Myelin basic protein and its autoantibody concentration in CSF and serum of patients with ischemic stroke

Taking into consideration the autoimmune sensitization to S100β protein found within the first hours of ischemic stroke, we thought that it would be interesting to elucidate if autoimmune reactions against other structural neurospecific proteins including those belonging to cerebral white matter were present.

Myelin basic protein (MBP) is one of the main protein components of CNS myelin. It constitutes over 30% of total protein content in myelin. MBP is not an integral membrane-associated protein. There exists its soluble fraction present in cytosol of oligodendrocytes before it is incorporated into the axonal myelin membrane, and the insoluble fraction of MBP which is one of the constituents of myelin. In the myelin itself MBP is located by the cytosolic side of the elementary membrane [17].

MBP plays a very important role in the arrangement and the support of the structural integrity of myelin. The portions of the protein that stimulate astroglial growth (residues 44–59 in the polypeptide chain) and that enhance synthesis of connective tissue elements (residues 44–166) have been isolated [94]. It was demonstrated that MBP inactivates a range of inhibitors of serine proteases, such as antitrypsin and microglobulin, and possesses lectin-like activity that is specific to galactose and galactosaminic residues [59].

Parenteral injection of MBP causes the development of allergic encephalitis accompanied by diffuse demyelination [7]. It was shown that various peptide fragments of the MBP molecule can cause encephalitis. The most probable protein carrier of the determinant responsible for the induction of encephalitis is a nonapeptide consisting of residues 114–122 [56].

Gene mutation leading to insufficient synthesis of MBP in experimental animals leads to the development of a demyelinating process in brain and to occurrence of hyperkinesis ("wobbler mice" model) [112]. Injection of the protein at the early stages of embryogenesis into the embryo of transgenic mice averts the development of the characteristic neurological symptoms and early fatality [112].

All the discovered properties of MBP connect the disturbances in its metabolism with the development of a demyelinating process. That is why the detection of MBP as well as of autoantibodies to MBP has been first of all conducted in patients with multiple sclerosis and other demyelinating disorders. White matter destruction is accompanied by release of MBP from the damaged tissue and by its accumulation in the CSF. At this sight the detection of this protein may appear a reliable indicator of how the pathological process is pronounced. Permeating through the BBB, MBP and its fragments stimulate synthesis of antibodies to myelin components that supports the course of the disease.

The role of MBP in cerebrovascular pathology has been studied fragmentarily. It was shown in experiments that *a priori* sensitization to MBP significantly worsens prognosis of stroke and enlarges the volume of infarction. At the same time, in animals (Lewis rats) with tolerance to MBP infarction size at 24 and 96 h after induced focal brain ischemia (occlusion of MCA) was significantly less than in control animals [5]. So, antigen-specific modulation of the immune response decreased infarction volume after focal cerebral ischemia and sensitization to the same antigen actually worsened outcome. In a clinical study performed by Palfreyman *et al.* [78], the elevation of MBP content in serum of stroke patients was shown. Such elevation remained until the 7th day from the stroke onset. They found a correlation between the serum level of MBP and stroke prognosis.

Together with colleagues from Anokhin Institute of Normal Physiology of the Russian Academy of Medical Sciences (V. V. Sherstnyov, M. A. Gruden, *et al.*) we studied MBP and its autoantibody content in CSF and serum of 25 patients with carotid ischemic stroke at the neurological clinic of the Russian State Medical University (the clinical description of the group is given in Section 9.1) [100, 101].

The content of MBP and its autoantibodies in CSF and serum was quantified by enzyme-linked immunosorbent assay (ELISA) repeatedly: on admission (before treatment was started) and on the 3rd day of stroke. Commercial immunoenzyme kits manufactured by DIAplus (Russia) were used for the detection of MBP.

To detect **MBP** content, CSF samples were injected into each well except control of a 96-well polystyrene plate for immunoenzyme essay (VNII Polymerbyt, Russia). The samples were previously diluted to the final volume of 100 µl in a buffer solution. The prepared plate was placed into the working chamber of a robot. Samples were thoroughly shaken and incubated for 2 h at room temperature. After incubation the plate was washed 5 times with distilled water (350 µl). After that 100 µl of conjugate was injected into each well, then the plate was thoroughly shaken and incubated for 1 h at room temperature. After that in all wells including control we added 100 µl of the substrate solution, which had been prepared 10 min before use in the following proportion: 1 volume of the substrate solution of TMB +

4 volumes of substrate H_2O_2 buffer. Samples were thoroughly shaken and incubated for 15 min at 37°C, constantly shaking in darkness. Then in each well we added 100 μl of sulfuric acid and the plate was thoroughly shaken. When the reaction was terminated we measured the optical density of the samples in relation to control using a photometer at 450 nm.

To detect **the level of autoantibodies to MBP** we injected 50 μl of S100β protein solution taken at the concentration of 25 μg/ml into each well except control of a 96-well polystyrene plate. Samples were incubated for 24 h at 4°C. After incubation the plate was washed thrice with PBS-Tween buffer at pH 7.2 and with distilled water. After that, to avoid nonspecific binding of antigen to polystyrene in each well except control 100 μl of 1% albumin solution was added. The plate was incubated for one hour at 4°C and then washed as described. In the following stage of immunoenzyme essay, we titrated serum samples of patients. In all rows except the first we injected 50 μl of Tris-HCl-buffer with pH 7.5. Then in the wells of the first and the second row, except control, we added 50 μl of baseline serum sample. Then using a multi-channel pipette of 50 μl volume the diluted sample was transposed from the second into the third row, etc., intensely shaking them. Then the mixtures were incubated for 24 h at 4°C. The plate with titrated samples was then washed 6 times with Tris-HCL-buffer with pH 7.5, Tween-20 solution, and distilled water. Then in each well except control we injected 50 μl of IgG goat conjugate against human IgG with horseradish peroxidase with standard 1:1000 titer. It was incubated for 2 h at 37°C. Then the plate was washed 6 times with Tris-HCl-buffer with pH 7.5, Tween-20 solution, and distilled water. Then in each well including control we injected 50 μl of the substrate: *o*-phenylenediamine in substrate buffer in 1 mg/ml content. Before being placed onto the plate, H_2O_2 was added in proportion of 3.5 μl for 10 ml of the substrate. The reaction was performed in darkness. Thirty minutes after the start of the reaction we analyzed the results using the automatic immunoenzyme reader according to the optical density of the colored substrate.

When we measured MBP content in CSF, we found a considerable variability of content on the 1st day after stroke onset. According to the results, all patients were subdivided onto 3 groups. In *the first group* MBP content registered on admission was normal and then elevated on the 3rd day of stroke on average by 238% ($p < 0.05$). In *the second group* we registered normal level of MBP on the 1st day and then abruptly elevated (by 1309%, $p < 0.004$) on the 3rd day. In *the third group* MBP content exceeded normal values 15- to 20-fold on admission and then continued to elevate until the 3rd day. Clinical and immunological analysis showed that the extent of MBP content elevation depended on the location as well as the distribution of the ischemic lesion. In all patients of the first group MRI revealed small cortical infarctions ($<10 cm^3$) (Fig. 9.6). In patients from the second group we found small subcortical infarctions ($<10 cm^3$). High MBP content on admission characterized patients with large subcortical lesions ($>25 cm^3$) (Fig. 9.6).

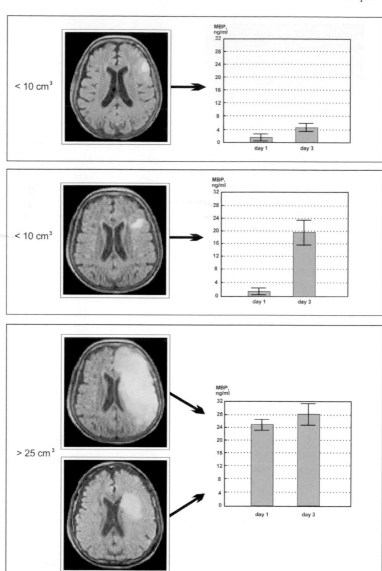

Figure 9.6. Relation of MBP content in CSF to the location and volume of ischemic lesion.

However, we found no correlation between MBP content and the extent of neurological deficit as well as its pathogenic variant ($r = -0.13$, $p > 0.05$ and $r = -0.18$, $p > 0.05$, respectively).

Thus, only by the 3rd day after stroke onset all patients exhibited elevated MBP content in CSF, whereas S100β content elevation has already found on admission. Such discrepancy possibly reflects the differences of metabolism inherent in these proteins. Being more labile and sensitive to calcium homeostasis changes, S100β responds to acute ischemia earlier, being released from its usual bounds. The chronology of its CSF content elevation roughly corresponds to the progression of the glutamate–calcium cascade reactions. MBP comes into CSF when the destruction of brain tissue has already occurred and appears to be a marker of morphological damage of white matter, as can be explained by its belated accumulation.

However, from the first hours of stroke all patients exhibited elevation of autoantibody titer to MBP. The ratio between highly (1:128, 1:256) and moderately (1:16, 1:32, 1:64) increased titers was identical to those for S100β: 16 (64%) and 9 (36%), respectively. The extent of elevation for titers of autoantibodies to S100β and MBP closely correlated with each other ($r = 0.9$, $p = 0.001$), possibly indicating *a priori* sensitization of stroke patients to various brain protein structures. The related study on the 3rd day did not reveal any significant change in titer of MBP autoantibodies versus the 1st day value ($p > 0.07$). We failed to find correlation between the protein content in CSF and titer of autoantibodies to MBP in serum.

It was of great interest to study simultaneously the intensity of autoimmune and inflammatory reactions in acute ischemic stroke.

An important role in the development of inflammation is played by neutrophils [41, 75]. They accumulate in the ischemized area already within the first hours of infarction (see Chapters 2 and 8) [6, 106]. Activated cells exhibiting various modes of reactivity (adhesion, "respiratory burst", degranulation), may damage not only vascular endothelium, but also cause significant tissue injury. Degranulation occurring in the last stages of neutrophil response is accompanied by release of intracellular proteolytic enzymes with wide substrate specificity, including leukocytic elastase (LE), into the extracellular space. This enzyme causes proteolytic fission of elastin and collagen fibers of basal vascular membrane that determines its pathogenic significance in the embodiment of inflammation and allows considering it a marker of pathological conditions complicated by inflammatory process [9, 47]. It has been shown that LE takes part in enhancement of BBB permeability [3, 92, 107] and brain tissue destruction during acute focal ischemia [18, 96]. Decrease in the BBB resistance creates conditions for release of brain neurospecific proteins into the blood flow and stimulation of autoimmune response onto beyond-barrier antigens.

Together with colleagues from the Mental Health Research Center (T. P. Klushnik *et al.*) and the Anokhin Institute of Normal Physiology (V. V. Sherstnyov, M. A. Gruden, *et al.*) of the Russian Academy of Medical Sciences, we studied activity of leukocytic elastase (LE) and its inhibitor α-1-antitrypsin (α-1-A) in blood serum of the same 25 patients with carotid ischemic stroke in whom we simultaneously investigated autoimmune reactions to neurospecific proteins (the clinical description of the group is given in the previous sections) [93]. The monitoring of the activity of LE and α-1-A in blood serum was performed on days 1, 3, 7, 14, and 21.

LE activity was measured as an index of degranulation neutrophilic activity by Visser and Blout [113] using specific substrate (XI-tert-BOC-Ala-Onp) by means of spectrophotometry [22].

Functional activity of α-1-A in serum was assessed by means of a universal enzymatic technique by the degree of inhibition of trypsin by the investigated serum using N-benzoyl-arginine-ethyl ester as a substrate using a spectrophotometer [76].

Normal values were obtained from serum of 19 healthy individuals of comparable age without clinical or laboratory signs of inflammatory and autoimmune diseases.

Statistical analysis was performed by nonparametric tests (Mann–Whitney U-test, Spearman's *r*-coefficient of rank correlation). Fatality was measured by Fisher's test. The differences between values were assumed to be significant at the 5% level in every comparison conducted.

The study of LE activity on admission showed its increase in serum of all patients versus controls (376 ± 21.8 and 258 ± 34 μM/min, respectively, $p <$ 0.01). At the same time, the α-1-A level in patients with moderately severe stroke was within normal range (30.1 ± 12.2 IU/ml, $p > 0.05$ versus control value 29.9 ± 1.2 IU/ml) and showed a tendency to decrease in patients with severe stroke in some cases down to zero (Fig. 9.7). Thus, even within the first day after stroke onset, the imbalance in the system of "LE-inhibitor" was obvious, which indicated marked degranulation activity of neutrophils in the ischemized area and of possible exhaustion of the LE inhibitor system.

In following days, the average LE activity tended to decrease: the mean value on day 21 was 280 ± 23.3 μM/min, while the level of α-1-antitrypsin insignificantly increased (by 12.2%). Only in patients with steady improvement and good functional recovery (Barthel Index more than 75) by the end of the acute stroke period, significant decrease in LE level to the normal value (246 ± 24.6, $p < 0.01$) was registered by day 21, which was accompanied by an increase in the inhibitor concentration by 66.6% (Fig. 9.8), this probably reflecting abatement of inflammatory processes.

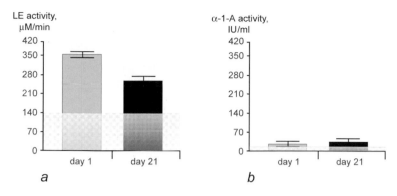

Figure 9.7. Dynamics of activity of leukocytic elastase (LE) (a) and its inhibitor α-1-antitrypsin (α-1-A) (b) in blood serum of patients with carotid ischemic stroke. Shaded area corresponds to normal range.

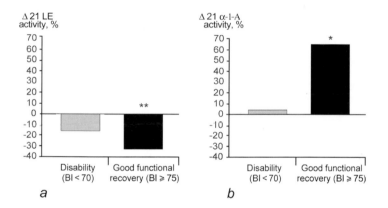

Figure 9.8. Dynamics (in %) of LE (a) and α-1-A (b) activity in blood serum of patients with carotid ischemic stroke by day 21 in dependence on stroke outcome. Significance of the differences between patients with disability (BI < 70) and good functional recovery (BI ≥ 75): * $p < 0.05$; ** $p < 0.001$.

Analysis of the interaction between the content of antibodies to neurospecific proteins and LE activity found a tight direct correlation between the reduction of serum LE level and the decrease in primary antibodies to S100β ($r = 0.8$, $p = 0.025$) and MBP ($r = 0.75$, $p = 0.04$) levels in serum (Fig. 9.9).

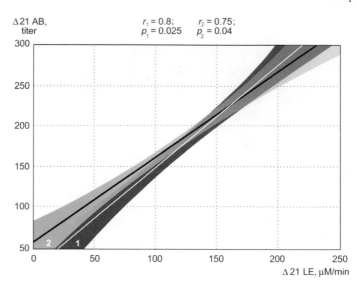

Figure 9.9. Correlation between the decrease of blood serum levels of leukocytic elastase (LE) and primary autoantibodies (AB) to S100β (1) and MBP (2) in patients with atherothrombotic ischemic stroke by day 21.

Thus, the abatement of inflammatory response averts the progression of BBB damage, which retards the permeation of neuroantigens into blood and, hence, decreases the intensity of additional autoimmune brain damage.

9.4.   Content of autoantibodies to NGF, S100β, and MBP in serum of patients with chronic brain ischemia

To elucidate whether the revealed autoimmune reaction had a "pre-stroke origin", we performed the study of autoantibody level to NGF, S100β, and MBP in serum of 40 patients (10 males and 30 females; average age 64 years) with different stages of chronic ischemic brain disease (CIBD) of Stages I–III [55, 100]. The diagnosis of CIBD and of its stage was established according to clinical criteria elaborated by Shmidt and Maksudov [97], to the results of neuropsychological testing by Luria [62, 63], the results of brain MRI, and ultrasound Doppler study of extra- and intracranial vessels and/or MR-angiography.

In all patients chronic ischemic brain disease developed on the background of atherosclerosis and arterial hypertension which was steady in

24 (60%) of patients and manifested as repeated crisis in 16 (40%) of patients. Ultrasound Doppler found atherosclerotic changes and stenosis in at least one main line head artery in all patients. In 6 patients (10%) we found stenosis significantly affecting blood flow (>70%) or arterial occlusions.

In 30 (75%) of patients we diagnosed CIBD of Stage I–II which clinically manifested by asthenia, non-systemic vertigo, headache, memory disorders, decrease of attention of mental working capacity, fatigue, sleep disorders, and soft focal neurological symptoms as eye convergence weakness, position nistagmus, oral reflexes, anisoreflexia, and moderate disturbances of coordination. In 10 (25%) patients we diagnosed CIBD of Stage III, which manifested as prominent intellectual and memory disorders (in 5 cases) and by hypokinesia and rigidity (also in 5 cases). To objectify the severity and dynamics of focal, cognitive and subjective symptoms and to standardize the processing of clinical data, we used three complimentary scales: Mini-Mental State Examination Scale [21], Sandoz Clinical Assessment Geriatric Scale [69, 79], and an original questionnaire of subjective symptoms.

CT scans and MR images revealed hydrocephalus of atrophic origin in all patients. In 4 patients we found prominent leukoareosis, and in 5 we found small post-ischemic cysts.

We performed the follow-up neurological assessment of patients with CIBD for 24 months and repeated clinical and immunological studies 6 times within this period: the first study of the "background" conducted before any therapeutic administrations; the second control study 2 weeks after the first to detect if the titer of antibodies was stable; the third following 2 weeks of neuroprotective therapy with glycine; the fourth 6 months after the third study; the fifth following 4 weeks of neuroprotective therapy with glycine, and the last 6 months after the fifth study (without therapy).

We failed to reveal significant elevation of titer of antibodies to NGF in serum of any patient. The average titer value was 0.7 ± 0.06 OD units versus normal level of 0.7 ± 0.15 OD units ($p > 0.05$). However, we detected significant differences in titer values of autoantibodies to S100β and MBP (Fig. 9.10). Levels of autoantibodies to S100β close to normal (1:4, 1:8) were registered in 8 (20%) patients and those to MBP in 10 (25%). Moderately elevated levels (1:16, 1:32, 1:64) of autoantibodies to S100β and MBP were registered in 24 (60%) and 9 (22.5%) patients, respectively; high titers (1:128, 1:256) in 8 (20%) and 21 (52.5%) of cases. Repeated study 2 weeks after the first examination revealed significant titer fluctuations only in 2 patients (not more than by 1 titer), showing that the changes were relatively stable.

We failed to find any correlation between the levels of autoantibodies to neurospecific proteins in serum and clinical manifestations assessed by all three clinical scales or CIBD stages as well as the variant of arterial hypertension course (stable or repeated crisis) and the extent of morphological change (according to CT and MRI data) ($r < 0.15$). The elevation of autoantibody titers was never accompanied by any signs of clinical worsening.

When we compared titers of autoantibodies to S100β and MBP it occurred that in the majority of patients (48%) MBP antibody titer exceeded those for S100β antibodies. In 30% of studied patients titer values for antibodies to both proteins were identical, whereas in 22% S100β titers prevailed. This possibly reflected the individual character of brain tissue morphological change on the background of chronic brain ischemia as well indicating the predominance of white matter damage in the majority of our patients. Along with this, comparing MRI and immunological data we failed to reveal any differences between the titers of autoantibodies to MBP in relation to the extent of leukoareosis and other white matter changes. Possibly, the MRI techniques and immunological essays used reflect events that cannot be chronologically compared.

We found certain conformities in ratios between titers of primary and secondary autoantibodies to S100β. In cases where primary titers were normal (1:2–1:8) the level of anti-idiotypic autoantibodies was high (1:32–1:128) and, vice versa, the higher was the level of primary autoantibodies (1:128–1:512) the closer to normal was the titer of anti-idiotypic autoantibodies. Taking into consideration the preliminary data acquired in patients with ischemic stroke, we suppose that the high index of primary to secondary autoantibody ratio is a criterion for "acuity" of autoimmune processes and, hence, of the pathologically altered BBB permeability.

All CIBD patients followed-up have undergone neuroprotective treatment with glycine (sublingually in daily dose of 600 mg) (see Section 14.5). Complex clinical (neurological and neuropsychological) study demonstrated that the condition being treated improved. The general state of health exhibited positive changes: the extent of non-systemic vertigo decreased, attention, working capacity, and general daily activities improved, whereas the soft focal neurological symptoms remained. Clinical and immunological analysis suggested the delayed character of the influence on autoimmune reactions produced by neuroprotective therapy: possibly a lapse of time is necessary for the generation of autoantibodies as well as for the decrease of previously synthesized autoantibodies. Only several weeks after the first course of glycine patients exhibited significant decrease in primary and secondary autoantibody titers to neurospecific proteins; in 65% down to 1:2–1:8, i.e. normal values; in 35% down to 1:16 (see Fig. 9.10). After the second course of neuroprotective treatment we registered further

Titer of AB

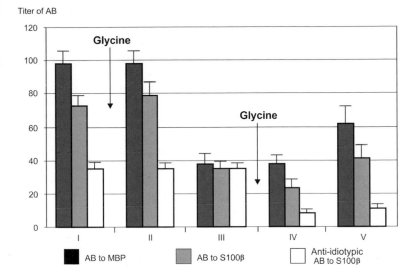

Figure 9.10. Dynamics of blood serum levels of autoantibodies (AB) to neurospecific proteins in patients with chronic brain ischemia. I – "background" study conducted before any therapeutic administration; II – following 2 weeks of neuroprotective therapy with glycine; III – 6 months after the II point; IV – following 4 weeks of neuroprotective therapy with glycine; V – 6 months after the IV point (without therapy).

decrease in titers, but 6 months after the cessation of glycine intake a moderate tendency towards the elevation of titers ensued again.

It is noteworthy that acute cerebrovascular events occurred in three patients during the follow-up period. These events varied in clinical course and in extension of focal morphological lesion according to MRI. In two of the patients acute cerebrovascular events developed during the repeated course of glycine and on the background of normal titers of autoantibodies to S100β and MBP, and manifested themselves as transitory ischemic attacks with no morphological lesion formed. In the third patient acute cerebrovascular event developed 4 weeks after the cessation of the repeated course of glycine, and manifested itself as severe ischemic stroke with grave neurological deficit and vast cortical and subcortical lesion. These differences may possibly emphasize the role of neuroprotection in the inhibiting of auto-aggression to neurospecific brain proteins and, so, in the decreasing of the background previous sensitization of brain tissue to its structural components. Taking into consideration the participation of autoimmune mechanisms in preparation and embodiment of brain tissue response to CBF decrease and focal ischemia, we suppose that preventive courses of neuroprotection are of great significance for prevention of severe morphological brain lesions in cases of sudden decrease of cerebral blood supply.

Thus, longitudinal clinical and immunological study in CIBD patients confirmed the main results acquired in the study of stroke patients. No case before clinical manifestation of stroke was accompanied by the elevation of autoantibody titer to NGF in serum, which possibly reflected the absence of "pre-stroke sensitization" of brain tissue to NGF and might have explained local character of intracerebral autoimmune response within the first days of stroke. Along with this, elevated level of primary antibodies to structural neurospecific proteins S100β and MBP was registered. The extent of primary titer elevation and the index of its relation to titer of anti-idiotypic autoantibodies was variable, this possibly reflecting that patients differed by "acuity" of autoimmune events within the group. Differences in ratios between autoantibodies to S100β and MBP reflected topographic peculiarities of brain tissue destruction.

Thus, being relatively functionally intact, CIBD patients when examined manifest generalized autoimmune reaction to structural components of neural tissue. Autoimmune process participates in the formation of the "background" cerebrovascular damage (encephalopathy) and prepares brain tissue for the development of infarction in response to acute CBF decrease.

As the titers of autoantibodies to S100β and MBP came to normal values after glycine administration along with clinical improvement, we suppose that such neuroprotective treatment is of preventive value. It may decrease the risk of large infarction development on the background of acute insufficiency of blood supply.

REFERENCES

1. Abraha, H. D., Butterworth, R. G., Sherwood, R. A., 1997, *Ann Clin Biochem.* **34** (4): 366–370.
2. Anderson, R. E., Hansson, L. O., Vaage, J., 1999, *Ann Thorac Surg.* **67** (6): 1721–1725.
3. Armao, D., Kornfeld, M., Estrada, E. Y., *et al.*, 1997, *Brain Res.* **767** (2): 259–264.
4. Aurell, A., Karlsson, B., Zbomikova, V., 1987, *Stroke.* **18**: 911–918.
5. Becker, K. J., McCarron, R. M., Ruetzler, C., *et al.*, 1997, *Proc Natl Acad Sci USA.* **94** (20): 10873-10878.
6. Benjelloun, N., Renolleau, A., Represa, A., *et al.*, 1999, *Stroke.* **30** (9): 1916–1924.
7. Berezin, V. A., 1990, *Neurochemistry.* **1** (1): 114–123 (in Russian).
8. Berezin. V. A., 1990, *Proteins Specific for Neural Tissue,* Kiev (in Russian).
9. Bless, N. M., Smith, D., Charlton, J., *et al.*, 1997, *Curr Biol.* **7** (11): 877–880.
10. Blomquist, S., Johnsson, P., Solem, J. O., 1997, *J Cardiothorac Vase Anesth.* **11**: 699–703.

11. Bono, F., Lamarche, I., Bornia, J., Savi, P., Della Valle, G., Herbert, J. M., 1999, *FEBS Lett.* **457** (1): 93-97.
12. Butterworth, R. J., Sherwood, R. A., Bath, P. V., 1998, *Stroke.* **29**: 730.
13. Chekhonin, V. P., Ryabukhin, I. A., Belopasov, V. V., *et al.*, 1995, *Monoclonal Antibodies in Neurobiology*, Novosibirsk, pp. 160–170 (in Russian).
14. Chen, S.-C., Holly, D., Soares, H. D., Morgan, J. I., 1996, *Adv Neurol.* **71**: 433–450.
15. Clarke, A. R., Purdie, C. A., Harrison, D. G., *et al.*, 1993, *Nature.* **362**: 849–852.
16. Davies, A. M., 1988, *Trends Genet.* **4**: 139–143.
17. Davies, L., Sonston, G., 1974, *J Neurochem.* **22**: 107–111.
18. Dawson, D. A., Ruetzler, C. A., Carlos, T. M., *et al.*, 1996, *Keio J Med.* **45** (3): 248–253.
19. Dawson, R., Elliot, D., Elliot, W., Jones, K., 1991, *Biochemists Manual.* Mir, Moscow (Russian translation).
20. Deckwerth, T. L., Johnson, E. M., Jr., 1993, *J Cell Biol.* **123**: 1207–1222.
21. Dick, J. P., Guiloff, R. J, Stewart, A., Blackstock, J., *et al*, 1984, *J Neurol Neurosurg Psychiatr.* **47** (5): 496–499.
22. Dotsenko, V. L., Neshkova, E. A., Jarovaja, G. A., 1994, *Probl Med Chem.* **40** (3): 20–25 (in Russian).
23. Dropcho, E. J., Chen, Y. T., Posner, J. B., Old, L. J., 1987, *Proc Natl Acad Sci USA.* **84**: 4552–4556.
24. Efremova, N. M., Skvortsova, V. I., Gruden, M. A., *et al.*, 2000, The study of S100β protein content and of primary and secondary antibodies to S100β in patients with acute focal brain ischemia in relation to pathogenetic variants of stroke. In *Modern Approaches to Diagnosis of Nervous and Psychiatric Disorders*, St. Petersburg, 294 (in Russian).
25. Einstain, E., 1988, *Cerebral Proteins and CSF in Normal and Pathological Conditions.* Mir, Moscow (Russian translation).
26. Fassbender, K., Hennerici, M., DoUman, M., *et al.*, 1997, *Lab Clin Med.* **130** (5): 535–539.
27. Festoff, B. W., Israel, R. S., Engel, W. K., 1977, *Neurol.* **27**: 963–970.
28. Freeman, R. S., Estus, S., Jonson, E. M., Jr., 1994, *Neuron.* **12**: 343–355.
29. Gannushkina, I. V., 1994, *Immunological Aspects of Brain Injury and Cerebrovascular Disorders.* Meditsina, Moscow, 198 (in Russian).
30. Gruden, M. A., Poletayev, A. B., 1987, *Biochemistry.* **52** (6): 915–917 (in Russian).
31. Gruden, M. A., Storozheva, M. I., Sherstnyov, V. V., 1999, Regulatory antibodies to neurotrophic factors: clinical and experimental study. In *Neuroimminopathology.* Moscow, pp. 19–20 (in Russian).
32. Gusev, E. I., Skvortsova, V. I., 1991, *Case Record Form for Examination and Treatment of Patients with Ischemic Stroke*, Moscow, pp. 1–44 (in Russian).
33. Gusev, E. I., Skvortsova, V. I., Burd, G. S., *et al.*, 1997, *J Neurol Sci.* **150**: 81–82.
34. Gusev, E. I., Skvortsova, V. I., Dambinova, S. A., *et al.*, 1996, *Cerebrovasc Dis.* **6** (2): 44.
35. Gusev, E. I., Skvortsova, V. I., Dambinova, S. A., *et al.*, 2000, *Cerebrovasc Dis.* **10**: 49–60.
36. Gusev, E. I., Skvortsova, V. I., Izykenova, G. A., *et al.*, 1996, *Korsakoff J Neurol Psychiatr.* **5**: 68–72 (in Russian).
37. Gusev, E. I., Skvortsova, V. I., Kovalenko, A. V., Sokolov, M. A., 1999, *Korsakoff J Neurol Psychiatr.* **2**: 71–76 (in Russian).
38. Gusev, E. I., Skvortsova, V. I., Raevsky, K. S., *et al.*, 1997, *Eur J Neurol.* **4** (1): 78.
39. Gwag, B. J., Canzoniero, L. M., Sensi, S. L., 1999, *Neuroscience.* **90** (4): 1339–1348.
40. Haglid, K., Carlsson, A., Stavron, D., 1973, *Acta Neuropathol.* **24**: 187–192.

41. Hofman, F. M., Chen, P., Jeyaseelan, R., *et al.*, 1998, *Blood.* **92** (9): 3064–3072.
42. Hohfield, A., 1989, *Ann Neurol.* **25**: 531–538.
43. Huber, M., Heiss, W. D., 1996, *Semin Thromb Hemost.* **22** (1): 53–60.
44. Isobe, T., Ishoka, N., 1983, *Biochem Int.* **6** (3): 419–426.
45. Jager, L., 1988, *Klin Immunol Allergol.* **1** (3): 3.1.4.
46. Jancovic, B. D., Jaculic, S., 1980, *Horvat J Clin Exp Immunol.* **40**: 598–602.
47. Jarovaja, G. A., Dotsenko, V. L., Neshkova, E. A., 1995, *Pathogenetic Role of Leukocytic Elastase*, Moscow, pp. 16–18 (in Russian).
48. Jerne, N., 1985, *Biosci Res.* **5** (4): 439–451.
49. Johansson, B. B., 1995, *Cerebrovasc Dis.* **5**: 278–281.
50. Johnson, E. M., Jr., Greenlund, L. J., Akins, P. T., Hsu, C. Y., 1995, *J Neurotrauma.* **12** (5): 843–852.
51. Johnsson, N., Liebiry, M., 1992, *Stroke.* **2**: 123–125.
52. Jonsson, H., Johnsson, P., Birch-Iensen, M., *et al.*, 2001, *Ann Thorac Surg.* **71** (5): 1433–1437.
53. Kang, H., Schuman, E., 1995, *Science.* **267**: 1658–1662.
54. Karas, I. Yu., 1988, *Candidate's dissertation*, Tomsk (in Russian).
55. Khadzhiyeva, M. Kh., Skvortsova, V. I., Sherstnyov, V. V., *et al.*, 2000, The study of neurotrophic factors and their autoantibodies in patients with chronic brain ischemia. In *Contemporary Approaches to the Diagnosis and Treatment of Nervous and Psychiatric Diseases.* St. Petersburg, 341 (in Russian).
56. Khokhlov, A. P., Baskayeva, T. S., 1986, *Bull Exp Biol Med.* **10**: 430–434 (in Russian).
57. Kogure, K., Yamasaki, Y., Malsuo, Y., *et al.*, 1996, *Acta Neurochir.* **66**: 40–43.
58. Kryzhanovsky, G. N., Lutsenko, V. K., 1995, *Adv Modern Biol.* **115** (1): 31–49 (in Russian).
59. Lisak, R. P., Zweiman, D., 1977, *New Engl J Med.* **297**: 850–858.
60. Lisyany, N. I., 1988, *Physiol J.* (Kiev). **34** (2): 112–118 (in Russian).
61. Lisyany, N. I., Cherenko, T. M., Komissarenko, S. V., *et al.*, 1993, *Korsakoff J Neurol Psychiatr.* **93**: 50–53 (in Russian).
62. Luria, A. R., 1973, *Proc Natl Acad Sci USA.* **70** (3): 959–964; **70** (4): 1278–1283.
63. Luria, A. R., 1999, *Neuropsychol Rev.* **9** (1): 9–22.
64. Malashkhia, Yu. A., 1986, *Immunological Brain Barrier.* Moscow (in Russian).
65. Malashkhia, Yu. A., 1990, *Immunology.* **3**: 12–15 (in Russian).
66. Malashkhia, Yu. A., 1995, *Int J Immunol Rehabilitol.* **1**: 10–17 (in Russian).
67. Marchetti, B., Gallo, F., Romeo, C., *et al.*, 1996, *Ann NY Acad Sci.* **784**: 209–236.
68. Martens, P., Raabe, A., Johnsson, P., 1998, *Stroke.* **29**: 2363–2366.
69. Maurer, W., Ferner, U., Patin, J., Hamot, H. B., 1982, *Z Gerontol.* **15** (1): 26–30.
70. Meisner, H., Czech, P., 1991, *Curr Biol.* **3**: 474–483.
71. Michetti, G., Miani, N., De Renzis, G., *et al.*, 1974, *J Neurochem.* **22** (2): 239–242.
72. Michetti, G., Missaro, A., Murazio, M., 1979, *Neurosci J.* **11** (2): 171–180.
73. Michetti, G., Missaro, A., Russo, G., Rigon, C., 1980, *J Neurol Sci.* **44**: 259–263.
74. Moister, D., 1984, *J Neurochem.* **42** (6): 1536–1541.
75. Nagy, Z., Kolev, K., Csonka, E., *et al.*, 1998, *Blood Coagul Fibrinolysis.* **9** (6): 471–478.
76. Nartikova, V. F., Paskhina, T. S., 1979, *Probl Med Chem.* **25** (4): 494–499 (in Russian).
77. Orgogozo, J. M., Dartigues, J. F., 1986, In *Acute Brain Ischemia: Medical and Surgical Therapy* (Battistini, N., Courbier, R., Fieschi, C., Fiorani, P., Plum, F., eds.) Raven Press, New York, pp. 282–289.
78. Palfreyman, J., 1979, *Clin Chim Acta.* **92** (3): 403–409.

79. Patin, J. R., Hamot, H. B., Singer, J. M., 1984, *Progr Neuropsychopharmacol Biol Psychiatr.* **8** (2): 293–306.
80. Paul, W., 1988, *Immunology.* Vol. 2. Meditsina, Moscow, p. 345 (Russian translation).
81. Perumol, A. S., 1976, *J Neurochem.* **27** (1): 173–177.
82. Pisa, E. K., 1999, *Stroke.* **30** (5): 1153–1154.
83. Poletayev, A. B., 1984, *Doctoral dissertation*, Moscow (in Russian).
84. Poletayev, A. B., Mcshcheryakova, O. D., 1982, *Biochemistry.* **47** (11): 1835–1838 (in Russian).
85. Poletayev, A. B., Sepiphanova, O. P., 1987, *Neurochemistry.* **6** (4): 572–580 (in Russian).
86. Poletayev, A. B., Sherstnyov, V. V., 1987, *Adv Modern Biol.* **103** (1): 124-132 (in Russian).
87. Ross, G. M., Shamovsky, I. L., Lawrance, G., Solc, M., Dostaler, S. M., Jimmo, S. L., Weaver, D. F., Riopelle, R. J., 1997, *Nat Med.* **3** (8): 872–878.
88. Sandalov, V. B., 1984, *Neurochemistry.* **3** (2): 116–123 (in Russian).
89. *Scandinavian Stroke Study Group Stroke,* 1985. **16**: 885–890.
90. Scotto, C., Deloulme, J. C., Rousseau, D., *et al.*, 1998, *Mol Cell Biol.* **18** (7): 4272–4281.
91. Scotto, C., Mrely, Y., Ohshima, H., *et al.*, 1998, *J Biol Chem.* **273** (3): 3901–3908.
92. Scherbakova, I. V., Klyushnik, T. P., Ermakova, S. A., Efremova, N. M., Skvortsova, V. I., 2001, *Korsakoff J Neurol Psychiatr., suppl. Insult.* **4**: 39–44 (in Russian).
93. Shcherbakova, I. V., Neshkova, E., Dotsenko, V., *et al.*, 1999, *Immunopharmacol.* **43**: 273–279.
94. Sheffild, W., Kim, S., 1977, *Brain Res.* **132**: 580.
95. Sherstnyov, V. V., 1983, *Doctoral dissertation,* Moscow (in Russian).
96. Shimakura, A., Kamanaka, Y., Ikeda, Y., *et al.*, 2000, *Brain Res.* **858** (1): 55–60.
97. Shmidt, E. A., Maksudov, G. A., 1971, *Korsakoff J Neurol Psychiatr.* **1**: 3–15 (in Russian).
98. Shtark, M. B., 1989, *Brain Specific Proteins (Antigens) and Neuronal Functions,* Moscow (in Russian).
99. Skvortsova, V. I., Gruden, M. A., Stakhovskaya, L. V., *et al.*, 1999, Structural neuro-specific proteins (S100 and basic myelin protein) and autoantibodies to them as markers of neuro–immune–pathological reactions in chronic brain ischemia. In *New Technologies in Neurology and Neurosurgery at the Turn of the Millenium*, Stupino, pp. 180–181 (in Russian).
100. Skvortsova, V. I., Gusev, E. I., Sherstnev, V. V., *et al.*, 2000, *Cerebrovasc Dis.* **12** (3).
101. Skvortsova, V. I., Klushnik, T. P., Stakhovskaya, L. V., *et al.*, 1999, The study of autoantibodies to NGF in patients with acute and chronic brain ischemia. In *New Technologies in Neurology and Neurosurgery at the Turn of Millennium*, Stupino, pp. 181–182 (in Russian).
102. Skvortsova, V. I., Myasoyedov, N. F., Klushnik, T. P., *et al.*, 2000, The study of NGF and its autoantibodies content in patients with acute brain ischemia. In *Contemporary Approaches to the Diagnosis and Treatment of Nervous and Psychiatric Diseases.* St. Petersburg, pp. 332–333 (in Russian).
103. Solimena, M., Folli, F., Denis-Doninis, D., 1988, *J Med.* **318** (16): 1012–1020.
104. Stabberod, P., Tomasevic, G., Kamme, F., Wieloch, T., 1994, *Abst Soc Neurosci.* **20**: 616.
105. Stanzani, L., Zoia, C., Sala, G., *et al.*, 2001, *Cerebrovasc Dis.* **12** (3): 240–244.
106. Stoll, G., Jander, S., Schroeter, M., 1998, *Progr Neurobiol.* **56** (2): 149–171.

107. Temesvari, P., Abraham, C. S., Gellen, J., Jr., *et al.*, 1995, *Biol Neonate.* **67** (1): 59–63.
108. Thompson, N., 1985, *Semin Immunopathol.* **8** (1): 57–70.
109. Turusov, V. S., 1990, *Arch Pathol.* **52** (1): 71–78 (in Russian).
110. Tuszynski, M. H., 1999, Neurotrophic factors. In *CNS Regeneration* (Tuszynski, M. H., Kordower, J. H., eds.) Academic Press, San Diego, pp. 109–158.
111. Twichell, T. E., 1951, *Brain.* **74**: 443–480.
112. Ulrich, J., 1993, *Acta Neuropathol.* **133**: 77–83.
113. Visser, L., Blout, E., 1972, *Biochim Biophys Acta.* **268**: 257–260.
114. Waters, C., 1997, *RBI Neurotransmissions, Newsletter for Neuroscientist.* **XIII** (2): 2–7.
115. Wong, C., Bonser, R. S., 1998, Stroke. **29** (11): 2446-2447.
116. Wunderlich, M. T., Ebert, A. D., Kratz, T., *et al.*, 1999, *Stroke.* **30**: 1190–1195.
117. Yamasaki, Y., Matsuo, Y., Matsuura, N., *et al.*, 1995, *Stroke.* **26**: 318–323.
118. Yamasaki, Y., Suzuki, T., Yamaya, H., *et al.*, 1992, *Neurosci Lett.* **142**: 45–47.
119. Yamasaki, Y., Yamaya, H., Watanabe, M., *et al.*, 1991, *J Cerebr Blood Flow Met.* **13**: 113.
120. Zhabotinsky, Yu. M., 1975, *Demyelinating Disorders of Nervous System in Experiment and in Clinic*, Minsk, pp. 21–26 (in Russian).
121. Zomezely-Neurath, C., Keller, A., 1982, *Development Neuroscience* (Karger, S., ed.) New York.

Chapter 10

Programmed Cell Death. Apoptosis in Focal Brain Ischemia

Until the 1980s it was believed that neuronal death in acute brain ischemia occurred only by the mechanism of necrosis. However, later studies elucidated the role of apoptosis, another pathway to infarction, which is a variant of programmed cell death.

The term "apoptosis" was proposed in 1972 by English scientists J. F. R. Kerr, A. H. Wyllie, and A. R. Currie. They coined this term from the Greek word meaning "falling off", as leaves do in autumn, to describe this natural, timely death of cells that is realized via cell fragmentation to "apoptotic bodies", which are then consumed by the adjacent phagocytosing cells of various types.

Studying normal and pathological tissues, Kerr et al. [67] found that the dying cells may be subdivided into two categories. In severely affected tissues the necrotic process predominated, which expanded as wide as cell fields and was characterized by passive cell degeneration with swelling and fragmentation of organelles, destruction of membranes, cell lysis, efflux of intracellular content into neighboring tissue and by the development of inflammatory response. As shown later, necrosis is always the result of severe pathology, its mechanisms never envisage energetic expenses, and it can be averted only by the removal of the pathogenic factor [20, 24, 124].

Apoptotic cells, in contrast to necrotic cells, are not found adjacent to each other, but lie solely or arranged in small groups and are scattered all over the affected tissue. They are smaller in size and possess unchanged membranes and organelles along with signs of cytosolic shrinkage and multiple cytosolic membrane-associated protrusions. Owing to intact membranes, no lytic enzymes come in contact with adjacent cells, thus no sign of their damage or local inflammation is observed, which is now regarded as one of the main morphological features distinguishing apoptosis from necrosis [140]. Shrunken cells as well as the apoptotic bodies contain intact masses of condensed chromatin. As a result of sequential destruction of DNA in the apoptotic cells, they lose the ability to replicate and participate in the intercellular interactions as these processes require the synthesis of new proteins. Dying cells are effectively removed from the tissue via phagocytosis occurring in the first minutes and hours after death. Later, it was shown that if the capacity for phagocytosis is limited and the apoptotic cells remained in the tissue for more than 1 or 2 days their membranes begin also to disintegrate and "secondary" necrosis occurs (the lysis of a cell that died via apoptosis) [140]. The basic differences in the features of necrosis and apoptosis are given in Table 10.1.

It is noteworthy that the morphological changes associated with apoptosis were described almost 100 years before the observation made by Kerr *et al.* In 1885 W. Flemming demonstrated semi-lunar picnotic chromatin typical for apoptosis. In 1886 the botanist G. Berthold showed that protein denaturation is responsible for certain forms of cell death. In 1951 Glucksmann classified different forms of cell death, including a form

Table 10.1. Differential features of necrosis and apoptosis

Feature	Necrosis	Apoptosis
1. Mode	Passive	Active
2. Energetic expenses	Absent	Present
3. RNA demand	Absent	Present
4. Demand for protein synthesis	Absent	Present
5. DNA degradation	Nonspecific	Specific fragmentation
6. Morphological features:		
distribution	Large cell fields	Selective cell loss
membranes	Destruction	Stay intact
organelles	Swelling and fragmentation	Stay intact
condensation of chromatin	In chaotic masses	In homogenous and dense mass
cytosolic membrane-associated protrusions	Absent	Present
7. End-point	Edema and lysis of cell	Cell shrinkage and formation of apoptotic bodies
8. Efflux of intracellular content into adjacent tissue	Present	Absent
9. Inflammatory reaction	Present	Absent
10. Phagocytosis	Intense	Moderate
11. Significance	Always pathological	Physiological (implicated in growth, development, homeostasis support); possibly, pathological
12. Intensity of triggering influence	Prominent	Moderate

specific for an embryo. In 1964 Lokshin and Williams studying insect metamorphosis noted the regularity inherent in the process of cell death which was evidently predestined and sequential for specific cells and proposed the term "programmed cell death". In 1966 it was proven that such programmed death requires protein synthesis, which emphasized the active role of the cell itself in the process of its death [133].

Kerr *et al.* [67] found apoptotic cells in pathological as well as within physiological conditions. Later it occurred that apoptosis appeared to be an inevitable part of development and homeostasis supporting events in a mature tissue. In the normal state the living organism uses this genetically programmed mechanism in embryogenesis to annihilate the "excessive" cellular material at early stages of tissue development, particularly in neurons that failed to establish contacts with target cells and, hence, bereft of trophic support from these cells [33]. In maturity the intensity of apoptosis in mammalian CNS significantly deceases, whereas it remains much higher in other tissues [96]. Apoptotic events underlie the process of normal cell replacement. Prominent inhibition of apoptotic death leads to the development of cancer diseases [57]. Necessary apoptotic reaction accompanies both the removal of cells affected by viruses and the development of the immune response.

Apoptosis is closely associated with programmed cell death [87, 88]. On balance, "programmed cell death" to a greater extent reflects the assignment of this process representing a natural part of multicellular organism life connected with metamorphosis and development [56, 117]. At the same time, the term "apoptosis" was proposed to define the type of cell death characterized by a certain set of morphological features [127]. At present some other types of programmed cell death are being discussed besides apoptosis.

Apoptotic mechanisms are triggered when the deleterious stimulus is not so strong as to cause necrosis [93, 140]. An organism responds to threatening cell death by a peculiar defense—by suicide of a relatively small number of cells to make the consequences of the pathological stimuli not so disastrous.

As already mentioned, morphological markers of apoptosis are the apoptotic bodies and shrunken neurons with intact membrane. DNA fragmentation became a biochemical marker, which appears practically identical to the concept of "apoptosis" [68, 95, 112, 135]. This process is activated by calcium and magnesium ions and inhibited by zinc ions. DNA fragmentation occurs as a result of calcium-magnesium-dependent endonuclease action. Wyllie [142] and Waters [140] showed that endonucleases cause DNA cleavage between histones, releasing fragments of regular length. Initially DNA is divided onto large fragments of 50 and

300 b.p., which then are fragmented onto 180 b.p. portions constituting a characteristic "ladder" when separated by gel electrophoresis [95, 112]. However, Tomei *et al.* [136] and Hockenbery [57] concluded that DNA fragmentation does not always correlate with the morphology typical for apoptosis and represents a conditional marker not equal to a morphological criterion. The most perfect technique to confirm the presence of apoptosis is a biological histochemical technique allowing not only registering the fact of DNA fragmentation, but also the presence of such important features as apoptotic bodies [42].

Apoptosis is an active process of self-destruction conducted by a cell and it requires protein synthesis [36, 73, 99]. This was confirmed in *in vivo* in experiments where attempts were made to avert neuronal death using inhibitors of protein synthesis [36]. Goto *et al.* [46] demonstrated the favorable effect of infusion of the protein synthesis inhibitor cycloheximide on the survival of CA_1 hippocampal sensory neurons after transient ischemia of rat brain. On the other hand, it has also been found that apoptosis can occur without *de novo* protein synthesis [140].

The program of apoptosis consists of three sequential stages:

- making a decision to die or to survive;
- proceeding in the mechanism of annihilation;
- removal of dead cells (degradation of cellular components and phagocytosis).

Genetic and molecular mechanisms of apoptosis have been being intensely studied since the 1980s. The first studies conducted in the nematode *Caenorhabditis elegans* [40, 56, 150] showed that 131 cells of a total of 1090 constituting in the worm's body are predetermined to die. Mutations in the DNA of nematodes allowed the creation of two interesting phenotypes: in the first one all the cells predetermined to death survived, leading to the formation of giant nematodes; in the second case small nematodes originate with total cell count less than 959. It occurred that those phenotypes were the results of mutations in *ced*-family genes, the expression of which leads to synthesis of the proteins determining cell death or survival. Protein products of two of these genes, *ced-3* and *ced-4*, named "assassin genes", are necessary for the embodiment of the apoptotic program. The loss of their function led to the formation of the first type of nematode. The protein product of the *ced-9* gene protects cells and averts apoptosis by inhibiting the activated *ced-3* and *ced-4*. The rest of the *ced* genes encode proteins implicated in packing and phagocytosis of dying cells. One of them (*nuc-1*) encodes a nuclease that causes DNA degradation in a dead cell. Studies directed toward the search for *ced*-gene analogs in mammals are of the utmost interest. Such homologs of assassin genes *ced-3* (and its protein products) are genes, encoding IL-1β-converting enzymes—caspases (cysteine aspartyl-proteases) that possess various substrate and inhibitory specificity. The role of caspases in apoptotic mechanisms has been confirmed by many experimental works [19, 34, 53, 110, 128, 144].

The linkage between apoptosis and IL-1β, the basic trigger of inflammatory reactions, is of pivotal importance and suggests the common induction of the delayed consequences of ischemia as well as for close connection between programmed cell death, local inflammatory reaction in the ischemized brain area, and necrosis.

As a result of these experiments, a virus-derived caspase inhibitor was found—a protein of varicella virus (CrmA) which averted premature death of neurons in culture lacking trophic support. In acute cerebral ischemia in mice with genetically inactivated caspase-1, the extent of ischemic brain damage of brain tissue was far less pronounced in comparison with control animals. These results were confirmed by parallel studies *in vitro*, which showed that isolated murine neurons with inactivated caspase-1 were more resistant to apoptosis caused by removal of trophic factors. Experiments with inactivation of caspase-3 led to mice with enlarged brain [140]. Thus, modern studies demonstrated the activating influence of caspase family representatives on apoptotic events in the period of development as well as in case there was a damaging influence to brain.

At present 14 variants of caspases are identified in a human, and they can be subdivided into 3 groups:

- cytokine activators (caspases 1, 4, 5, 13);
- inducers of effector caspases activation (caspases 2, 8, 9, 10);
- effector caspases—executors of apoptosis (caspases 3, 6, 7).

Procaspases (inactive precursors of caspases) are present in all cells. The activation of caspases is produced by their proteolytic fission in sites where asparagine residues are located with following dimerization of the generated active subunits. This process may proceed by autostimulation during rapprochement of two caspases, as well as via direct action of other proteases and special adapter proteins during apoptotic signals transduction.

The activation of mammalian procaspases is mediated by the analog of the *ced-4* gene, which is an activating factor of apoptotic protease-1 (Apaf-1). It is noteworthy that Apaf-1 possesses a binding site for ATP, which emphasizes the importance of the level of energy support for choosing of death mechanism (energy-dependent apoptosis or energy-independent necrosis).

Activated caspases initiate the chain of proteolytic events aimed at apoptotic "dismantling" of a cell. There are several directions of their destructive action [138]. Caspases may inactivate inhibitors of those proteins which cause apoptotic changes. Thus, nuclease which leads to DNA fragmentation (CAD—caspase-activated deoxyribonuclease), is usually inactive being bound to an inhibitory protein (ICAD). Activated caspases 3 and 7 destroy this protein and trigger DNAase action [41, 86, 125]. Along with this caspases produce direct fission of cellular structural proteins. For instance, activated caspase 6 destroys nuclear lamin A, a protein strictly bound to nuclear membrane and organizing the structure of chromatin [132, 118]. Caspases also break regulation of protein synthesis. Indirect destruction of cytoskeleton via fission of proteins regulating their structure can be a consequence of this. Thus, effector caspases destroy gelsolin, which regulates tension of actin fibers. That leads to their destruction [72].

Destruction of cytosolic proteins causes a cell to loose water and decrease in size, decreasing the cytosolic pH. The cellular membrane looses its properties, the cell shrinks, and apoptotic bodies form. The activation of sphingomyelinase underlies the reconstitution of cellular membranes. This enzyme causes cleavage of membrane sphingomyelin resulting in generation of the second messenger ceramide, which activates phospholipase A_2 [149]. The products of arachidonic acid metabolism accumulate. The proteins expressed during apoptosis, such as phosphatidyl serine and vitronectin, come to the cell surface and produce signals for macrophages, which proceed to phagocytosis of the apoptotic bodies.

Activated caspases may also inactivate proteins involved in DNA reparation (for instance, DNA-PK_{sc}), mRNA formation (e.g., U1-70K), and DNA replication (e.g., replication factor C) [29, 122].

It is noteworthy that although caspases are basic effectors of apoptosis, apoptotic process may proceed in several types of cells even in cases when caspases are inactivated. This is possible due to activity of other effectors, such as calcium-activated proteases, serine and lysosomal proteases, and endonucleases [9]. However, the significance of these effectors in the embodiment of apoptosis is not considerable and possesses additional compensatory character.

Homologs of the nematode *ced-9* gene determining cell survival in mammals are the family of protooncogenes *bcl-2*. *Bcl-2* genes were initially revealed in B-cell lymphoma where their non-regulated production blocked mechanisms of cell death, thus maintaining oncogenesis. Bcl-2 and its kindred protein Bcl-x-l are present in mammalian brain. They protect neurons from apoptosis when ischemia occurred, trophic factors are removed, or neurotoxic influences are present *in vivo* and *in vitro*. Molecular genetic studies conducted during the last decade showed that in the so-called *bcl-2* gene family, mapped to chromosome 18 in humans, 16 genes are included that express proteins with opposite function (see Table 10.2). Six of them cause anti-apoptotic effects, like the family founder, bcl-2, and the

Table 10.2. The *Bcl-2* gene family [139]

Antiapoptotic genes	Proapoptotic genes
**** *bcl-2* [#]	*** *bax* [#]
**** *bcl-xl* [#]	*** *bak* [#]
**** *bcl-w* [#]	*** *bok/mtd* [#]
**** *boo* [#]	** *bcl-xs* [#]
*** *al*	* *bad*
*** *mcl-1* [#]	* *bik/nbk* [#]
	* *bid*
	* *hrk/dp5* [#]
	* *blk* [#]
	* *bim/bod* [#]

Note: Asterisks designate the number of conservative amino acid sequences typical for *bcl-2* gene: **** 4 regions (BH1-BH4); *** 3 regions (BH1-BH3); ** 2 regions (BH3, BH4); * 1 region (BH3).
[#] COOH terminal hydrophobic domain that is responsible for attachment to the outer membrane of mitochondria.

other 10 family members support apoptosis. Protein derivates of these genes have resembling morphology—each of them has not less than one of 4 conservative amino acid sequences typical for the *bcl-2* gene (BH1–BH4) [119]. Bcl-2-related proteins include anti-apoptotic (Bcl-2, Bcl-x-l, etc.), as well as pro-apoptotic (Bcl-x-s, Bax, Bad, Bag, etc.) proteins [107].

Regulation of the activity of Bcl-2-related proteins is conducted on the gene level (thus, P53 protein increases *bax* gene expression), as well as on posttranscriptional protein level (influence of cytokines). Along with this, complex interactions exist between the proteins themselves, such as formation of homo- and heterodimers within their protein group or with proteins of the oppositely directed action [116]. Bax and Bad proteins possess a homologous sequence and form heterodimers with Bcl-2 and Bcl-x-l *in vitro*. To reach activity enabling them to inhibit cell death, Bcl-2 and Bcl-x-l should form dimers with Bax protein, whereas dimers formed with Bad enhance death. This indicated that Bcl-2 and related molecules are crucial determinants of cell death or cell survival in the CNS [70, 145].

Pro- and anti-apoptotic actions of activated Bcl-2 family proteins are realized mainly via modulation of mitochondrial activity. This is because mitochondria are the source of cytochrome *c*, ATP, Ca^{2+}, apoptosis-inducing factor (AIF)—components required for the further ongoing of apoptotic signal. Release of these factors from mitochondria proceeds only when its membrane interacts with activated proteins of Bcl-2 family, which play a pivotal role in regulation of mitochondrial permeability [85]. Activated Bcl-2 family proteins are attached by their carboxy terminal hydrophobic groups to the outer membrane of mitochondria, in those sites where the outer and inner membranes come into proximity and where obviously the permeability transition (PT) pores are located. The PT pore represents a multiprotein complex built up at the contact site between the inner and the outer mitochondrial membranes [65, 103, 111, 138]. The PT pores are mega-channels with diameter not larger than 2 nm which are permeable for ions, active oxygen species, but never for larger molecules (cytochrome c, ATP, AIF), required for induction of apoptosis. It was demonstrated that pro-apoptotic proteins of Bcl-2 family, attached to the outer membrane, contact with adenine-nucleotide-translocator (ANT), incorporated into the inner membrane in these loci, and temporarily form larger channels of 2.4–3 nm in diameter [14]. Cytochrome c, ATP and AIF can percolate through them from mitochondria to cytosol. Anti-apoptotic proteins of the Bcl-2 family are not able to increase mitochondrial membrane permeability, and according to Marzo [103], can close pre-existing channels, thus interrupting the ongoing of apoptotic signal and thus defending the cell from apoptosis.

Thus, in response to decrease in *bcl-2* gene expression or superexpression of pro-apoptotic genes (e.g., *bax*) the permeability of mitochondrial membrane increases and cytochrome c, ATP, and AIF are released [65, 111]. Cytochrome c in complex with Apaf-1, procaspase-9, and ADP form the apoptosome, which activates effector caspases 9 and 3 [13], which are pivotal enzymes for initiation of apoptosis [78]. Thus is realized **the first pathway of apoptosis induction**—the mitochondrial pathway [121] (Fig. 10.1).

Along with this, release of AIF can trigger caspase-independent mechanisms of apoptosis. Being released from mitochondria, AIF comes into the nucleus where it causes DNA fragmentation resembling apoptosis, i.e. this is an additional "assassin-factor", which doubles the action of cytochrome c and caspases when they are blocked. Interestingly, translocation of AIF from the mitochondria to the nucleus is dependent on the activation of a nuclear enzyme, poly(ADP-ribose) polymerase (PARP) [150], which in its turn is activated by $NO^\bullet$ and peroxynitrite (see Section 4.3). At the same time, AIF is necessary for PARP-dependent cell death [150], which goes via catalyzing attachment of ADP ribose units from NAD to nuclear proteins following DNA damage. So, excessive activation of PARP by nitric oxide not only causes energy depletion-dependent cell death, but also activates apoptosis induction [39, 47, 97] (Fig. 10.2).

It is important to note that during apoptosis mitochondria do not loose their integrity and are not destroyed. According to some authors [63], exactly the mitochondrion is a pivotal figure of the apoptotic process. Others [59, 134] regard activation of caspase cascade to be a pivotal apoptotic event, where cytochrome c is not the initial factor, but just plays an amplifying role.

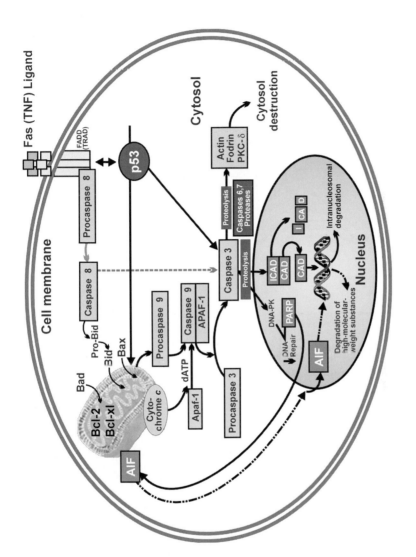

Figure '0.1. Main pathways of apoptosis induction.

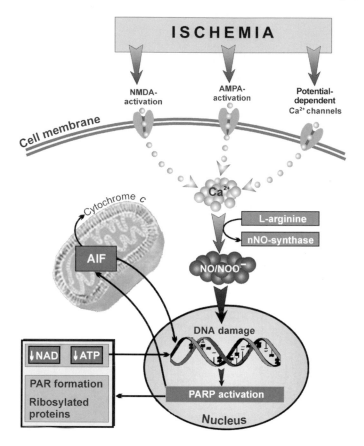

Figure 10.2. Participation of poly(ADP-ribose) polymerase (PARP) activation in mechanisms of ischemic cell damage.

The second pathway of apoptosis induction is connected with activation of superficial receptors (Fas, TNF) after extracellular "death-ligands" bind to them with following excitation of caspase-8 and triggering of the caspase cascade [139]. Transduction of pro-apoptotic signal when ligands are bound with "death domain" receptors is produced via adapter proteins FADD/MORT1 (FADD, Fas-associated death domain), whose N-terminal region (DED—death effector domain) in its turn binds to the analogous regions of procaspase-8, causing its autocatalytic activation [109]. Activation of several TNF-receptors (including TNF-R1) is produced via additional adapter protein TRADD (TNF-R-associated death domain).

Activated caspase-8 is able to directly activate effector caspases. At the same time, caspase-8 also activates Bid protein, which intensifies the release of cytochrome c from mitochondria. Thus, the receptor and mitochondrial pathways of apoptosis induction are tightly interlinked (see Fig. 10.1).

It is interesting that the receptor pathway with involvement of other adapter proteins is used for realization of other cellular programs. Thus, the signal through TNF-R1 can activate transcription factors NF-kappaB and AP-1 (see Chapter 7), this leading to activation of genes promoting production of pro-inflammatory factors [3]. Thus, when TNF-R1 is activated the development of both apoptosis and inflammation pathways is possible [139]. It has been demonstrated that inhibition of TNF and TNF-R1 in mice 30 min after induction of ischemic stroke blocks stroke related damage at two levels, the primary ischemic and the secondary inflammatory injury [102], which is accompanied by a marked decrease in both infarction volume and mortality.

DNA damage can also serve to trigger programmed cell death (**the third pathway of apoptosis induction**). Caused by oxidative stress or calcium-activated endonucleases, DNA damage initiates expression of transcriptional factor p53 [92], which in turn can alter transcription of a number of genes additionally inducing apoptosis [108]. It was shown that natural p53 causes apoptosis in tumor cell lines [146] and *in vivo* [126]. Transformation of p53 from its natural type into a mutant leads to cancer in many organs due to the inhibition of apoptosis.

The *p53* "tumor suppressor" gene, located on *p17*, encodes a nuclear protein which consists of 393 amino acids, with 53 kDa molecular weight. This transcriptional factor in latent state is situated in cytosol. Its activation occurs as a response to very early signs of DNA damage, as well as to many other intracellular processes (hypoxia, aging, trophic deficiency, etc.) [52]. It is not an accident that this protein was called "guardian of the genome" [75], as it regulates genomic stability and the cellular response to any impending metabolic changes [92]. Activated P53 protein may induce two programs: temporary suspension of cellular cycle in G1/S phase with the help of P21(WAF1) protein, which inhibits cyclin-dependent kinases [24, 130, 143] and induction of apoptosis via activation of the *bax* gene and/or activation of the generation of reactive oxygen species, those promoting cytochrome c release from mitochondria [91].

There are inhibitors of effector caspases such as proteins of the IAP family, which suppress activity of caspases 3 and 9. It is interesting that one of them, Survin, was found in the majority of various cancer cells [1]. Neuroprotective action of HSP72 was shown, which is connected with its blocking of translocation of apoptosis-inducing factor (AIF) from the mitochondria to the nucleus and inhibition of activated caspase 3 [129] (Scheme 10.1).

The role of apoptosis in both neuronal death and formation of brain infarction in acute focal cerebral ischemia is now being actively studied [95, 114].

Morphological features of apoptosis in ischemia can be found in neurons as well in glial cells [82, 95]. In models of global ischemia, it was demonstrated that apoptosis predominantly affects oligodendrocytes.

In focal ischemia the majority of apoptotic cells are registered along the internal border of the ischemic core [15, 16, 79, 82, 124], which may possibly reflect the participation of apoptotic processes in the enlargement of the infarction area. The number of apoptotic cells increases in relation to the duration of acute focal ischemia.

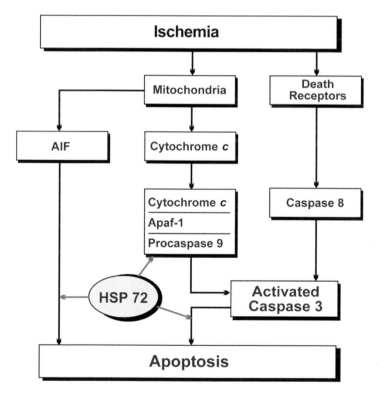

Scheme 10.1. Protective anti-apoptotic effects of HSP72.

According to Chopp and Li [24], 2 h after the constant occlusion of the rat middle cerebral artery and 24 h after the reperfusion began one looking through the ischemized striatum can observe cells that underwent apoptosis. Among them 90–95% are neurons, 5–10% astrocytes, and not more than 1% endothelial cells. The number of apoptotic cells increases up to a peak 24 48 h after the induced ischemia and tended later to decrease, however remaining significantly elevated versus control animals for approximately 4 weeks after the occlusion. The lengthy functioning of apoptosis confirms the dynamic character of the cell death process and gives evidence that any procedures carried out even weeks after the ischemic stroke onset might protect the ischemized brain tissue.

The DNA fragmentation phenomenon taking place in neurons that suffered transient global ischemia has been demonstrated in a number of works [7, 68, 95, 135, 141]. High doses of BDNF neurotrophin rescued CA_1 hippocampal neurons from apoptotic destruction during ischemia [7].

Direct connection between early response gene expression (*fos, c-jun, mkp-1*), the synthesis of stress-proteins (HSP72), and the induction of apoptosis was shown [141]. According to Harrison *et al.* [55], there is a certain genetic program that is triggered in response to brain ischemia and determines "turning on" of various signaling pathways leading to cell death. It was demonstrated that *in vitro* microinjection of antibodies to Fos and c-Jun protein families can block apoptosis.

Chen *et al.* [17] reported both the increase of *bcl-2* gene expression and HSP72 synthesis in neurons surviving acute focal cerebral ischemia. Isenmann *et al.* [62] confirmed that that elevation of Bcl-2 and Bcl-x-l contents is registered in surviving neurons located in the peri-infarction zone, whereas the increase of Bax proteins content is connected with DNA fragmentation and cell death occurring in the ischemic core. Identical results were acquired in the experimental work of Gillardon [43], who showed that 6 h after the occlusion of the middle cerebral artery the expression of *bax* mRNA significantly increases as well as the synthesis of Bax proteins in severely ischemized cortical and thalamic neurons. Asahi *et al.* [2] detected moderate increase in *bcl-x* gene transcription level in cells located in the ischemized brain area. In a study by Cao *et al.* [14], the role of intracellular Bax translocation in ischemic brain injury was investigated in a rat model of transient focal ischemia (30 min) and reperfusion (1 to 72 h). Immuno-chemical investigations revealed that transient ischemia induced a rapid translocation of Bax from cytosol to mitochondria in caudate neurons, with a temporal profile and regional distribution coinciding with the mitochondrial release of cytochrome *c* and caspase-9. In post-ischemic caudate putamen *in vivo* and in isolated brain mitochondria *in vitro*, the authors found enhanced heterodimerization between Bax and the mitochondrial membrane permeabilization-related proteins adenine nucleotide translocator (ANT) and potential-dependent channel. These results suggest that the Bax-mediated mitochondrial apoptotic signaling pathway may play an important role in the development of ischemic neuronal injury.

Protective effects of genetic and pharmacological interventions directed to the selective inhibition of apoptosis are the best proof of its participation in cell death during brain ischemia. It was shown that experimental transgenic hyperexpression of *bcl-2* [101] or introduction of the gene via herpes virus vector [76, 82] decrease brain infarction size in mice and improve neuronal survival of CA_1 hippocampal area in global ischemia [69]. Decrease of infarction size was also demonstrated in transient focal ischemia in mice null for the *bax* gene [77].

De Bilbao *et al.* [35] did not confirm previous data showing a strong protective role of bcl-2 in neocortical infracted areas, but demonstrated in *bcl-2* transgenic mice that anti-apoptotic therapies may constitute a possible treatment for areas of the brain remote from those directly affected by ischemia.

It was shown that within a few hours after experimental transient ischemic attack caspase-3 is activated in cortical and striatal neurons. This enzyme is the most homologous to the lethal gene *ced-3* of *Caenorhabditis elegans* [110, 144]. In the model of ischemic stroke induced by occlusion of the middle cerebral artery (MCAO) in mice, Guegan *et al.* [50] showed by immunohistochemistry a constitutive expression of caspase-3 that is enhanced after MCAO in neurons localized within the infarcted zone. As a function of time intervals after MCAO, the cytochrome *c* amount increased in the cytosolic fraction of ischemic cortical extracts. The kinetics of the release was in concordance with the expression of caspase-3 and the subsequent cleavage of poly(ADP-ribose) polymerase (PARP) appearing before the internucleosomal fragmentation of DNA, the ultimate step of apoptosis.

After intraventricular injection of the caspase inhibitor z-DEVD.fmk in focal ischemia the size of rat brain infarction decreases [53], and neurological functional recovery improves [18], in transient global ischemia the number of deceased neurons of the CA_1 hippocampal area declines [19].

At the same time, Love *et al.* [90] examined brain tissue from patients who had experienced a cardiac arrest with resuscitation or an atherothrombotic ischemic stroke and died 12 h and 9 days later. They concluded that in humans the activation of caspase-3 plays little part in neuronal death in atherothrombotic ischemic stroke but may contribute to delayed death of neurons after cardiac arrest. Sections, which were immunostained for activated caspase-3 or the 89 kDa caspase-3-mediated cleavage product of poly(ADP-ribose) polymerase, showed that brain ischemia caused activation of caspase-3 in macrophages/microglia. Some neurons demonstrated delayed activation of caspase-3 after cardiac arrest, but very few in atherothrombotic infarctions.

Investigations performed *in vitro* have shown that the culture of murine cortical neuron temporarily lacking oxygen and glucose dies via "excitotoxic" necrosis. However, after the blockage of excitotoxicity by NMDA- and AMPA/kainate antagonists (when the deprivation period is elongated to avert the therapeutic effects of those medications) neurons die via apoptosis sensitive to z-VAD.fmk effects [48]. Neuroprotective effects of caspase inhibitors in ischemic stroke *in vivo* may be partially connected with the inhibition of both IL-1β-converting enzyme and IL-1β induced inflammation [123]. However, we cannot explain that in cell culture referring to this mechanism because when higher concentration of IL-1β is injected the neuroprotective effect was stable [48].

The participation of p53 and p21 in the development of apoptotic change in neurons of the CA_1 hippocampal area of the forebrain has been proven [24, 26, 130]. Middle cerebral artery occlusion is accompanied by significant elevation of p53 level in the ischemic lesion [25, 80]. The infarction area decreases in transgenic mice lacking p53 [87] as well as under protection of P53-inhibitor, pifithrin-alpha, decreasing P53 DNA-binding activity and expression of the p53 target gene *bax* [31]. This emphasizes p53 participation in apoptosis during ischemic stroke.

Experimental studies [10, 37, 77, 104] demonstrated that when necrosis and apoptosis are coherently involved in the ischemic process in brain, the choice of predominant death pathway is determined by several factors including the severity of ischemia, the degree of neuronal maturity, the accessibility of trophic support, the intracellular content of calcium, and levels of certain cytokines. Low concentrations of "excitotoxins" *in vitro* induce neuronal apoptosis, whereas higher concentrations induce necrosis [10]. Following soft transient focal ischemia (30-min long middle cerebral artery occlusion) infarction develops in a reduced way within more than 3 h after stroke onset. It is noteworthy that morphological study and sensitivity of delayed cell death mechanisms to such inhibitor of protein synthesis as cycloheximide confirms the significant role of apoptosis [37].

The maturity of cells is a very important factor that determines the mechanism of their death. After a ligature was placed onto carotid arteries and hypoxia ensued in newborn rats, the caspase inhibitor BAF decreased the size of infarction by more than 50%, indicating a significant role of apoptosis in the damage of immature tissue [21]. Such a conclusion was also made in the study by McDonald *et al.* [106], which showed that the predisposition of cultured neurons to apoptosis decreases proportionally to their maturation *in vitro* and, coherently, their susceptibility to NMDA-induced "excitotoxic" necrosis is enhanced.

Insufficient brain trophic support plays a crucial role in the development of apoptosis. When cortical neurons are exposed to low content of calcium ions, morphological damage to axons and following "triggered" apoptosis were induced [51]. When exposed to "excitotoxins" neuronal axons were also damaged, predominantly in the area of the central infarction. Penumbral neurons continued to resist this glutamate damage and later underwent apoptosis due to the loss of trophic factor delivery via synapses [8].

The state of calcium homeostasis in cells undergoing apoptosis remains to be further elucidated [77]. In some cases excessive content of calcium ions are observed in apoptotic cells as these ions induce mechanisms of apoptotic cascade (for instance, the activation of endonucleases that leads to fragmentation of nuclear DNA) [105]. Along with this, according to Balaz *et al.* [5], Johnson and Deckwerth [64], and Lampe *et al.* [74], apoptosis can be suspended by increasing intracellular content of calcium ions via activation of potential-dependent calcium channels, inhibition of calcium sequestration, or adding of low doses of NMDA into the culture media. Intracellular calcium contents inhibiting apoptosis in sympathetic neurons are significantly lower (180–240 nmol) [71] than those levels mediating glutamate–calcium "excitotoxicity" in neurons (>5 µmol) [60]. Possibly, neuronal survival depends to a considerable extent on the optimal content of intracellular calcium ions [64] and the paradoxical, at first sight, hypothesis about calcium neuroprotection may turn out to be real [77].

It is noteworthy that young sympathetic neurons strongly depending on trophic support have lower optimal level of intracellular calcium than mature sympathetic neurons less dependent on trophic support [84].

Thus, calcium homeostasis appears to be a powerful modulating system [71, 77]: when intracellular level of calcium is inadequately low, neurons are at risk of apoptosis and strongly depend on trophic support; when these concentrations are intermediate the conditions for survival are optimal and cell demands for trophic support are minimal; when calcium concentrations are increased the cytotoxic events of necrosis ensue. It is logical to suggest that the excess of intracellular calcium participates in cell damage predominantly in the ischemic core within the first minutes after ischemia started, whereas their lack plays a negative role later and especially in the penumbral area. Obviously, some cells undergo both stages.

Johnson and Deckwerth [64] found that increase in intracellular potassium ions *in vitro* inhibits apoptosis. Their neuroprotective effect is not eliminated when the potential-dependent calcium channels are blocked and, hence, intracellular calcium accumulation is inhibited [147]. This led the authors to conclude that such variants of apoptosis may exist that are connected with the early enhancement of potassium efflux out of cells and the decrease in their intracellular level [147]. It was shown that the injection of potassium channel blocker 4-aminopyridine inhibits apoptotic shrinkage of human eosinophils caused by removal of cytokines [6]. Along with this, when potassium ion content is decreased to a level lower than physiological, transformation of caspase-3, which induces apoptotic processes, is observed and the intensity of this transformation also depends on the level of potassium [11].

Thus, it was established that genetically programmed cell death participated in the formation of brain infarction. Apoptotic mechanisms that allow cells to die without both prominent signs of inflammation and release of genetic material are triggered later than swift necrotic cascade events are—at least 1 h after the ischemia onset, then they begin to manifest fully 12 h later and reach a peak on the 2nd–3rd day of stroke. Thus, apoptosis along with other delayed consequences of ischemia takes part in the "up-formation" of the infarction, additionally damaging the area of the penumbra.

All three basic mechanisms of apoptosis induction participate in the common process and are triggered simultaneously, functioning in tight interaction. Of course, many details of regulation of apoptosis still remain to be elucidated. Several facts were demonstrated showing the possibility of the development of apoptosis in cells with no nucleus [4], under conditions of suspension of protein synthesis [137], as well as in isolated nuclei, located outside of cells [27]. Cells can undergo apoptosis in every stage of the cell cycle; their sensitivity to apoptotic stimuli is different [32]. These findings suggest that different cellular compartments are autonomic in relation to apoptosis, and apoptotic effectors are constitutionally expressed in every cell. Along with that their activity can be controlled by both intra- and extracellular signals [131].

Experimental investigations of recent years [45, 152] have shown the close connection between molecular mechanisms of necrotic and apoptotic cell death. One of the main necrotic markers PARP plays an important role in AIF translocation to the nucleus and supports apoptotic processes [150]. Overexpression of *bcl-2* has been found to protect against both apoptotic and necrotic cell death [66]. Changes in the expression of anti-apoptotic genes appear to be extremely relevant to the evolution of ischemic injury regardless of its precise mechanism of morphology.

NMDA-mediated increase in cytoplasmic calcium and following nitric oxide generation have been shown to be the main links leading to long-lasting neuronal nucleus responses such as differentiation, survival, and synaptic plasticity. It has been demonstrated that NO activates two kinase

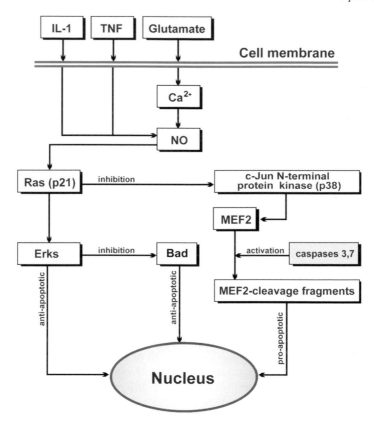

Scheme 10.2. Protein kinase signaling pathways involved in mechanisms of cell death and survival.

signaling pathways involving Ras (p21) with following induction of the mitogen-activated protein kinases (MAPKs)/extracellular signal-regulated kinases (Erks) cascade and c-Jun N-terminal protein kinases (JNK)/p38/SAPK2 cascade [61, 113, 153]. Extracellular signal-regulated kinases pathway was shown to promote cell survival and proliferation, while JNK/p38 pathway induces apoptosis in general (Scheme 10.2). P38 mitogen-activated protein kinase activates myocyte enhancer factor-2 (MEF2), which triggers pro-apoptotic cascade under ischemic conditions [115]. During focal ischemic brain damage, activated caspases 3 and 7 cleave MEF2 molecules into fragments, which block MEF2 transcriptional activity and cause pro-apoptotic effects [115]. Erks activation can result in the prevention of JNK/p38 activation and thus be involved in the protective responses in ischemic tolerance [49]. At the same time, the roles of MAPK cascades in neuronal death and survival seem to be complicated and altered by the type of cells and the severity and timing of stroke [113].

It is interesting that in many cases along with the apoptotic features one can detect additional necrotic features in brain ischemized cells [89, 99]. Such miscellaneous morphology of damage with fragmented intranucleosomal DNA was found in necrotic cells [77]. Programmed cell death and necrotic mechanisms connected with glutamate "excitotoxicity" are triggered in the ischemized brain area coherently, so the concomitance of their morphological features and biochemical markers is rather predictable [22, 77, 99]. Taking into account the common intracellular signaling system and the "cross-talks" between all the molecular cascades, we suppose that necrosis and apoptosis are interrelated and dependent on each other parts of the single ischemic mechanism.

Anti-apoptotic signaling pathways are activated by neurotrophic factors, certain cytokines, and some stress factors [30, 104]. Such protective pathways include: activation of the transcription factors (e.g., NF-kappa B), which induce expression of stress proteins, antioxidant enzymes, and calcium-regulating proteins; activation of extracellular signal-regulated kinases; phosphorylation-mediated modulation of ion channels and membrane transporters; cytoskeletal alterations, which modulate calcium homeostasis; as well as modulation of proteins that stabilize mitochondrial function (e.g., Bcl-2) [83, 104].

Intervention studies in experimental stroke models have identified a battery of approaches of potential benefit in reducing neuronal death in stroke patients, including administration of antioxidants, calcium-stabilizing agents, caspase inhibitors, HSPs, MAPK regulators, and agents that influence NF-kappaB [30, 38, 98, 104, 113].

REFERENCES

1. Ambrosini, G., Adida, C., Altieri, D., 1997, *Nat Med.* **3**: 917–921.
2. Asahi, M., Hoshimaru, M., Uemura, Y., *et al.*, 1997, *J Cerebr Blood Flow Metab.* **17**: 11–18.
3. Ashkenazi, A., Dixit, V., 1998, *Science.* **281**: 1305–1308.
4. Bakhshi, A. J., Jensen, P., Goldman, P., *et al.*, 1985, *Cell.* **41**: 889–906.
5. Balazs, R., Jorgens, O. S., Hack, N., 1988, *Neurosci.* **27**: 437–451.
6. Beauvais, F., Michel, L., Dubertret, L., 1995, *J Cell Physiol.* **57**: 851–855.
7. Beck, T., Lindholm, D., Casiren, E., Wree, A., 1994, *J Cerebr Blood Flow Metab.* **14**: 689–692.
8. Bindokas, V. P., Miller, R. J., 1995, *J Neurosci.* **15**: 6999–7011.
9. Blagosklonny, M., 2000, *Leukemia.* **14**: 1502–1518.

10. Bonfoco, E., Krainc, D., Ankarcrona, M., *et al*., 1995, *Proc Natl Acad Sci USA*. **92**: 7162–7166.
11. Bortner, C. D., Hughes, F. M., Jr., Cidlowski, J. A., 1997, *J Biol Chem.* **272**: 32436–32442.
12. Brenner, C., Zamzami, N., *et al*., 1998, *J Exp Med.* **187** (8): 1261–1271.
13. Budihardjo, I., Oliver, H., Lutter, M., *et al*., 1999, *Ann Rev Cell Develop Biol.* **15**: 269–290.
14. Cao, G., Minami, M., Pei, W., *et al*., 2001, *J Cerebr Blood Flow Metab.* **21** (4): 321–333.
15. Charriaut-Marlangue, C., Margaill, I., Plotkine, M., Ben-Ari, Y., 1995, *J Cerebr Blood Flow Metab.* **15**: 385–388.
16. Charriaut-Marlangue, C., Margaill, I., Represa, A., 1996, *J Cerebr Blood Flow Metab.* **16**: 186–194.
17. Chen, J., Graham, S. H., Chan, P. H., *et al*., 1995, *Neuroreport.* **6**: 394–398.
18. Chen, J., Li, Y., Wang, L., *et al*., 2002, *J Neurol Sci.* **199** (1–2): 17–24.
19. Chen, J., Nagayama, T., Jin, K., *et al*., 1998, *J. Neurosci.* **18** (13): 4914–4928.
20. Chen, S. C., Soares, H. D., Morgan, J. I., 1996, *Adv Neurol.* **71**: 433–450.
21. Cheng, Y., Deshmukh, M., D'Costa, A., *et al*., 1998, *J Clin Invest.* **101** (9): 1992–1999.
22. Choi, D. W., 1996, *Curr Opin Neurobiol.* **6**: 667–672.
23. Chopp, M., Chan, P. H., Hsu, C. Y., *et al*., 1996, *Stroke.* **27** (3): 363–369.
24. Chopp, M., Li, Y., 1996, *Acta Neurochir.* **66**: 21–26.
25. Chopp, M., Li, Y., Dereski, M. O., *et al*., 1992, *Stroke.* **23**: 104–107.
26. Clarke, A. R., Purdie, C. A., Harrison, D. G., *et al*., 1993, *Nature.* **362**: 849–852.
27. Cleary, M., Smith, S., Sclar, J., 1986, *Cell.* **47**: 19.
28. Crumrine, R. C., Thomas, A. L., Morgan, P. F., 1994, *J Cerebr Blood Flow Metab.* **14**: 887–891.
29. Cryns, V., Yuan, J., 1998, *Genes Dev.* **12**: 1551.
30. Culmsee, C., Zhu, Y., Krieglstein, J., Mattson, M. P., 2001, *J Cerebr Blood Flow Metab.* **21** (4): 334–343.
31. Culmsee, C., Zhu, Y., Yu, Q. S., *et al*., 2001, *J Neurochem.* **77** (1): 220–228.
32. Danmeade, S., *et al*., 2000, *Cancer Control J.* **3** (4): 1–10.
33. Davies, A. M., 1988, *Trends Genet.* **4**: 139–143.
34. Davies, A. M., 1995, *Trends Neurosci.* **18**: 355–358.
35. De Bilbao, F., Guarin, E., Nef, P., *et al*., 2000, *Eur J Neurosci.* **12** (3): 921–934.
36. Deckwerth, T. L., Johnson, E. M., Jr., 1993, *J Cell Biol.* **123**: 1207–1222.
37. Du, C., Hu, R., Csernansky, C. A., *et al*., 1996, *J Cerebr Blood Flow Metab.* **16**: 195–201.
38. Eldadah, B. A., Faden, A. I., 2000, *Neurotrauma.* **17** (10): 811–829.
39. Eliasson, M. J., Sampei, K., Mandir, A. S., *et al*., 1997, *Nat Med.* **3** (10): 1089–1095.
40. Ellis, H. M., Horvitz, H. R., 1986, *Cell.* **44**: 817–829.
41. Enari, M., Sakahira, H., Yokoyama, H., *et al*., 1998, *Nature.* **391** (6662): 43–50.
42. Gavrieli, Y., Sherman, Y., Ben-Sasson, S. A., 1992, *J Cell Biol.* **119**: 493–501.
43. Gillardon, F., Lenz, C., Waschke, K. F., 1996, *Mol Brain Res.* **40**: 254–260.
44. Glucksmann, P., Klempt, N., Guan, J., *et al*., 1992, *Biochem Biophys Res Commun.* **182**: 593–599.
45. Gonzalez-Zulueta, M., Feldman, A. B., Klesse, L. J., *et al*., 2000, *Proc Natl Acad Sci USA.* **97** (1): 436–441.
46. Goto, K., Ishige, A., Sekiguch, K., *et al*., 1990, *Brain Res.* **534**: 299–302.
47. Goto, S., Xue, R., Sugo N., *et al*., 2002, *Stroke.* **33** (4): 1101–1106.
48. Gottron, F. J., Ying, H. S., Choi, D. W., *et al*., 1997, *Mol Cell Neurosci.* **9** (3): 159–169.

49. Gu, Z., Jiang, Q., Zhang, G., 2001, *Neuroreport.* **12** (16): 3487–3491.
50. Guegan, C., Sola, B., 2000, *Brain Res.* **856** (1-2): 93–100.
51. Gwag, B. J., Canzoniero, L. M., Sensi, S. L., 1999, *Neurosci.* **90** (4): 1339–1348.
52. Hansen, R., Oren, M., 1997, *Curr Opin Genet.* **7**: 46–51.
53. Hara, H., Fink, K., Endres, M., *et al.*, 1997, *J Cerebr Blood Flow Metab.* **17**: 370–375.
54. Hara, H., Friedlander, R. M., Gagliardini, V., *et al.*, 1997, *Proc Natl Acad Sci USA.* **94**: 2007–2012.
55. Harrison, D. C., Davis, R. P., Bond, B. C., *et al.*, 2001, *Brain Res Mol Brain Res.* **89** (1–2): 133–146.
56. Hedgecock, E. M., Salston, J. E., Thomson, J. N., 1983, *Science.* **220**: 1277–1279.
57. Hockenbery, D. M., 1995, *Am J Pathol.* **146**: 16–19.
58. Hockenbery, D. M., Nunez, G., Milliman, C., 1990, *Nature.* **348**: 334–336.
59. Hu, Y., Benedict, M., Wu, D., *et al.*, 1998, *Proc Natl Acad Sci USA.* **95** (8): 4386–4391.
60. Hyrc, K., Handran, S. D., Rothman, S. M., Goldberg, M. P., 1997, *J Neurosci.* **17**: 6669–6677.
61. Irving, E. A., Barone, F. C., Reith, A. D., *et al*, 2000, *Brain Res Mol Brain Res.* **77** (1): 65–75.
62. Isenmann, S., Stoll, G., Schroeter, M., *et al.*, 1998, *Brain Pathol.* **8** (1): 49–62.
63. Jacotot, E., Constantini, P., *et al.*, 1999, *Ann NY Acad Sci.* **887**: 18–30.
64. Johnson, E. M., Jr., Deckwerth, T. L., 1993, *Annu Rev Neurosci.* **16**: 31–46.
65. Jurgensmeier, J. M., Xie, Z., Deveraux, Q., *et al.*, 1998, *Proc Natl Acad Sci USA.* **95** (9): 4997–5002.
66. Kane, D. J., Sarafian, T. A., Anton, R., *et al.*, 1993, *Science.* **262**: 1274–1277.
67. Kerr, J. F. R., Wyllie, A. H., Currie, A. R., 1972, *J Cancer.* **26**: 239–257.
68. Kihara, S., Shiraishi, T., Nakagawa, S., *et al.*, 1994, *Neurosci Lett.* **175**: 133–136.
69. Kitagawa, K., Hayashi, T., Mitsumoto, Y., *et al.*, 1998, *Stroke.* **29** (7): 1417–1422.
70. Kluck, R. M., Bossy-Wetzel, E., Green, D. R., Newmeyer, D. D., 1997, *Science.* **275** (5303): 1132–1136.
71. Koike, T., Tanaka, S., 1991, *Proc Natl Acad Sci USA.* **88**: 3892–3896.
72. Kothakota, S., 1997, *Science.* **278**: 294.
73. Kryzhanovsky, G. N., Lutsenko, V. K., 1995, *Adv Modern Biol.* **115** (1): 31–49 (in Russian).
74. Lampe, P. A., Cornbrooks, E. B., Juhasz, A., *et al.*, 1995, *J Neurobiol.* **26**: 205–212.
75. Lane, D. P., 1992, *Nature.* 358: 15–16.
76. Lawrence, M. S., Ho, D. Y., Sun, G. H., 1996, *J Neurosci.* **16** (486): 44–96.
77. Lee, J. M., Zipfel, G. J., Choi, D. W., 1999, *Nature.* **399** (6738): 7–14.
78. Li, P., Nijhawan, D., Budihardjo, I., *et al.*, 1997, *Cell.* **91** (4): 479–489.
79. Li, Y., Chopp, M., Jiang, N., 1995, *Stroke.* **26**: 1252–1258.
80. Li, Y., Chopp, M., Zhang, Z. G., *et al.*, 1994, *Stroke.* **25**: 849–856.
81. Linnik, M. D., Zahos, P., Geschwind, M. D., Federoff, H. J., 1995, *Stroke.* **26**: 1670–1675.
82. Linnik, M. D., Zobrist, R. H., Hatfield, M. D., 1993, *Stroke.* **24**: 2002–2008.
83. Lipton, P., 1999, *Physiol Rev.* **79** (4): 1431–1568.
84. Lipton, S. A., Kater, S. B., 1989, *Trends Neurosci.* **12**: 265–270.
85. Lithgow, T., van Driel, R., Bertram, J. F., Strasser, A., 1994, *Cell Growth Differ.* **5** (4): 411–417.
86. Liu, X. S., Zou, H., Slaughter, C., Wang, X., *et al.*, 1997, *Cell.* **89** (2): 175–184.
87. Lockshin, R. A., Williams, C. M., 1964, *J Insect Physiol.* **10**: 643–649.
88. Lockshin, R. A., Williams, C. M., 1965, *J Insect Physiol.* **11**: 123–133.

89. Van Lookeren Campagne, M., Gill, R., 1996, *Neurosci Lett.* **213**: 111–114.
90. Love, S., Barber, R., Srinivasan, A., *et al.*, 2000, *Neuroreport.* **11** (11): 2495–2499.
91. Lowe, S., Lin, A., 2000, *Cancerogenesis.* **21**: 485–495.
92. Lowe, S. W., Schmitt, E. M., Smith, S. W., *et al.*, 1993, *Nature.* **362** (6423): 847–849.
93. Lushnikov, E. F., Zagrebin, V. M., 1987, *Arch Pathol.* **2**: 84–89 (in Russian).
94. MacManus, J. P., Buchan, A. M., Hill, J. E., *et al.*, 1993, *Neurosci Lett.* **164**: 89–92.
95. MacManus, J. P., Linnik, M. D., 1997, *J Cerebr Blood Flow Metab.* **17**: 815–832.
96. Majno, G., Joris, I., 1995, *Am J Pathol.* **146**: 3–15.
97. Mandir, A. S., Poitras, M. F., Berliner, A. R., *et al.*, 2000, *J Neurosci.* **20** (21): 8005–8011.
98. Margulis, B., Guzhova, I., 2000, *Cytology.* **42** (4): 323–342 (in Russian).
99. Martin, D. P., Schmidt, R. E., DiStefano, P. S., *et al.*, 1988, *J Cell Biol.* **106**: 829–844.
100. Martin, L. J., *et al.*, 1998, *Brain Res Bull.* **46**: 281–309.
101. Martinou, J. C., Dubois-Dauphin, M., Staple, J. K., *et al.*, 1994, *Neuron.* **13**: 1017–1030.
102. Martin-Villalba, A., Hahne, M., Kleber, S., *et al.*, 2001, *Cell Death Differ.* **8** (7): 659–661.
103. Marzo, I., Brenner, C., Zamzami, N., *et al.*, 1998, *J Exp Med.* **187**: 1261–1271.
104. Mattson, M. P., Culmsee, C., Yu, Z. F., 2000, *Cell Tissue Res.* **301** (1): 173–187.
105. McConnkey, D. J., Orrenius, S. J., 1996, *Leukoc Biol.* **59**: 775–783.
106. McDonald, J. W., Behrens, M. I., Chung, C., *et al.*, 1997, *Brain Res.* **759**: 228–232.
107. Merry, D. E., Korsmeyer, S. J., 1996, *Annu Rev Neurosci.* **20**: 245–267.
108. Miyashita, T., Krajewski, S., Krajewska, M., *et al.*, 1994, *Oncogene.* **9** (6): 1799–1805.
109. Musio, M., Stockwell, B., Stennicke, H. R., *et al.*, 1998, *J Biol Chem.* **273** (5): 2926–2930.
110. Namura, S., *et al.*, 1998, *J Neurosci.* **18**: 3659–3668.
111. Narita, M., Shimuzu, S., Ito, T., *et al.*, 1998, *Proc Natl Acad Sci USA.* **95** (25): 14681–14686.
112. Nitatori, T., Sato, N., Waguri, S., *et al.*, 1995, *J Neurosci.* **15**: 1001–1011.
113. Nozaki, K., Nishimura, M., Hashimoto, N., 2001, *Mol Neurobiol.* **23** (1): 1–19.
114. Nowak, T. S., Jr., Osbome, O. C., Suga, S., 1993, Stress protein and proto-oncogene expression as indicators of neuronal pathophysiology after ischemia. In *Neurobiology of Ischemic Brain Damage. Progress in Brain Research* (Kogure, K., Hossmann, K.-A., Siesjo, B.-K., eds.) Vol. 96, Elsevier Science Publishers, Amsterdam, pp. 195–208.
115. Okamoto, S., Li, Z., Ju, C., *et al.*, 2002, *Proc Natl Acad Sci USA.* **99** (6): 3974–3979.
116. Oltvai, Z., Korsmeyer, S., 1994, *Cell.* **79**: 189–192.
117. Oppenheim, R. W., 1991, *Ann Rev Neurosci.* **14**: 453–501.
118. Orth, K., Chinnaiyan, A., Garg, M., *et al.*, 1996, *J Biol Chem.* **271** (28): 16443–16446.
119. Pellegrini, M., Strasser, A., 1999, *J Clin Immunol.* **19**: 365–375.
120. Petito, C. K., Olarte, J.-P., Roberts, B., *et al.*, 1998, *J Neuropathol Exp Neurol.* **57**: 231–238.
121. Reed, J. C., *et al.*, 1998, *Biochim Biophys Acta.* **1366**: 127–137.
122. Rheaume, E., Cohen, L. Y., Uhlmann, F., *et al.*, 1997, *EMBO J.* **16** (21): 6346–6354.
123. Rothwell, N., Allan, S., Toulmond, S., 1997, *J Clin Invest.* **100**: 2648–2652.
124. Sadoul, R., Dubois-Dauphin, M., Fernandel, P. A., *et al.*, 1996, *Adv Neurol.* **71**: 419–424.
125. Sakahira, H., Enari, M., Ohsawa, Y., *et al.*, 1999, *Curr Biol.* **9** (10): 543–546.
126. Schimke, R. T., Mihich, E., 1994, *Cancer Res.* **54**: 302–305.
127. Schwartz, L. M., Milligan, C. E., 1996, *Trends Neurosci.* **19**: 555–562.

128. Schwartz, L. M., Osborne, B. A., 1993, *Immunol Today*. **14**: 582–590.
129. Sharp, F. R., Massa, S. M., Swanson, R. A., 1999, *Trends Neurosci.* **22** (3): 97–99.
130. Stabberod, P., Tomasevic, G., Kamme, F., Wieloch, T., 1994, *Abst Soc Neurosci*. **20**: 616.
131. Steller, H., Grether, M., 1994, *Neuron.* **13**: 1269–1275.
132. Takahashi, A., Alnemri, E. S., Lazebnik, Y. A., *et al.*, 1996, *Proc Natl Acad Sci USA.* **93** (16): 8395–8400.
133. Tata, J. R., 1966, *Dev Biol*. **13**: 77–94.
134. Thorberry, N. A., Lazebnik, Y., 1998, *Science*. **281**: 12–16.
135. Tobita, M., Nagano, I., Nakamura, S., *et al.*, 1995, *Neurosci Lett*. **200**: 129–132.
136. Tomei, L. D., Shapiro, J. P., Cope, F. O., 1993, *Proc Natl Acad Sci USA.* **90**: 853–857.
137. Tsujimoto, Y., Gorham, J., Cossman, J., *et al.*, 1985, *Science*. **229**: 1390–1393.
138. Vladimirskaja, E. B., 2001, *Biological Basis for Anti-tumor Therapy*, Agat-Med, Moscow, p. 110 (in Russian).
139. Wallach, D., Varfolomeev, E. E., Malinin, N. L., *et al.*, 1999, *Annu Rev Immunol*. **17**: 331–367.
140. Waters, C., 1997, *RBI Neurotransmissions, Newsletter for Neuroscientist*. **XIII** (2): 2–7.
141. Wiessner, C., Vogel, P., *et al.*, 1996, *Mechanisms of Secondary Brain Damage in Cerebral Ischemia and Trauma.* New York, pp. 1–7.
142. Wyllie, A. H., 1980, *Nature*. **284**: 555–556.
143. Xiong, Y., Hannon, G., *et al.*, 1993, *Nature*. **366**: 701–704.
144. Xue, D., Shaham, S., Horvitz, H. R., 1996, *Genes Dev*. **10**: 1073–1082.
145. Yang, J., Liu, X., Bhalla, K., *et al.*, 1997, *Science*. **275** (5303): 1129–1132.
146. Yonish-Rouach, E., Resnitzky, D., Lotem, J., *et al.*, 1991, *Nature*. **353**: 345–347.
147. Yu, S. P., Yeh, C. H., Sensi, S. L., *et al.*, 1997, *Science.* **278** (5335): 114–117.
148. Yu, S. W., Wang, H., Poitras, M. F., *et al.*, 2002, *Science.* **297** (5579): 259–263.
149. Yu, Z. F., Nikolova-Karakashian, M., Zhou, D., *et al.*, 2000, *J Mol Neurosci.* **15** (2): 85–97.
150. Yuan, J., Horvitz, H. R., 1990, *Dev Biol*. **138**: 33–41.
151. Yun, H. Y., Dawson, V. L., Dawson, T. M., 1999, *Diabetes Res Clin Pract.* **45** (2–3): 113–115.
152. Yun, H. Y., Dawson, V. L., Dawson, T. M., 1999, *Diabetes Res Clin Pract.* **45** (2–3): 113–115.
153. Zhu, Y., Yang, G. Y., Ahlemeyer, B., *et al.*, 2002, *J Neurosci.* **22** (10): 3898–3909.

Chapter 11

Reaction of the Stress-Mediating Endocrine System in Response to Acute Brain Ischemia

Modern molecular studies suggest that the hormonal response to acute brain ischemia is a component of a common response of the neuro–immune–endocrine system caused by the expression of immediate-early genes—by the synthesis of heat-shock (or stress) proteins (HSP) and cytokines, which are immune mediators [5, 25, 26, 41] (see Chapters 7 and 8; Fig. 8.1). The stress-mediating response provides the metabolic basis for compensatory and adaptive changes that to a considerable extent determine the features of the course of disease [7, 11, 27]. This response is mediated by the sympathoadrenal, hypothalamic–pituitary–adrenal, renin–angiotensin, and thyroid hormonal axes [13].

The results of hormonal profile studies in patients with different forms of urgent somatic pathology (sepsis, burns, various forms of shock, cranio-cerebral injury) demonstrated that the type of endocrine system response does not depend on the cause of the stress [1, 19, 31]. When hormonal interactions were studied, it was shown that the response of the sympathoadrenal system occurs simultaneously with the stimulation of the hypothalamic–pituitary–adrenal and renin–angiotensin axes [2, 8, 12]. Corticoliberin plays a crucial role in the simultaneous activation of these systems [15, 18]. It, on one hand, stimulates the production of adrenocorticotropic hormone (ACTH) in the pituitary, which leads to activation of the hypothalamic–pituitary–adrenal axis and to inhibition of local inflammatory and autoimmune processes [6, 9, 31, 32], and, on the other hand, induces the sympathoadrenal system [16, 37, 48], mediates excessive production of renin [14, 39, 50] and angiotensin-II, as well as the release of catecholamines from axons of sympathetic neurons and secondary increase of ACTH secretion [33, 34] (Fig. 11.1). This leads to additional sympathetic activation and enhances the cytotoxic effect of catecholamines, steadily disturbing the permeability of cellular membranes [7] and enhancing the development of oxidative stress. Prominent hypercatecholaminemia aggravates the present hyperglycemia, which in the first days of stroke is often insulin-resistant, aggravates the disease course due to worsening the gas-transporting properties of hemoglobin along with both enhancement of hypoxia and accumulation of anaerobic glycolysis products [30]. The increased content of renin additionally activates the expression of immediate-early genes inducing the apoptotic program [22].

Thus, stress reaction, being initially adaptive, begins to take part in the pathogenic mechanisms; a basically compensatory act becomes a "disease of adaptation".

Chiolero [12], Hack *et al.* [21], and Wang [49] demonstrated in their experimental and clinical studies that the influence of any acute stress is accompanied by a special combination of parameters of thyroid status, which is characterized by decrease in triiodothyronine (T_3) level with concomitant normal or even elevated thyrotropic hormone (TTH) level. This phenomenon has been called the "low T_3-syndrome". The disruption of thyroxine (T_4) transformation into its active metabolic form triiodothyronine, caused by free radicals and free fatty acids [4, 10, 23], plays an important role in the development of this syndrome. The other proposed reason for "low T_3-syndrome" development is the decrease of TTH production in response to stimulation of the sympathoadrenal and hypothalamic–pituitary–adrenal axes [3, 18, 50] (Plate 2). Yu and Koenig [52] showed that low T_3-syndrome can be caused by decreased activity of type I iodothyronine 5'-deiodinase (5' D-I), the hepatic enzyme that converts thyroxine to T_3. It was demonstrated that cytokines decrease T_3 induction of 5' D-I, resulting in decreased T_3 production and hence a further decrease in 5' D-I. The proposed mechanism of this process is competition for limiting amounts of nuclear receptor coactivators between the 5' D-I promoter and the promoters of cytokine-induced genes.

We studied concentrations of stress-mediating system hormones in plasma of 25 patients with carotid ischemic stroke on days 1, 2, 3, and 7 of their illness [43, 44]. On the background of unified complex therapy without neuroprotection, we studied TTH level, free triiodothyronine (fT_3), free thyroxine (fT_4), ACTH, and renin by radioimmune techniques based on the detection of antigens or antibodies labeled by iodine isotope that bind to complementary portions of the hormone molecules. We used standard test kits (IRMA TTH CT, Belarus; Immunotech, Czech Republic; ACTH-PH CIS BIO International, France).

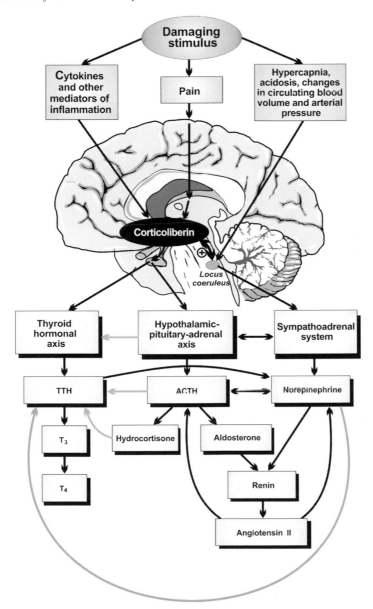

Figure 11.1. Interaction between hormonal axes of the stress-mediating endocrine system. Black arrows show activating, grey – inhibitory effects.

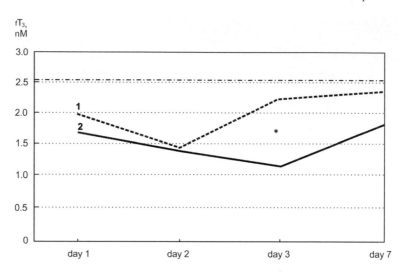

Figure 11.2. Dynamics of fT3 concentration in plasma within the first week after carotid ischemic stroke onset. Here and in Figs. 11.3–11.4, 11.6: 1 – moderately severe stroke (OSS > 40); 2 – severe stroke (OSS < 39). The dash-dot line shows the lower value of normal range (the upper level is 5.8 nM). Significance of the differences between patients with moderately severe and severe stroke: * $p < 0.05$.

In all patients, starting with day 1 from stroke onset and during all the following period of study, we registered the decrease in fT_3 serum level, more prominent in severe strokes that were assessed on admission as less than 30 points on the Scandinavian scale [38], less than 40 points on the Orgogozo scale [40], and less than 36 points on the original scale [42] (Fig. 11.2). The lowest level of fT_3 was registered in a patient who died. By day 2 after stroke onset in patients with moderately severe disease, fT_3 titer, being higher on admission, was significantly decreased ($p < 0.05$), being comparable to those values in patients with severe neurological deficit. However, already by day 3 fT_3 concentration in patients with moderately severe stroke became significantly higher than in those with severe deficit ($p < 0.05$) (see Fig. 11.2). The elevation of fT_3 level positively correlated with the degree of neurological recovery estimated by the change in OSS score ($r = 0.65$, $p = 0.047$).

TTH content remained within normal levels in all patients during the whole period of immune–biochemical monitoring. Along with this, on day 2 all patients exhibited a tendency towards increase in TTH content in comparison with values obtained on day 1. This tendency reached statistically significant difference only in patients with severe stroke

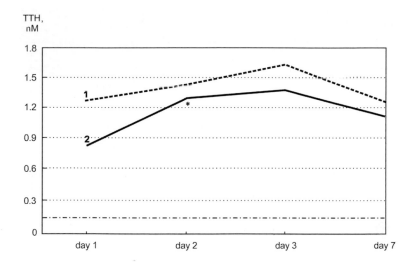

Figure 11.3. Dynamics of TTH concentration in plasma within the first week after carotid ischemic stroke onset. The dash-dot line shows the lower value of normal range (the upper level is 4.05 nM). Significance of the differences versus the value on day 1: * $p < 0.03$.

($p < 0.03$) (Fig. 11.3). TTH concentration fluctuations were not accompanied by elevation of fT_3 and exhibited a reverse correlation with the degree of neurological improvement estimated by the change in the OSS score ($r = -0.5, p = 0.05$).

Plasma fT_4 concentration was also within normal levels during the first days of stroke. However, on day 7 we registered its elevation, statistically significant in patients with severe stroke ($p < 0.01$) (Fig. 11.4). The increase in fT_4 level exhibited positive correlation with elevation of TTH concentration ($r = 0.68, p = 0.035$) and reverse correlation with fT_3 level ($r = -0.59, p = 0.03$) as well as with the dynamics of the OSS score ($r = -0.75, p = 0.01$) irrespective of stroke severity (Fig. 11.5). Thus, the lower was fT_3 level in acute focal brain ischemia, the higher was TTH and fT_4 concentrations in serum and more severe was the stroke course.

Experimental studies performed by Boelen *et al.* [4], Chan [10], and Huang *et al.* [23] suggested that the disturbances of thyroid metabolism seen in acute ischemic stroke were a result of both retardation of deiodination of T_4 and inhibition of its transformation to the active metabolic form (T_3) under conditions of high contents of free radicals and free fatty acids. The more prominent the blocking of the T_4 to T_3 transformation, the larger the

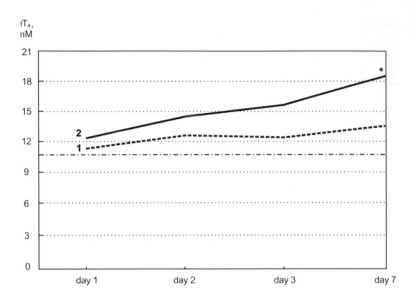

Figure 11.4. Dynamics of fT$_4$ concentration in plasma within the first week after carotid ischemic stroke onset. The dash-dot line shows the lower value of normal range (the upper level is 23.5 nM). Significance of the differences versus the value on day 1: * $p < 0.01$.

ischemized area and, hence, the more pronounced the processes of free radical oxidation [53]. In response to T$_3$ decrease, the pituitary increases production of TTH, which leads to increased secretion of T$_4$ (see Plate 2).

Normal ACTH concentration, seen in all studied patients on day 1 after stroke onset, was significantly elevated (by 35-fold) by day 2 ($p < 0.04$), and

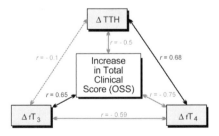

Figure 11.5. Correlation between the dynamics of thyroid hormone concentrations and the shift of total clinical score estimated by the Orgogozo Stroke Scale (OSS) within the first week after carotid ischemic stroke onset. Black arrows show positive correlation, grey – negative.

this was more pronounced in patients with severe neurological deficit. From day 3 all the patients began to exhibit the decrease in initially elevated ACTH concentration. This decrease became significant ($p < 0.05$) on day 7 (Fig. 11.6). The degree of ACTH decrease oppositely correlated with the degree of neurological recovery estimated by the OSS score ($r = -0.77$, $p = 0.03$). In severe strokes, the ACTH decrease was less pronounced and on day 7 the level of this hormone was higher than that in strokes with moderate neurological deficit.

Of interest is the fact that changes in ACTH concentration closely correlated ($r = 0.87$, $p = 0.006$) with fluctuations of TTH level (Fig. 11.7) and the concentrations of both hormones positively correlated with the severity of neurological deficit at the time of plasma examination ($r = 0.8$, $p = 0.017$ and $r = 0.65$, $p = 0.045$, respectively). Findings of Krachkevich and Protas [29], Shutova and Chudinova [42], and Zieger *et al.* [53] suggest that the activity of oxidant stress reactions determines the extent of hypothalamic–pituitary system activation, thus this may explain the relation of pituitary hormone concentration with the severity of ischemic stroke.

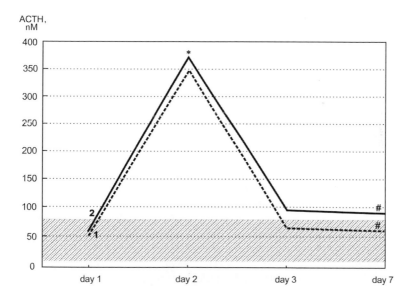

Figure 11.6. Dynamics of ACTH concentration in plasma within the first week after carotid ischemic stroke onset. Shaded area corresponds to normal range. Significance of the differences versus the value on day 1: * $p < 0.03$. Significance of the differences versus the value on day 2: # $p < 0.05$.

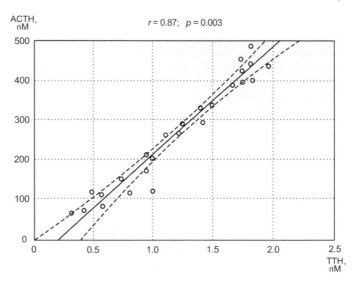

Figure 11.7. Correlation between concentrations of ACTH and TTH in plasma of patients with carotid ischemic stroke.

The concentration of renin in plasma of stroke patients remained steadily normal over the whole observation period. Only on day 7 a tendency towards its elevation was registered, but the level of renin did not exceed the upper border of the normal range.

We failed to find any significant differences in the levels of hormones detected in relation to right or left hemispheric location of the ischemized area as well as in relation to whether the lesion was small cortical (less than 5 cm^3), vast involving cortical and subcortical structures (more than 25 cm^3), or separately damaging only white brain matter (5–15 cm^3) ($p > 0.10$).

From retrospective immune–biochemical analysis, we elucidated that in patients with good neurological recovery (Barthel Index about 75–100) by the end of the acute period (on day 21 after stroke onset), significantly higher level of fT$_3$ on day 3 ($p < 0.04$) and lower levels of fT$_4$ and ACTH on day 7 ($p < 0.03$ and $p < 0.05$, respectively) are characteristic in comparison with disabled patients (Barthel Index less than 70) (Fig. 11.8). Renin concentration within the first days of stroke was lower in patients with good outcome of the disease; significant changes in comparison with disabled patients were noted on day 2 ($p < 0.01$).

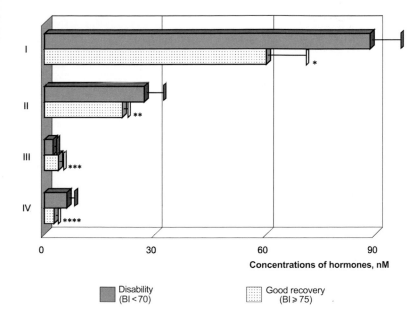

Figure 11.8. Retrospective analysis of concentrations of hormones of stress-mediating endocrine system in relation to the stroke outcome. I – ACTH on day 7; II – fT_4 on day 7; III – fT_3 on day 3; IV – active renin on day 2. Significance of the differences versus disabled patients: * $p < 0.05$, ** $p < 0.04$, *** $p < 0.03$, **** $p < 0.01$.

Close correlation between plasma levels of fT_3, fT_4, ACTH, and renin and the degree of neurological recovery estimated by the OSS score were registered over the entire observation period, thus these values can be used for prediction of the disease course and outcome. The most valid predictive criterion of good recovery is a high level of fT_3. Its participation in reparative events after ischemic stroke may be due to the neurotransmitter mechanisms of its action [24, 35].

Thus, in urgent somatic pathology [13, 28, 36, 45–47] as well as in acute focal brain ischemia stereotype response of stress-mediating endocrine system develops showing activation of both hypothalamic–pituitary–adrenal and renin–angiotensin hormonal axes as well as changes in thyroid metabolism when "low T_3-syndrome" forms. The extent of stress neuro–hormonal reorganization reflects the severity of brain ischemia, influences the course of acute period of stroke, and has prognostic value.

REFERENCES

1. Arem, R., Ghust, H., Ellerhorst, J., Comstock, J. P., 1997, *Clin Biochem.* **30** (5): 419–424.
2. Barton, R., 1987, *J Clin Endocrinol Metab.* **1**: 355–374.
3. Boado, R. J., Romeo, H. E., Chuluyan, H. E., *et al.*, 1994, *Neuroendocrinology.* **53** (4): 360–364.
4. Boelen, A., *et al.*, 1993, *J Clin Endocrinol Metab.* **77** (6): 1695–1699.
5. Brebner, K., Hayley, S., Zacharko, R., Merali, Z., Anisman, H., 2000, *Neuropsychopharmacology.* **22** (6): 566–580.
6. Brunetti, L., 1994, *Clin Ther.* **144** (2): 147–153.
7. Burd, G. S., 1983, *Respiratory Insufficiency in Patients with Acute Disorders of Blood Supply.* Doctoral dissertation. Moscow, 355 (in Russian).
8. Calogero, A., Gallued, W., Chrousos, G., *et al.*, 1968, *J Clin Invest.* **82**: 839–846.
9. Catanina, A., Lipton, J. M., 1998, *Ann NY Acad Sci.* **856**: 62–68.
10. Chan, P., 1992, *Neurotrauma.* **9**: 417–423.
11. Chikhikshvili, Ts. Sh., Antadze, Z. I., Sturua, I. T., 1990, *Travelling Section of the Presidium of All-Russian Scientific Societies of Therapeutists, Cardiologists and Neurologists "Essential Hypertension and Cerebrovascular Diseases".* Perm, 207 (in Russian).
12. Chiolero, R., 1994, *New Horizons*, **2** (4): 432–442.
13. Chiolero, R., Lemarchand-Beraud, T., Schut, Y., *et al.*, 1988, *J Trauma.* **28**: 1368–1374.
14. Clifton, G., Ziegler, M., Groesman, R., 1981, *J Neurosurg.* **8**: 10–14.
15. DeKeyser, F. G., Leker, R. R., 2000, *Weidenfeld J Neuroimmunomodulation.* **7** (4): 182–188.
16. Delgado, R., Carlin, A., Airaghi, L., *et al.*, 1998, *J Leukoc Biol.* **63** (6): 740–745.
17. Faber, J., Kirkegaard, C., Ramsmussem, B., *et al.*, 1987, *J Clin Endocrinol Metab.* **65** (2): 315–320.
18. Filcher, L., Brown, M., 1991, *J Clin Endocrinol Metab.* **5**: 35–50.
19. Fliers, E., *et al.*, 1997, *J Clin Endocrinol Metab.* **82** (12): 4032–4036.
20. Gusev, E. I., Skvortsova, V. I., 1990, *Case Record Form for the Examination and Treatment of Patients with Ischemic Stroke*, Moscow, pp. 1–44 (in Russian).
21. Hack, J., Gottardis, M., Wiesser, C., *et al.*, 1991, *Intensive Care Med.* **17**: 25–29.
22. Huang, C., Jeffrey, J. J., 1998, *Mol Cell Endocrinol.* **139** (1-2): 79–87.
23. Huang, T., Boado, R., Chopra, I., *et al.*, 1987, *J Endocrinol.* **121**: 498–503.
24. Iniguez, M. A., De Lecea, L., *et al.*, 1996, *Endocrinol.* **137** (3): 1032–1041.
25. Johansson, A., Ahren, B., Nasman, B., Carlstrom, K., Olsson, T., 2000, *J Int Med.* **247** (2): 179–187.
26. Judd, A. M., Call, G. B., Barney, M., McIlmoil, C. J., Balls, A. G., Adams, A., Oliveira, G. K., 2000, *Ann NY Acad Sci.* **917**: 628–637.
27. Kalinin, A. A., Neretin, V. Ya., Kotov, S. V., 1991, *Korsakoff J Neurol Psychiatr.* **91**: 134 (in Russian).
28. Krachkevich, N. G., Protas, P. N., 1980, *J Probl Neurosurg.* **3**: 571–522 (in Russian).
29. Krachkevich, N. G., Protas, P. N., 1986, *J Probl Neurosurg.* **3**: 35–38 (in Russian).
30. Kuzin, V. M., 1989, *Ischemic Stoke. Clinical and Pathogenic Aspects and Differential Treatment.* Doctoral dissertation. Moscow (in Russian).
31. Lipton, J. M., Catanina, A., Delgado, R., 1998, *Ann NY Acad Sci.* **840**: 373–380.
32. Lipton, J. M., Catanina, A., Delgado, R., 1998, *Neuroimmunomodulation.* **5** (3–4): 178–183.

33. Litvitsky, P. F., 1995, *Pathophysiology*. Meditsina, Moscow, pp. 490–495 (in Russian).
34. Lopau, K., Mark, J., Schramm, L., Heidbreder, E., Wanner, C., 2000, *Transplant Int.* **13**: 282–285.
35. Martin, J. V., Williams, D. B., Fitzgerald, R. M., *et al.*, 1996, *Neurosci.* **73**: 705–713.
36. McLarty, D., Ratcliffe, W., McColl, K., *et al.*, 1975, *Lancet,* 275–276.
37. Meglic, B., Kobal, J., Osredkar, J., Pogacnik, T., 2001, *Cerebrovasc Dis.* **11** (1): 2–8.
38. Orgogozo, J. M., Dartigues, J. F., 1986, Clinical trials in acute brain infarction: the question of assessment criteria. In *Acute Brain Ischemia: Medical and Surgical Therapy* (Battistini, N., Courbier, R., Fieschi, C., Fiorani, P., Plum, F., eds.) Raven Press, New York, pp. 282–289.
39. Rosner, M., Newsome, H., Becker, D., 1984, *J Neurosurg.* **61**: 76–86.
40. *Scandinavian Stroke Study Group. Stroke*, 1985. **16**: 885–890.
41. Schmidt, E. D., Schoffelmeer, A. N., De Vries, T. J., Wardeh, G., Dogterom, G., Bol, J. G., Binnekade, R., Tilders, F. J., 2001, *Eur J Neurosci.* **13** (10): 1923–1930.
42. Shutova, A. A., Chudinova, A. A., 1987, *J Probl Neurosurg.* **3**: 7–10 (in Russian).
43. Skvortsova, V. I., Ostrovtsev, I. V., Platonova, I. A., Efremova, N. M., 2000, Laboratory diagnosis of changes in stress-mediating endocrine system contents in acute period of ischemic stroke. In *Biomedpribor 2000, Abst Symp.*, Moscow. **1**: 239–241 (in Russian).
44. Skvortsova, V. I., Platonova, I. A., Ostrovtsev, I. V., *et al.*, 2000, *Korsakoff J Neurol Psychiatr.* **4**: 22–27 (in Russian).
45. Slag, M. F., Morly, J. E., Elson, M. K., *et al.*, 1961, *JAMA.* **245**: 43–45.
46. Song, Y. M., Ho, W. M., Tsou, C. T., *et al.*, 1991, *Chung Hua I Hsuen Tsa Chih.* **47** (4): 242–248.
47. Spratt, D., Cox, P., Orav, J., *et al.*, 1993, *J Clin Care Metab.* **76**: 1548–1554.
48. Vale, W., Spiees, J., Hivier, C., *et al.*, 1981, *Science.* **213**: 1394–1397.
49. Wang, G. L., 1991, *J Clin Med.* **104**: 764.
50. Woolf, P., 1992, *Crit Care Med.* **20**: 216–226.
51. Woolf, P., Harnffel, R., Lee, L., *et al.*, 1987, *J Neurosurg.* **66**: 875–882.
52. Yu, J., Koenig, R. J., 2000, *J Biol Chem.* **275** (49): 38296–38301.
53. Ziegler, M., Morrissey, B., Marshall, L., 1990, *Crit Care Med.* **18**: 253–258.

Chapter 12

Molecular Mechanisms of Post-Ischemic Reparation Events

When effective blood flow is recovered before irreversible necrotic damage forms, a stereotype reparative process will have been started in the brain tissue.

Intermediate metabolites of oxidant stress, being second messengers, initiate the cascade of life-rescuing reactions in the ischemized tissue [32]. Their phosphorylation is a signal for expression of immediate-early, or early response, genes in affected neuronal and glial cell pools (see Scheme 7.1 and Fig. 7.1), which occurs on the background of the inhibited (due to trans-membrane amino acid transport dysfunction and destruction of ribosomes) protein synthesis. Expression of early response genes leads to generation of such transcription factors as nerve growth factor inducer A (NGFI-A) containing zinc-finger domain; fos, jun, c-myc family proteins containing leucine zipper and helix–loop–helix domains; NF-kappaB, transcription factor for the receptor of steroid hormone NGFI-B (see Chapter 7). Transcription factors may play an important role in the production of acute phase stress proteins. Thus, the expression of zinc-finger genes promotes the synthesis of HSP72 stress protein, presumably possessing neuroprotective properties. At present such enzymes as superoxide dismutase, catalase, and ornithine decarboxylase are also related to stress proteins, but their role needs to be further elucidated [32].

Besides HSPs, other signaling molecules also participate in post-ischemic life-rescuing reactions. Extracellular signal-regulated kinases (Erks) switch on pathways that promote cell survival and proliferation [48], causing protection of neurons and oligodendrocytes [27]. Activation of the Ras/Erks cascade resulted in the development of oxygen–glucose deprivation tolerance in cortical neurons [18, 20], preventing JNK/p38 activation.

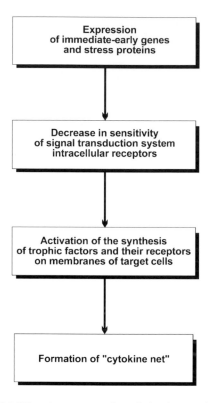

Scheme 12.1. Triggering sequence of post-ischemic reparative events.

Early response gene expression and induction and generation of transcription factors and stress proteins may enhance resistive capacity of the affected brain cells to induced ischemic changes. One of the basic mechanisms underlying the acquisition of tolerability to secondary ischemia is a **decrease in signal transduction system receptor sensitivity.** However, if this arranged response fails to avert threatening cell death, the second level of defense is switched on—the **synthesis of trophic factors and their receptors is activated** on surfaces of target cells (Scheme 12.1).

Experimental data [33, 40, 63] suggest that short-term brain ischemia in rats and mice, which leads to no cell damage, is accompanied by the enhanced activity of mRNA encoding neurotrophins—NGF and brain derived neurotrophic factor (BDNF). Neurotrophin-3 (NT-3) expression was shortly decreased. More prolonged focal ischemia caused the sequential elevation of neurotrophin synthesis reaching a peak 12 h after stroke onset and then coming back to the normal range by the 2nd day [62]. Kokaia *et al.* [33] showed that the extent of BDNF synthesis elevation correlated with the intensity of early response *c-foc* gene expression. However, when BDNF content increased only within the area encircling ischemized tissue, *c-foc* expression was found in the ischemic core as well. Activation of BDNF synthesis was inhibited by hyperglycemia and hypercapnia [68]. In transgenic mouse lines that overexpress the dominant-negative truncated splice variant of BDNF receptor trkB (trkB.T1) in postnatal cortical and hippocampal neurons, it was shown that endogenously expressed BDNF is neuroprotective and that BDNF signaling may have an important role in preventing brain damage after transient ischemia and limiting its spread [13, 52]. Our clinical studies [54–56] confirmed the fact of NGF content increase in CSF within the first 12 h after carotid ischemic stroke, and the extent of this increase had predictive value (see Section 9.1).

It was proven that activated astrocytes and possibly microglial cells adjacent to affected neurons support reparative events that take place in those cells [32]. Thus, already 10 min after the induction of global cerebral ischemia the synthesis of basic fibroblastic growth factor (bFGF) has been found as well as of its receptors in astrocytes [61, 62] and neurons [12], and it should be mentioned that the main source of this factor is glial cells. In focal ischemic rat brain after middle cerebral artery occlusion, bFGF accumulated predominantly in acutely reacting astrocytes [15, 70]. In the region adjacent to infarction, some neurons also showed an upregulation of bFGF expression [36, 70].

Erythropoietin (EPO) has been shown to be also produced in ischemized brain tissue [11]. It is up-regulated by hypoxia-inducible factor-1 (HIF-1) and protects neurons from products of oxidative and nitrosative stress. Activation of EPO receptors (EPORs) prevented apoptosis induced by NMDA or NO by triggering a cross-talk between the signaling pathways of Janus kinase-2 and NF-kappaB [11]. EPOR-mediated activation of Janus kinase-2 led to phosphorylation of the inhibitor of NF-kappaB, subsequent nuclear translocation of the transcription factor NF-kappaB, and NF-kappaB-dependent transcription of neuroprotective genes.

Induction of vascular endothelial growth factor (VEGF) was found in acute focal brain ischemia in neurons, astrocytes, microglial, and endothelial cells 18 h after stroke onset. It remained for 2 weeks. In transient focal ischemia the early elevation of VEGF synthesis was found (with peak 13 h after the beginning of re-circulation) with following normalization by the end of the 1st day [22].

Increased production of insulin-dependent growth factor II (IGF-II) and its receptors in ischemized neurons including the core zone [57] was registered within the first days of experimental ischemic stroke as well as the elevation of IGF-I level, that correlated with the extent of reactive gliosis [17, 37].

The expression of genes involved in post-ischemic reparation led to the synthesis of growth-associated phosphoproteins (for instanse, GAP-43) [46, 60]. It was shown that increased production of GAP-43 contributed to the molecular basis for brain plasticity, neurite out-growth, and synaptogenesis in post-ischemic brain, correlated with both neurological recovery and reduction of behavioral disturbances in rats that underwent stroke [53, 58].

Morphological and functional plasticity of ischemized brain tissue are also determined by the production of microtubule-associated protein 2 (MAP2) [10, 26]. The extent of post-ischemic expression of MAP2 reflects brain tolerance to ischemia and its capacity for repair [49].

Thus, trophic factors along with the autocrine stimulation of their endogenous receptors induce further signaling cascades, which lead to the synthesis of regulatory proteins averting induction of apoptosis and promoting cell viability. When ischemic brain damage is massive and accompanied by severe energy deficit, the enhancement of trophic support appears to be insufficient to rescue affected cells. Additional protective capacities are searched via induction of cytokines and other signaling molecules (e.g., tyrosine kinases) and the formation of so-called **"cytokine net"** [5, 8, 24, 30, 41, 42].

It was demonstrated that such anti-inflammatory cytokines as TGF-β_1 and IL-10 as well as neuromodulator TNF-α also participate in reparative events [7, 30, 42, 59]. Experimental studies confirmed their expression in rescued cells of the penumbra, whereas in the area of infarction no expression was detected [23, 25]. TNF-α and TGF-β_1 mRNA contents increased within the first 2–4 days after stroke [6, 46, 65, 66]. Glial expression was retarded and was elevated during days 3–5 of stroke [14, 30, 31, 38, 42, 67]. Benveniste [5], Lindholm *et al.* [39], Muzino *et al.* [47], Hu *et al.* [25], and Lodge and Sriram [43] showed that TGF-β_1 inhibits the expression of pro-inflammatory cytokines by astrocytes, decreases proliferation of astrocytes and microglia, and mediates the release of other trophic factors. The role of TNF-α in post-ischemic process as mentioned in Chapter 8 is multifaceted. It stimulates astroglial and microglial proliferation, induces local inflammatory response, but increasing NGF [16, 19, 28] and TGF-β_1 [29, 45] expression, thus mediating trophic support to neurons being under conditions of inflammatory cascade development.

Prolonged induction of anti-inflammatory cytokines, revealed in experiment, may, according to Ulich [69], relect their participation in modulation of glial and inflammatory reactions after neuronal damage. Nevertheless, the clinical study that we performed in patients with carotid ischemic stroke showed the integrative result of multi-directional degenerative and reparative events: in the first hours of stroke the content of anti-inflammatory cytokines (IL-10, TGF-β_1) decreased in CSF on the background of prominent elevation of pro-inflammatory cytokine levels [54–56]. This suggests both the insufficiency of protective mechanisms and the development of local inflammatory reaction in the ischemic focus (see Section 8.1).

Thus, when the ischemic damage is relatively "soft", intermediate metabolites of membrane-associated phospholipids acting as second messengers enhance progammed events of reconstitution, which allow a cell to restore its structural and functional integrity. If the ischemic damage is more severe, not only neurons, but also glial cells take part in the reparative process. When energy metabolism is greatly affected the damaged tissue undergoes inflammatory reactions [32], which lead to irreversible destruction of cellular structures.

One of the most theoretically important and complex question is what is to be considered a "critical point" of irreversible cell damage?

Up to the present, irreversible cell damage is predominantly associated with both anoxic depolarization of neuronal membranes and cell destruction. On the other hand, studies by Abe *et al.* [2, 3] and Aoki *et al.* [4] showed that even in short-term and "soft" brain ischemia dysfunction of shuttle mitochondrial neuronal systems may occur. Mitochondria can no longer receive information from the cell nucleus, usually delivered via newly synthesized proteins encoded in the nucleus. As a result, severe energy failure of cellular metabolism occurs. It is clear that the formation of a cytokine net as well as trophic factor and phosphoprotein synthesis are possible only when a cell preserves a capability to read out information encoded by DNA and ribosomes preserve their ability to synthesize proteins. So, it is logical to suggest that the "critical point" of damage, which determines the choice between life and death made by a cell, must take place earlier than the irreversible dysfunction of energy metabolism and membrane depolarization occur, i.e. the cell's fate is predetermined before the formation of structural changes typical for necrosis.

According to Chopp and Li [9], taking in regard the morphological features it is hard to assess whether a cell has passed the brink of irreversible damage or death. On one hand, a cell may maintain its functions and be metabolically active even after irreversible structural damage has occurred. On the other hand, it could be mortally damaged and doomed even showing no morphological changes typical for necrosis or apoptosis, considering death as a functional concept.

Neurons highly sensitive to ischemia are, obviously, the first to constitute the cell pool resistant to therapy, where no functional recovery is possible, yet at the same time no morphological signs of death are present [44, 50].

These cells are characterized by the inhibition of signal transduction excited by growth factors as the activities of protein kinases are inhibited after ischemia and those of phosphatases increased. The inhibition of signals from the growth factor receptors leads to changes in gene expression and programmed cell death, accompanied by the activation of P53-mediated events. Besides, it was proven that in cell pools highly sensitive to ischemia the release of activated ligands to low affinity receptors of p75 growth factor (structurally similar to TNF-α receptors) occurs. This may accelerate cell death via pathways similar to TNF-α-mediated cytotoxicity: activation of sphingomyelin cycle and ceramide formation, as well as inflammation [21] (see Chapter 10).

REFERENCES

1. Abe, K., Kawagoe, J., Aoki, M., Kogure, K., 1993, *J Cerebr Blood Flow Metab.* **13**: 773–780.
2. Abe, K., Kawagoe, J., Aoki, M., Kogure, K., 1993, *Mol Brain Res.* **19**: 69–75.
3. Abe, K., Kawagoe, J., Kogure, K., 1993, *Neurosci Lett.* **153**: 173–176.
4. Aoki, M., Abe, K., Yoshida, T., *et al.*, 1995, *Brain Res.* **669**: 189–196.
5. Benveniste, E. N., 1992, *Chem Immunol Basel Karger.* **52**: 106–153.
6. Bruce, A. J., Baling, W., Kindy, M. S., *et al.*, 1996, *Nat Med.* **2**: 788–794.
7. Buttini, M., Sauter, A., Boddeke, H. W., 1994, *Brain Res.* **25**: 126–134.
8. Campos-Gonzalez, R., Kindy, M., 1992, *J Neurochem.* **59**: 1955–1958.
9. Chopp, M., Li, Y., 1996, *Acta Neurochir.* **66**: 21–26.
10. Dawson, D. A., Hallenbeck, J. M., 1996, *J Cerebr Blood Flow Metab.* **16**: 170–174.
11. Digicaylioglu, M., Lipton, S. A., 2001, *Nature.* **412** (6847): 601–602.
12. Endoh, M., Pulsinelli, W. A., Wagner, J. A., 1994, *Mol. Brain Res.* **22**: 76–88.
13. Endres, M., Fan, G., Hirt, L., *et al.*, 2000, *J Cerebr Blood Flow Metab.* **20** (1): 139–144.
14. Finch, C. F., Laping, N. J., Morgan, T. E., *et al.*, 1993, *J Cell Biochem.* **53**: 314–322.
15. Finklestein, S. P., Apostolides, P. J., Caday, C. G., *et al.*, 1988, *Brain Res.* **460**: 253–259.
16. Gadient, R. A., Cron, K. C., Otten, U., 1990, *Neurosci Lett.* **117**: 335–340.
17. Gluckman, P., Klempt, N., Guan, J., *et al.*, 1992, *Biochem Biophys Res Commun.* **182**: 593–599.
18. Gonzalez-Zulueta, M., Feldman, A. B., Klesse, L. J., *et al.*, 2000, *Proc Natl Acad Sci USA.* **97** (1): 436–441.
19. Goossens, V., Grooten, J., De Vos, K., Fiers, W., 1995, *Proc Natl Acad Sci USA.* **92**: 8115–8119.
20. Gu, Z., Jiang, Q., Zhang, G., 2001, *Neuroreport.* **12** (16): 3487–3491.
21. Hannum, Y. A., Obeid, L. M., Wolff, R. A., 1992, *Adv Lipid Res.* **25**: 43–64.
22. Hayashi, T., Abe, K., Suzuki, H., Itoyama, Y., 1997, *Stroke.* **28**: 2039–2044.
23. Henrick-Noack, P., Prehn, J. H. M., Kreiglestein, J., 1996, *Stroke.* **27**: 1609–1615.
24. Hu, B. R., Wieloch, T., 1994, *J Neurochem.* **62**: 1357–1367.
25. Hu, S., Sheng, W. S., Peterson, P. K., Chao, C. C., 1995, *Glia.* **13**: 45–50.
26. Inuzuka, T., Tamura, A., Sato, S., *et al.*, 1990, *Stroke.* **21**: 917–922.
27. Irving, E. A., Barone, F. C., Reith, A. D., *et al*, 2000, *Brain Res Mol Brain Res.* **77** (1): 65–75.

28. Jaattela, M., 1991, *Lab Invest.* **64**: 724–742.
29. Kalthoff, H., Roeder, C., Brockhaust, M., *et al.*, 1993, *J Biol Chem.* **26**: 2762–2766.
30. Kim, J .S., 1996, *J Neural Sci.* **137**: 69–78.
31. Klempt, M., Singh, K., Williams, C., Nikolics, K., 1992, *Biochem Biophys Res Commun.* **182**: 593–599.
32. Kogure, K., Yamasaki, Y., Matsuo, Y., *et al.*, 1996, *Acta Neurochir.* **66**: 40–43.
33. Kokaia, Z., Zhao, Q., Kokaia, M., *et al.*, 1995, *Exp Neurol.* **136**: 73–88.
34. Kovacs, Z., Ikezaki, K., Samoto, K., *et al.*, 1996, *Stroke.* **27**: 1865–1873.
35. Krupinski, J., Kumar, P., Kumar, S., *et al.*, 1996, *Stroke.* **27**: 852–857.
36. Kumon, Y., Sakaki, S., Kadota, O., 1993, *Brain Res.* **605**: 169–174.
37. Lee, W. H., Clemens, J. A., Bondy, C. A., 1992, *Mol Cell Neurosci.* **3**: 36–43.
38. Lehrmann, E., Kiefer, R., Finsen, B., *et al.*, 1995, *Exp Neurol.* **131**: 114–123.
39. Lindholm, D., Castren, E., Keifer, R., *et al.*, 1992, *Brain Res.* **117**: 395–400.
40. Lindvall, O., Ernfors, P., Bengzon, J., *et al.*, 1992, *Proc Natl Acad Sci USA.* **89**: 648–652.
41. Liu, T., Dark, R. K., McDonnell, P. C., *et al.*, 1994, *Stroke.* **25**: 1481–1488.
42. Liu, T., McDonnell, P. C., Young, P. R., *et al.*, 1993, *Stroke.* **24**: 1746–1751.
43. Lodge, P. A., Sriram, S. J., 1996, *Leuk Biol.* **60**: 502–508.
44. Memezawa, H., Smith, M. L., Siesjo, B. K., 1992, *Stroke.* **23**: 552–559.
45. Merrill, J. E., Koyanagi, Y., Zack, J., l992, *J Virol.* **66**: 2217–2225.
46. Minami, M., Kuraishi, Y., Satoh, M., 1991, *Biochem Biophys Res Commun.* **176**: 593–598.
47. Mizuno, T., Sawada, M., Suzumura, A., Marunouchi, T., 1994, *Brain Res.* **656**: 141–146.
48. Nozaki, K., Nishimura, M., Hashimoto, N., 2001, *Mol Neurobiol.* **23** (1): 1–19.
49. Okawa, M., Halaby, I., Pulsinelli, W. A., Nowak, T. S., Jr., 1995, Attenuated hsp72 and MAP2c mRNA expression in rat hippocampus in a model of induced ischemic tolerance. In *Pharmacology of Cerebral Ischemia 1994* (Kriegistein, J., Oberpichkr-Schwenk, H., eds.) Stuttgart, Germany, pp. 467–472.
50. Park, C. K., Nehls, D. G., Graham, D., *et al.*, 1988, *J Cerebr Blood Flow Metab.* **8**: 757–762.
51. Pettigrew, L. C., Holtz, M. L., Craddock, S. D., 1996, *J Cerebr Blood Flow Metab.* **16**: 1189–1202.
52. Saarelainen, T., Lukkarinen, J. A., Koronen, S., *et al.*, 2000, *Mol Cell Neurosci.* **16** (2): 87–96.
53. Schmidt-Kastner, R., Bedtird, A., Hakim, A., 1997, *Cell Tissue Res.* **288**: 225–238.
54. Skvortsova, V. I., Gusev, E. I., Sherstnev, V. V., *et al.*, 2000, *Cerebrovasc Dis.* **12** (3).
55. Skvortsova, V. I., Klushnik, T. P., Stakhovskaya, L. V., *et al.*, 1999, The study of autoantibodies to NGF in patients with acute and chronic brain ischemia. In *New Technologies in Neurology and Neurosurgery at the Turn of Millenium.* Stupino, pp. 181–182 (in Russian).
56. Skvortsova, V. I., Myasoyedov, N. F., Klushnik, T. P., *et al.*, 2000, The study of NGF and its autoantibodies content in patients with acute brain ischemia. In *Contemporary Approaches to the Diagnosis and Treatment of Nervous and Psychiatric Diseases.* St. Petersburg, pp. 332–333 (in Russian).
57. Stephenson, D. T., Rash, K., Clemens, J. A., 1995, *J Cerebr Blood Flow Metab.* **15**: 1022–1031.
58. Stroemer, R. P., Kent, T. A., Hulsebosch, C. E., 1995, *Sinid.* **26**: 2135–2144.
59. Szaflarski, J., Burtrum, D., Silverstein, F. S., 1995, *Stroke.* **26**: 1093–1100.

60. Tagaya, M., Matsuyama, T., Nakamura, H., *et al.*, 1995, *J Cerebr Blood Flow Metab.* **15**: 1132–1136.
61. Takami, K., Iwane, M., Kiyota, Y., *et al.*, 1992, *Exp Brain Res.* **90**: 1–10.
62. Takami, K., Kiyota, Y., Iwane, M., *et al.*, 1993, *Exp Brain Res.* **97**: 185–194.
63. Takeda, A., Onodera, H., Sugimoto, A., *et al.*, 1993, *Neuroscience.* **55**: 23–31.
64. Takeda, A., Onodera, H., Yamasaki, Y., *et al.*, 1992, *Brain Res.* **569**: 177–180.
65. Taupin, V., Toulmond, S., Serrano, A., *et al.*, 1993, *J Neuroimmunol.* **42**: 177–186.
66. Tchelingerian, J. L., Quinonero, J., Booss, J., *et al.*, 1993, *Neuron.* **10**: 213–224.
67. Thompson, N. L., Flanders, K. C., Smith, J. M., *et al.*, 1989, *J Cell Biol.* 661–669.
68. Uchino, H., Lindvall, O., Siesjo, B. K., Kokaia, Z., 1997, *J Cerebr Blood Flow Metab.* **17**: 1303–1308.
69. Ulich, T. R., Songemi, Y., Guo, K., *et al.*, 1991, *Am J Pathol.* **138**: 1099–1101.
70. Wei, O. Y., Huang, Y. I., Da, C. D., Cheng, J. S., 2000, *Acta Pharmacol Sin.* **21** (4): 296–300.

PART II

NEUROPROTECTION IN BRAIN ISCHEMIA

Chapter 13

Modern Therapeutic Approaches to Acute Focal Brain Ischemia. Basic Strategies for Neuroprotection

Modern scientific and technological advances give reason to revise the previous pessimistic opinion about the therapeutic possibilities for acute focal brain ischemia, i.e. ischemic stroke.

Modern technologies when put into practice significantly improved the diagnostic procedures in ischemic stroke. They allow not only the visualization of the area of structural damage, but also features of blood supply and functional state of brain tissue. They also can be use to objectify vascular factors leading to disturbances of blood supply.

Fundamental neurosciences and clinical neurology have united their efforts, and this has promoted the discovery of basic developmental mechanisms of cerebrovascular pathology and ischemic brain damage. It has been proven that irreversible brain damage occurs later after the onset of the manifestations of stroke. This evidence obliged physicians to treat stroke as an **urgent condition** that requires fast and pathogenically substantiated medical aids desirably within the first 3–6 h of its appearance. Thus, much attention is now given to specialized neurological emergency teams responsible for both the earliest beginning of therapeutic interventions already at the pre-admission stage and fast delivery of patients to hospital.

Early admission of stroke patients is one of the basic factors determining the success of treatment. This has no medical limitations. Traditionally considered conditional medical and social limitations to the delivery from home to hospital are terminal comas, as well as cases when stroke develops on the background of terminal stages of cancer and other chronic disorders, and prominent psychic disturbances in senile patients. The most effective intensive treatment of stroke is provided in specialized Intensive Stroke Units (ISU) within neurological or neuro-resuscitation departments. Many multi-center trials and systemic meta-analyses have proven the advantages of specialized ISU over conventional neurological and therapeutic departments. The most valid arguments are the significant mortality and disability decrease in stroke patients, as well as life quality improvement of surviving patients in the remote period after stroke [3, 25, 26]. These advantages are connected with peculiarities of their structural arrangement as well as with high qualification level of their multi-disciplinary personnel. Specialized ISUs are arranged in multi-field hospitals with constant access to neuroimaging service (CT/MRI), laboratory (with access to a wide range of assays, including detection of glucose, electrolytes, blood gases, rheological, and coagulation properties) and different specialists including neurosurgeon (or having the possibility of urgent consultation). The personnel of ISUs consists of 24 h specialized stroke teams including a neurologist trained in resuscitation and intensive treatment and a cardiologist. ISUs are always equipped with devices for cardiovascular monitoring (pulse, ECG, BP), blood oxygen saturation, skin temperature; for monitoring of cerebral functional state (computerized EEG with compressed spectrum analysis (CSA), mapping EEG, and three-dimensional location (Loretta, Brain Loc); evoked potentials, etc.); ultra sound extra- and transcranial Doppler [36].

The more severe the stroke, the more necessary is so-called **basic therapy** included in the whole treatment complex. This is directed to support of all vital functions. That is why the selection of patients for ISUs is required. Such patients are preferably those as having consciousness disturbances (coma I–II), systemic depression of blood supply (due to myocardial infarction, cardiac arrhythmia, and conduction disturbances), respiratory deficiency (obstruction of airways, lung edema, severe pneumonia), and paroxysmal states. The main directions of basic therapy are the correction of respiratory and cardiovascular disturbances, water electrolyte imbalance, prevention and treatment of intracranial hypertension and brain edema, autonomic and trophic disorders, and complications of ischemic stroke [2, 4, 10, 14, 15, 19, 20, 24, 28, 32, 40, 41].

Extending our knowledge about ischemic brain damage formation changed our views on pathogenic treatment strategies. Analyzing the dynamics of molecular and biochemical mechanisms triggered by acute focal ischemia, one can see the chronological sequence of their occurrence. Within the first 3 h of stroke the energy failure is maximally pronounced; within 3–6 h glutamate "excitotoxicity", calcium homeostasis disorders and lactic acidosis arc marked, decreasing by the end of the 3rd day. Delayed consequences of ischemia begin to manifest on the 2nd–3rd h, the activity of oxidant stress and local inflammation reaches a peak 12–36 h after onset, "peak" activity of apoptosis is registered on the 2nd–3rd day. Delayed consequences of ischemia last for several months after stroke onset mediating the progression of atherogenesis and diffuse brain tissue damage (encephalopathy) in the post-stroke period (see Figs. 6.1 and 7.2). Taking into consideration the elucidated conformities, to systematize complex reactions of ischemic cascade the following conditional and simplified **scheme of sequential ischemic cascade events** can be proposed:

1) CBF decrease;
2) ion pump failure and glutamate "excitotoxicity";
3) intracellular accumulation of calcium;
4) activation of intracellular enzymes;
5) increased NO synthesis and the development of oxidative stress;
6) expression of early response genes;
7) reactions of activated glia (local inflammation, microcirculation disturbances, BBB damage);
8) apoptosis.

Each stage of the ischemic cascade appears to be a potential target for therapeutic interventions. The earlier the cascade is disrupted, the more prominent effect should be expected. **Two basic directions** of pathogenesis-substantiated therapy of ischemic stroke can be proposed:

- improvement of brain tissue perfusion, i.e. therapeutic reperfusion (influence on the 1st stage of the cascade);
- neuroprotective (cytoprotective) therapy (influence on the 2nd–8th stages of the cascade) (Fig. 13.1).

The problem of **therapeutic reperfusion** is rather complex. Reperfusion is most effective within the first minutes of acute focal ischemia. Even occurring after 5 min, the massive blood re-flow into the ischemized area via switched on collateral vessels or restituted lumen of intrinsic vessel does not restitute blood supply in the ischemized area completely. Sequential disturbances of perfusion occur in brain tissue: hyperemia (or "luxuriant perfusion") in the first minutes; then post-ischemic hypoperfusion as a result of severe microcirculation disturbances caused by release of pro-inflammatory and vasoactive metabolites in the ischemized tissue (see Section 8.2). The longer is the period that precedes reperfusion, the less is the chance to normalize microcirculation in the ischemized area and the higher the risk of additional reperfusion damage of brain tissue. Such damage is oxidative, mediated by oxygen involved in free radical damage, and osmotic, caused by escalation of cytotoxic edema due to excess of water and osmotically active substances [27, 29]. According to experimental data, the expediency of therapeutic reperfusion remains within the first 3–6 h.

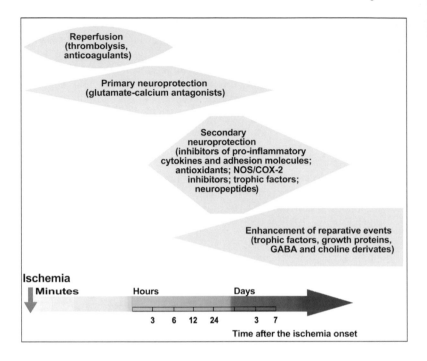

Figure 13.1. Therapeutic strategy in the acute period of ischemic stroke.

Later enhances the risk of both reperfusion damage and hemorrhagic complications. Thus, reperfusion should be early, active, and desirably short-term.

The character of reperfusion therapy is determined by the pathogenic variant of stroke development. In thrombotic or embolic damage of vessels of large and middle diameter, thrombolysis can be a method of treatment.

The NINDS study [30] where 624 patients were enrolled for the first time proved the efficacy of tissue plasminogen activator (rtPA) administered within the first 3 h after stroke onset, especially within the first 90 min, in dose 0.9 mg/kg (maximum 90 mg). The drug provided significant lumen restitution in the affected vessel. By the 90th day 11–13% of patients who received rtPA exhibited highly significant neurological improvement, this prominently exceeding those in the placebo group. However, in 6.4% of cases within 36 h after the drug administration intracerebral hemorrhages occurred, whereas only 0.6% of placebo-treated patients developed such a condition. Despite the risk of early hemorrhagic complications, 90-day mortality in rtPA group was significantly lower than in placebo group (17 and 21%, respectively). Later it was shown that in extremely large ischemized territory (early occurrence of large edema or hypodensity zone on CT) as well in patients with diabetes mellitus, thrombolysis neither normalized cerebral blood supply nor improved the disease prognosis [17]. That is because the risk of hemorrhagic complications significantly elevates and severe microcirculation disturbances are present in these patients [17]. When rtPA is administered in higher dose of 1.1 mg/kg (the ECASS study) or injected within later period (the ATLANTIS, the ECASS, the ECASS II studies), the risk of secondary hemorrhagic complications increases [1, 6, 7, 11, 18, 21–23, 35, 43]. On the other hand, Furlan *et al.* [16] showed that in middle cerebral artery occlusion confirmed by angiography the intra-arterial injection of recombinant pro-urokinase administered with low doses of IV heparin improved the outcome of the disease (within 3 months of follow-up) as well as in cases when used 6 h after stroke onset. This suggests the possible efficacy of thrombolytic therapy beyond the 3 h span in case patients were selected thoroughly.

Fibrinogen-depleting drug, ancrod, based on the poison of the Malaysian pit viper, also improved clinical outcome of stroke within the first 3 h after stroke onset, but to less extent than rtPA administered within the 3 h span and pro-urokinase, administered within 6 h span [33].

Of course, besides enumerated contraindications to thrombolysis, it is also contraindicated in lacunar strokes and strokes following systemic depression of blood supply. According to Hacke [24], at present only 1–2% of stroke patients receive thrombolytic therapy, although even in cases if the drug is administered within 3 h span, this percent may be higher (up to 5–10%).

Multi-center trials failed to confirm the efficacy of other reperfusion methods. However, subgroup post-hoc analysis showed that it is expedient to administer heparin or low molecular weight heparinoids within the first day of progressive atherothrombotic stroke or stroke following proven cardiac embolism or angiosurgical intervention. Anticoagulants are contraindicated in patients with steady elevated BP >180 mm Hg or, vice versa, significantly decreased BP, as well as with comas, epileptic seizures, severe liver and kidney insufficiency, gastric ulcer, and various hemorrhagic manifestations.

Heparin is preferable to use within the first 2–5 day of stroke in daily dose of 10,000–15,000 units. Laboratory control of heparin effect is compulsory: bleeding time should be elongated 1.5–2-fold, activated partial thromboplastin time should not be elevated more than 2-fold. For 1–2 days before the withdrawal of heparin, gradual lowering of its dose is expedient as well as parallel administration of indirect anticoagulants for 3–4 weeks. It is known that in some patients with progressive basilar or internal carotid thrombosis, there is a deficit of antithrombin-3. Taking into account antithrombin-3-dependent efficacy of heparin, plasma should be administered to these patients in dose of 100 ml 1–2 times a day along with heparin.

Two major trials of aspirin in acute ischemic stroke (the CAST and the IST) showed that aspirin started early (within the first 48 h after stroke onset) in hospital produces a small but definite net benefit [37, 38].

The efficacy of hemodilution in the first hours after an acute hemispheric stroke has not been confirmed. Many trials have yielded strong evidence that reduction of the hematocrit from 43 to 37% does not improve short- or long-term stroke outcome [39].

Active therapeutic reperfusion is possible only in hospital after neuroimaging is performed (brain CT or MRI) to rule out the hemorrhagic component of brain damage and to access the volume of ischemic area and the pathogenic variant of stroke. This emphasizes advantages of another therapeutic approach—**neuroprotection** (cytoprotection)—which may be resorted to already at the pre-admission stage when first stroke manifestations originate even if a hemorrhagic component is possible.

To compare with reperfusion therapy, neuroprotection is far more complex and varied, which reflects the variety of ischemic brain damage mechanisms. Practically for each stage of ischemic cascade at least one neuroprotector has been elaborated or undergone clinical trial. At present over 30 multi-center international trials in different phases are being conducted. Drugs of various pharmacological groups with potential neuroprotective properties are being tested. According to experimental studies [5, 9, 12, 13, 31, 42], early administration of neuroprotectors allows: 1) increasing the quota of transient ischemic attacks and "mild" strokes within acute ischemic cerebrovascular events; 2) significantly decreasing the volume of infarction; 3) lengthening the period of the therapeutic window, expanding the possibilities of metabolic therapy; 4) establishing protection from reperfusion damage.

The variety of mechanisms of brain infarction formation allows two main strategies of neuroprotection to be distinguished.

Primary neuroprotection is aimed at disrupting the fast mechanisms of necrotic cell death—reactions of the glutamate-calcium cascade (see Fig. 13.1). This type of neuroprotection should be started from the first minutes of ischemia and continue for the first 3 days of stroke, most intensely within the first 12 h.

Secondary neuroprotection aimed to decrease the extent of delayed consequences of ischemia: to block pro-inflammatory cytokines, cell adhesion molecules, to inhibit pro-oxidant enzymes, to enhance trophic support and temporarily to disrupt apoptosis. It can be started 3–6 h after stroke and should be maintained for at least 7 days.

Starting with the first days of stroke, after infarction begins to form, more and more importance is gained by **reparative therapy** directed to improvement of healthy brain tissue plasticity in areas adjacent to the infarction, into activation of polysynaptic contacts, and increase of receptor density. Secondary neuroprotectors possessing trophic and modulatory effects as well as GABA and choline derivates enhance reparation and regeneration and promotes neurological recovery.

It is noteworthy that acute ischemic stroke therapy includes components of its **secondary prevention** as well. Secondary prevention becomes very important from the 2nd week of the disease when the risk of repeated vascular events greatly increases. The most important preventive measures are BP, glucose, and lipid control and correction as well as administration of basic antiplatelets (aspirin, clopidogrel, ticlopidine, dipyridamol).

Thus, the result of recent clinical and experimental studies shows the necessity of early combined pathogenesis-substantiated treatment (within the therapeutic window) of ischemic stroke, which includes reperfusion, combined neuroprotection, stimulation of reparative processes and elements of secondary prevention (to avert re-embolism or secondary vascular or tissue damage). Modern principles of ischemic stroke therapy being put in practice allow significantly improved stroke outcome: decreasing mortality and increasing efficacy of neurological recovery.

REFERENCES

1. Albers, G. W., Bates, V. E., Clark, W. M., *et al.*, 2000, *JAMA.* **283**: 1145–1150.
2. Allain, H., Decombe, R., Saiag, B., *et al.*, 1991, *Cerebrovasc Dis.* **1**: 83–92.
3. Barer, D., 1999, *Int J Med Pract.* **5**: 26.
4. Bogolepov, N. K., 1971, *Cerebral Crisis and Stroke.* Meditsina, Moscow, 392 (in Russian).
5. Bowers, M. P., Rothlein, R., Fagan, S. C., Zivin, J. A., 1995, *Neurology.* **45**: 815–819.
6. Chiu, D., Krieger, D., Villar-Cordova, C., *et al.*, 1998, *Stroke.* **29**:18–22.
7. Clark, W. M., Wissman, S., Albers, G. W., *et al.*, 1999, *JAMA.* **282**: 2019–2026.
8. *Collaborative Systematic Review of the Randomized Trials of Organized In-patient (Stroke Unit) Care After Stroke.* 1997, Stroke Unit Trialists Collaboration, BMJ. **314**: 1151–1159.
9. Devuyst, G., Bogousslavsky, J., 1999, *J Neurol Neurosurg Psych.* **67** (4): 419–425.
10. EFNS Task Force on Neurological Stroke Care. 1997, *Eur J Neurol.* **4** (5): 435–441.
11. Egan, R., Lutsep, H. L., Clark, W. R., *et al.*, 1999, *J Stroke Cerebrovasc Dis.* **8**: 298–290.
12. Fisher, M., 1997, *Stroke.* **28**: 866–872.
13. Fisher, M., 1999, *The Neuroscientist.* **6**: 392–401.
14. Fisher, M., Bogousslavsky, J., 1993, *JAMA.* 270–360.
15. Fisher, M., Bogousslavsky, J., 1998, *JAMA.* **279**: 1298–1303.

16. Furlan, A. J., Higashida, R., Wechsler, L., *et al.*, 1999, PROACT II: Recombinant prourokinase (r-ProUK) in acute cerebral thromboembolism. Initial trial results. *Stroke.* **30**: 234.
17. Generalized efficacy of t-PA for acute stroke. Subgroup analysis of the NINDS t-PA stroke trial. 1997, *Stroke.* **28**: 2119–2125.
18. Grond, M., Stenzel, C., Schmulling, S., *et al.*, 1998, *Stroke.* **29**: 1544–1549.
19. Gusev, E. I., 1992, *Ischemic Brain Disease. Assembly Speech.* Moscow (in Russian).
20. Gusev, E. I., Burd, G. S., Bogolepov, N. N., 1979, *Cerebrovascular Diseases.* Meditsina, Moscow, 142 (in Russian).
21. Hacke, W., Brott, T., Caplan, L. R., *et al.*, 1999, *Neurology.* **53**: 3–14.
22. Hacke, W., Kaste, M., Fieschi, C., *et al.*, 1995, *JAMA.* **274**: 1017–1025.
23. Hacke, W., Kaste, M., Fieschi, C., *et al.*, 1998, *Lancet.* **352**: 1245–1251.
24. Hacke, W., Krieger, D., Hirschberg, H., 1991, *Cerebrovasc Dis.* **1**: 93–99.
25. Indredavik, B., Bakke, F., Slordahl, S. A., *et al.*, 1998, *Stroke.* **29**: 895–899.
26. Juby, L. C., Lincoln, N. B., Berman, P., 1996, *Cerebrovasc Dis.* **6**: 106–110.
27. Kidwell, C. S., Saver, J. C., Mattiello, J., *et al.*, 1999, *Neurology.* **52**: 536.
28. Krieger, D., Hacke, W., 1997, Intensive care treatment of ischemic stroke. In *Acute Stroke Treatment* (Bogousslavsky, J., ed.) Martin Dunitz Ltd, pp. 79–108.
29. Li, F., Hsu, S., Tatlisumak, T., *et al.*, 1999, *Ann Neurol.* **46**: 333–342.
30. The National Institute of Neurological Disorders and Stroke rt-PA Stroke Study Group. Tissue plasminogen activator for acute ischemic stroke. 1995, *N Engl J Med.* **333**: 1581–1587.
31. Onal, M. Z., Fisher, M., 1996, *Drugs of Today.* **32** (7): 573–592.
32. Popova, L. M., 1983, *Neuro-resuscitation.* Meditsina, Moscow, 175 (in Russian).
33. Sherman, D. G., for the STAT writers group. 1999, *Stroke.* **30**: 234.
34. Skvortsova, V. I., 1993, *Clinical and Neurophysiological Monitoring and Neuroprotective Therapy in Acute Ischemic Stroke.* Doctoral dissertation. Moscow, 379 (in Russian).
35. Tanne, D., Mansbach, H. H., Verro, P., *et al.*, 1998, *Stroke.* **29**: 288.
36. The European Ad Hoc Consensus Group. 1997, *Cerebrovasc Dis.* **7**: 113–128.
37. The CAST Collaborative Group. 1997, *Lancet.* **349**: 1641–1649.
38. The IST Collaborative Group. 1997, *Lancet.* **349**: 1569–1581.
39. The Italian Acute Stroke Study Group. 1988, *Lancet.* **1**: 318–321.
40. Toole, J., 1995, *Management of Acute Ischemic Stroke.* Winston-Salem.
41. Vereshchagin, N. V., 1980, *Vertebral Basilary System Pathology and Cerebral Strokes.* Meditsina, Moscow, 310 (in Russian).
42. Wahlgren, N. G., 1997, A review of earlier clinical studies on neuroprotective agents and current approaches. In *Neuroprotective Agents and Cerebral Ischemia* (Green, A. R., Cross, A. J., eds.) Academic Press, pp. 337–363.
43. Wang, D. Z., McLean, J. M., Rose, J. A., *et al.*, 1998, *Neurology.* **50**: 436.

Chapter 14

Primary Neuroprotection

Primary neuroprotection is aimed at disruption of the earliest events of the ischemic cascade occurring within the therapeutic window and underlying the fast necrotic damage of brain tissue (Table 14.1).

Table 14.1. Main directions of the primary neuroprotection

Direction	Basic drug groups	Representatives	Present state of study
Potential-dependent calcium channel antagonists	Dihydropiridines	Nimodipine	Efficacy is not proven
		Darodipine (PY 108-068)	Efficacy is not proven, **X**
		Isradipine	Severe adverse effects, **X**
		Flunarizine	Efficacy is not proven, **X**
		Cerebrokrast	Efficacy is not proven, **X**
Glutamate receptors antagonists	*NMDA receptor antagonists* non-competitive	Dizopcipine, MK-801	Severe adverse effects, **X**
		Dextrorphan	Severe adverse effects, **X**
		Dextromethorphan	Severe adverse effects, **X**
		Cerestat (CNS-1102, aptiganel hydrochloride)	Severe adverse effects, **X**

(*continued*)

Table 14.1 (continued)

Direction	Basic drug groups	Representatives	Present state of study
		Remacemide hydrochloride	Studies are in progress
		Magnesium sulfate	Studies are in progress
	competitive	Selfotel (CGS-19755)	Severe adverse effects, **X**
	selective: – polyamines site blockers	Eliprodil (SL-82.0715)	Efficacy is not proven, **X**
	– glycine site blockers	Gavestinel (GV-150526A)	Efficacy is not proven, **X**
		Licostinel (ACEA-1021)	Efficacy is not proven
	AMPA receptors antagonists	NBQX	Severe adverse effects, **X**
		ZK 200775	Severe adverse effects, **X**
Inhibitors of synthesis and pre-synaptic release of glutamate		BW-619C89	Severe adverse effects, **X**
		Propentofylline	Severe adverse effects, **X**
		Phenytoin	Efficacy is not proven
		Fos-Phenytion	Efficacy is not proven
		Lubeluzole	Efficacy is not proven, **X**
GABA agonists		Chlomethiazole	Studies are in progress
Glycine		Glycine	Studies are in progress

Notes: **X** – trials are stopped.

14.1. Potential-dependent calcium channel antagonists

Calcium-induced toxicity was discovered in the end of 70s. The concept of "calcium cell death" in brain ischemia was widely accepted [13, 19, 43, 171, 177, 178] and for many years (until the 90s) a mechanism of intracellular calcium content was regarded as crucial in the pathological biochemical ischemic cascade. It was supposed that calcium ions are delivered into a cell via only potential-dependent (potential-regulated) ion channels. In this connection, elaboration of potential-dependent calcium channel antagonists became the first substantiated neuroprotective strategy. Since the 80s over 20 randomized placebo-controlled studies of L-type potential-dependent calcium channels blockers, dihydropyridines, have been conducted.

The dihydropyridine derivate *nimodipine* is better studied. It is a lipophilic drug easily permeating through the blood brain barrier [5, 57, 62, 118, 209, 210]. It was proven that its basic property is prevention of vascular spasm after subarachnoid hemorrhage when administered orally or intravenously [5, 31, 142–152, 154–160]. The first clinical trials of oral [53, 54] and intravenous [78] nimodipine in acute ischemic stroke provided reassuring results. However, randomized, double-blind, placebo-controlled trials (the American Nimodipine Study, ANSG that enrolled 1064 stroke patients [199], and the Trust study that included 1215 patients [200]), failed to confirm that nimodipine favorably influenced stroke outcome when it was administered in daily doses of 60 and 120 mg orally, starting within 48 h of the stroke. Mohr *et al.* [128] published results of meta-analysis based on nine independent studies which enrolled 3719 patients. Significant decrease in the extent of neurological deficit after oral administration of nimodipine (by 38% versus placebo) was demonstrated only when therapy was started within the first 12 h after stroke onset. In contrast, significant deterioration of stroke prognosis was registered when therapy had a delayed start, 24 h after onset or later. These results became a prerequisite to study efficacy of oral nimodipine forms in the very early period of stroke, within 6 h after the onset [115]. But, this study again failed to prove any benefit of the drug in regard to stroke outcome [82].

Many trials have studied efficacy of intravenous nimodipine in acute ischemic stroke [14, 77, 135, 213]. Nimodipine was administered in dose of 2 mg/h for 5 days and then orally in dose of 120 mg/day for 21 day. Placebo and a lower dose of intravenous nimodipine (1 mg/h) were used to control the results [16, 215]. In patients who received 2 mg/h of nimodipine, significant worsening of stroke outcome versus patients receiving placebo was registered. There were no differences in the 30-day mortality and functional recovery in patients treated with the lower nimodipine dose versus placebo. Bridgers *et al.* [16] registered neurological improvement only in patients with moderately severe stroke when nimodipine was administered within the first 12 h. Norris *et al.* [137] performed the study of intravenous and oral nimodipine in 164 stroke patients. The drug was administered intravenously in dose of 2 mg/h for 10 days, and then orally in daily dose of 180 mg for 6 months. Statistically significant differences in mortality, neurological, or functional outcome were not found between patients who received nimodipine or placebo. Besides, Wahlgren *et al*. [215] showed that intravenous nimodipine could cause such serious adverse effect as dose-dependent decrease of systolic and diastolic blood pressure that directly correlated with frequency of unfavorable outcomes and increasing mortality.

Another dihydropyridine derivate tried in ischemic stroke is *darodipine* (PY 108-068). This drug has been traditionally administered for treatment of arterial hypertension and angina pectoris. In animal models of acute focal ischemia, darodipine increased the extent of oxygen saturation in brain tissue, improved neurological recovery, and decreased mortality [219]. The first clinical trials held in stroke showed that low doses of the drug improve blood flow in the ischemized brain area, whereas high doses may worsen blood flow in peri-infarction area [210]. A pilot study performed by Oczkowsky *et al*. [141], when darodipine had been administered orally in daily dose of 150 mg within the first 48 h of the stroke, proved relative safety of the drug and revealed a trend towards functional improvement. However, there were no significant differences in stroke outcomes and mortality rate in patients treated with darodipine versus placebo. Transient arterial hypotension is an adverse effect of the drug.

Isradipine is also a potential-dependent calcium channel antagonist of the dihydropyridine group; it decreases infarction size in experimental ischemic stroke [169]. Clinical trials of isradipine were ceased with the appearance of prominent adverse effects of the drug [7, 106].

Flunarizine is a dihydropyridine derivate that inhibits T-type of potential-dependent calcium channels. The drug is widely administered in clinic for treatment of vestibular disorders and prevention of migraine. Randomized trials of flunarizine in ischemic stroke [116, 163] have not confirmed its neuroprotective effects as well as its influence on blood pressure level.

In the Neurology Clinic of the Russian State Medical University, a pilot, simple blind, trial of novel dihydropyridine derivate *Cerebrokrast* ("Grindex", Latvia) was performed in 40 patients with carotid ischemic stroke admitted not later then 12 h after stroke onset [70]. Cerebrokrast (2,6-dimethyl-3,5-bis[2-(propoxyethoxycarbonyl-4-(2-difluoromethoxyphenyl)]-1,4-dihydroxypyridine) differs from nimodipine and other known dihydro-pyridine derivates in the presence in its structure of a difluoromethoxy-group, which enhances lipophilic properties of the drug and cerebral selectivity of vasodilating action. In an experimental model on isolated brain vessels, Cerebrokrast exhibited spasmolytic action in concentrations 1000-fold lower than those producing effect on peripheral limb arteries. High efficacy of the drug was observed in spasms of brain vessels that were caused by spasmogenic action of prostaglandin F_2 and KCl as well as by developed subarachnoid hemorrhage.

Cerebrokrast was administered in daily dose of 1 mg (10 ml of 0.01% solution) intravenously beginning from the first 12 h of the stroke for 5 days. The relative safety and good tolerability of the drug was confirmed. No marked influence on blood pressure or rheological and coagulation properties of blood was found. A trend towards accelerated and more complete improvement of neurological status (according to the Orgogozo and the Scandinavian Stroke Scales, $p < 0.07–0.09$) as well as to improvement of functional recovery (according to the Barthel Index; $p < 0.07$) was showed only in patients with mild stroke (on admission the total clinical score more than 60 by the Orgogozo Stroke Scale) treated with Cerebrokrast. In patients with severe stroke (total clinical score by the Orgogozo Stroke scale less than 40) we found no advantages of Cerebrokrast treatment; moreover, a trend towards retardation of neurological improvement and functional recovery was registered. Neurophysiological monitoring (EEG with mapping) held right after the first administration of Cerebrokrast and within 3 h after that revealed negative influence of Cerebrokrast on EEG pattern in patients with severe stroke. After the short-term (within first 15–30 min after the infusion) activation of basic fast rhythms (α, β), we registered a prominent "exhaustive" effect of the drug which comprised the significant decrease in power of all frequency ranges in the ischemized hemisphere, especially in the projection of the ischemic focus. This effect was accompanied by retardation of clinical improvement, and in 27.8% cases, by an increase in focal neurological deficit. On day 5 significantly more prominent EEG alterations persisted in patients with severe stroke treated with Cerebrokrast versus patients from the placebo group comparable in severity ($p < 0.05$). Possibly, negative clinical and neurophysiological effects of Cerebrokrast were connected with the development of the "steal syndrome" in the ischemic zone due to vasodilatation in the peri-infarction area.

Thus, an overview of all trials involving potential-dependent calcium channel antagonists found that they did not improve outcome after ischemic stroke. At present, this fact does not provoke any surprise, as it has been shown that intracellular accumulation of calcium is a rather complex process, under which many mechanisms lie, and only one of them is realized through potential-dependent calcium channels (see Chapter 4). Besides, adverse effects of these drugs, connected with vasodilation and transient arterial hypotension, impose limitations on their clinical application within the first hours of acute focal brain ischemia.

14.2.   Glutamate receptor antagonists

The role of agonist-dependent (receptor-regulated) calcium channels incorporated into glutamate receptors in the formation of ischemic brain damage became obvious after the phenomenon of glutamate excitotoxicity had been discovered and the concept of the glutamate–calcium cascade had developed. Changes in neuroprotective strategies followed this, and such drugs as NMDA and AMPA receptors antagonists were created.

NMDA receptor antagonists decrease calcium ion influx into cells via agonist-dependent calcium channels. They were the first neuroprotective drugs that significantly decreased (by 40–70%) infarction volume in experimental models owing to survival of the penumbra zone [23, 150]. Receptor blockade may be embodied via competitive or non-competitive antagonists.

Non-competitive antagonists of NMDA receptors are phencyclidine (PCP), ketamine, dizolcipine maleate (MK-801), dextromethorphan, dextrorphan, cerestat (CNS-1102, aptiganel hydrochloride), remacemide hydrochloride (FPL-12495), and magnesium. These compounds bind phencyclidine recognizing site on NMDA-associated ion channels [18, 179, 211, 222].

Dizolcipine exhibits significant neuroprotective effect under experimental conditions, significantly decreasing the infarction size in acute focal brain ischemia. Positive effect of this drug persists in various animal models as well as in different administration schedules (both before induction of experimental stroke and after) [18, 99, 149]. Clinical trials of this drug were stopped due to a wide range of its serious adverse effects (cataplexy, locomotor disturbances, transient arterial hypotension, dose-dependent consciousness disturbances) [17, 100, 152]. Besides, studies performed by Olney *et al.* [143] showed that even relatively low doses of dizolcipine caused vacuolization of brain tissue, and Leung and Desborough [114] revealed that the drug induced severe disturbances in brain functional activity according to EEG.

Dextrorphan and its derivate *dextromethorphan* are non-competitive antagonists of NMDA receptors with more complicated mechanisms of action in comparison with dizolcipine. Along with blocking of NMDA-associated ion channels, they inhibit L-type of potential-dependent calcium channels [4, 132]. Like dizolcipine, they, under experimental conditions, decrease the infarction size when administered before and after occlusion of middle cerebral artery [55, 190–192]. A pilot study conducted by Albers *et al.* [4] in 10 patients with ischemic stroke to whom dextromethorphan was administered orally in daily dose of 240 mg for 3 weeks, did not reveal any marked adverse effects of the drug. However, intravenous injection of dextromethorphan caused nystagmus, dizziness, vomiting, hallucinations, somnolence, confusion, motor anxiety, and arterial hypotension. These severe adverse effects accrued as the dose increased [2]. Clinical trials of dextromethorphan were stopped.

Cerestat is a non-competitive NMDA receptor antagonist with strong neuroprotective effects manifesting in experimental focal brain ischemia. Fifteen minutes after performed permanent occlusion of middle cerebral artery, its administration to experimental animals decreased the infarction size by 66% [126]. Delayed administration caused a reduction of the infarction volume by 50% [122]. The analysis of results of cerestat treatment in 94 patients with ischemic stroke established relative safety of the drug in cases of its use in middle doses, which are sufficient for manifestation of its neuroprotective properties [46]. Adverse effects were moderately expressed and included transient catatonia, moderate agitation, blood pressure increase (in higher doses). Phase II of clinical trial of cerestat that enrolled 120 patients who received the drug within the first 6 h of the stroke, showed significant neuroprotective effect of the drug when administered in maximal dose of 110 μg/kg [39]. However, the extent of adverse effects led to trial cessation.

Remacemide hydrochloride and its principal active desglycinyl metabolite are low-affinity non-competitive N-methyl-D-aspartate (NMDA)-receptor channel blockers having the property of inhibiting potential-dependent calcium channels [144]. Remacemide hydrochloride has demonstrated neuroprotection in animal models of hypoxia and ischemic stroke. The first clinical trials revealed slight adverse effects such as diplopia and somnolence [131]. At Phase II of clinical trials of intravenous and oral remacemide in dose of 400 mg each 12 h, there were no adverse effects registered [132–134]. Dyker and Lees [38] published results of the special study performed for assessment of the safety, tolerability, and pharmacokinetics of ascending doses of remacemide hydrochloride in patients with recent onset (within 12 h) ischemic stroke. This was a placebo-controlled, dose escalating, parallel group study. Groups of 8 patients (6 active, 2 placebo) were planned to receive twice-daily treatment, with 100, 200, 300, 400, 500, or 600 mg remacemide hydrochloride given as two intravenous infusions followed by 6 days of oral treatment. Neurological and functional outcome data were collected, but the study was not powered to demonstrate drug efficacy. It was confirmed that the most common adverse events of the treatment were related to the central nervous system, and these events appeared to increase with dose. Infusion site reactions and gastrointestinal upset were also reported and considered to be treatment related. On the evidence of this study, the maximum well-tolerated dose for remacemide hydrochloride in acute stroke is 400 mg BID. Doses of 200 mg

BID or higher attained the putative neuroprotective plasma concentrations of remacemide predicted from animal models (250 to 600 ng/ml). The authors concluded that the expected gradual accumulation of active metabolite suggest that optimal neuroprotective concentrations are unlikely to be achieved within the early hours of treatment at this dose. Further clinical trials are now in progress.

Magnesium ions block NMDA-associated channels in a potential-dependent way, but in terms of electrophysiology extracellular magnesium behaves like non-competitive NMDA-receptor antagonist [76, 138, 144]. Experimental studies held on rat models of focal brain ischemia showed that magnesium sulfate administration in doses 25-90 mg/kg significantly reduced brain infarction volume ($p < 0.01$) [84, 96, 121, 223], significantly increased survival rate in rats starting treatment at 2 h after middle cerebral artery occlusion, and improved neurological outcome in all animals treated with magnesium (at 2, 6, or 8 h after ischemic stroke onset) [223]. It was concluded that the therapeutic window for neuroprotection by magnesium can be extended up to 6 h in the focal cerebral ischemia model. The relatively long window of opportunity for effective dosing may be explained by the proposed multiple mechanisms of actions for magnesium [223]. Intravenous infusion of magnesium sulfate (25 mg/kg) also decreased up-regulation of c-fos and COX-2 mRNAs in the ischemized zone of rat brain 2 h after the occlusion [96]. In the global brain ischemia model, the application of magnesium chloride (50 mM), administered directly to the CA_1 sector of the rat hippocampus before and at various intervals following 20 min of ischemia, significantly normalized neuronal density, even when magnesium chloride was administered 24 h, but not 48 h, after the ischemic episode [206].

A pilot clinical trial of magnesium sulfate in 13 patients within the first hours of ischemic stroke demonstrated safety of its administration; besides, the number of patients with good outcome increased, and disability rate reduced [193, 211]. The results of treatment of 60 stroke patients with magnesium sulfate [134] not only confirmed safety of the drug, but also showed significant decrease of poor outcomes (death or severe disability) frequency (by 10%) versus patients from the placebo group (30 and 40%, respectively). A special study was undertaken to optimize the magnesium sulfate infusion regimen for a multi-center trial [136]. Within 24 h after the onset of clinically diagnosed stroke, patients were randomized to receive placebo or one of three intravenous infusions of magnesium sulfate: a loading infusion of 8, 12, or 16 mM, followed by 65 mM over 24 h. It was demonstrated that magnesium sulfate infusions rapidly elevate the serum magnesium concentration to potentially therapeutic levels, and they are well tolerated and have no major hemodynamic effects in patients with acute stroke. No effects of magnesium on heart rate, blood pressure, or blood glucose were evident. The 16-mmol loading infusion achieved target serum concentrations most rapidly and was chosen for further trials. A randomized, placebo-controlled, double-blind study [104] was performed as a pilot study to examine the benefit of the administration of magnesium sulfate given intravenously as a protective substance during the first 24 h following a stroke. Forty-one patients who had cortical infarction in the middle cerebral artery territory with moderate to severe neurological deficit lasting for more than 15 min with onset less than 24 h were included. Significant positive effect of magnesium sulfate on the outcome in patients by day 30 after the stroke onset was shown. Improvement of neurological deficit (by the Orgogozo Stroke Scale $p < 0.001$ versus placebo) and functional recovery (by the Rankin Disability Score $p < 0.05$ versus placebo) was demonstrated. Further studies on a larger scale are needed to confirm these findings [135].

Competitive antagonists of NMDA receptors (APH, CPP, CGS-19755, MDL 100, 453) directly block the glutamate-recognizing site of NMDA-receptors [217]. Efficacy of these drugs in experimental acute focal brain ischemia was found, and the infarction volume significantly decreased even in cases when the drug was administered after induction of ischemia [14, 77].

Selfotel (CGS-19755) is the most studied drug of this group. Selfotel showed neuroprotective properties in animal models when it was administered within 2 h after the stroke onset. Clinical trial of dose-dependent efficacy and safety of selfotel [63] showed that its administration within the first 12 h of the stroke was accompanied by such adverse effects as motor excitation, hallucinations, confusion, paranoid, delirium (the dose 2 mg/kg resulted in 100% of cases with complications; 1.75 mg/kg for 66%; 1.5 mg/kg for 50%). Adverse effects appeared on average 1–3 h after the first dose was administered (i.e. before 22 h) and remained for 2–60 h. The extent of neurological improvement assessed by the National Institute of Health Score (the NIH Score) on the background of selfotel treatment comprised 71 versus 36% in placebo group. Good recovery (Barthel Index more than 70) was registered in 95% of patients treated with selfotel versus 29% in the placebo group. Viewing the greatest safety of the 1.5 mg/kg dose, a special multi-center randomized placebo-controlled trial of this dose was held. The single dose 1.5 mg/kg was administered within the first 6 h of symptom onset [25]. Adverse effects (agitation, confusion, and hallucinations) were registered in 57% of patients from the selfotel group (13% of them were under severe condition) and in 18% of placebo-treated individuals ($p = 0.0001$). It is worth mentioning that 4% of patients treated with selfotel experienced respiratory disturbances which were absent in placebo-treated patients. Randomized, double-blind, placebo-controlled trial

[30] has shown that selfotel was not an effective treatment for acute ischemic stroke. Moreover, a trend towards increased mortality, particularly within first 30 days and in patients with severe stroke ($p = 0.05$) suggests that the drug might have a neurotoxic effect in brain ischemia.

Thus, the pre-clinical trials on various animal models demonstrated prominent neuroprotective effects of both competitive and non-competitive NMDA antagonists, such as considerable reduction of the infarction volume, significant increase in survival rate, and improvement of neurological outcome. However, the majority of these drugs proved inapplicable under clinical conditions due to a wide range of serious adverse effects (general toxic, psychic, locomotor, etc.). Adverse effects of several drugs developed after administration of low and average doses, before they reached plasma level sufficient to express neuroprotective properties. Only two drugs are being currently tried. They are the less toxic non-competitive NMDA antagonists, remacemide, and magnesium sulfate. The efficacy of these drugs within the first hours after the stroke onset requires further confirmation in larger clinical groups.

Taking into account high toxicity of basic NMDA receptor site (glutamate and phencyclidine) blockers, an attempt to **selectively block NMDA receptors** has been made by influencing their modulatory (the polyamine and the glycine) sites [48, 49].

Williams *et al.* [220] revealed the property of endogenous polyamines (spermine and spermidine) to regulate NMDA receptor activity, mediate their binding to non-competitive antagonists (dizolcipine, MK-801), and increase ion flows via NMDA-associated channels, which confirmed the existence of the modulatory polyamine site on NMDA receptors. Gotti *et al.* [58] and Poignet *et al.* [162] showed that *eliprodil* (SL-82.0715), which blocks both the modulatory polyamine site of the NMDA receptor and neuronal potential-dependent calcium channels, reduced the infarction size in experimental ischemic stroke models. Repeated intraperitoneal injection of the drug (5 min, 6 and 18 h after middle cerebral artery occlusion) decreased the size of infarction area by 60–70% versus controls [220]. Lekieffre *et al.* [112] used eliprodil alone or in combination with a thrombolytic agent, rt-PA, in a rat embolic stroke model (embolization was induced by intracarotid injection of an arterial blood clot). Eliprodil, administered intravenously at the dose of 1 mg/kg, 10 min and 2 h 30 min after embolization, reduced the neurological deficit by 54% ($p < 0.01$) and the total volume of the brain lesion by 49%. Thrombolysis with rt-PA (2.5 mg/kg, as a 30 min intravenous infusion beginning 1 h after embolization) decreased the neurological deficit by 48% ($p < 0.05$) and the size of the total infarct by 55% ($p < 0.05$). Combined therapy greatly improved the degree of neuroprotection as assessed by neurological and histological outcomes (70%, $p < 0.001$, and 89%, $p < 0.01$, respectively). So, the administration of eliprodil or thrombolytic agent (rt-PA) similarly reduced the volume of brain damage and the neurological deficit in a rat embolic stroke model. At the same time, combined cytoprotective therapy and thrombolysis markedly improved the degree of neuroprotection and may, thus, represent a valuable approach for the treatment of stroke in humans.

A pilot clinical trial of eliprodil in 114 ischemic stroke patients revealed minimal adverse effects, such as reversible elongation of QT interval on ECG and vertigo [47]. Phase III of a multi-center clinical trial, performed in 483 patients [195], failed to confirm the efficacy of eliprodil in ischemic stroke, so the trial was suspended.

Glycine is a co-agonist of NMDA receptors. In submicromolecular concentration (0.1 μM), it saturates the glycine recognition site and increases the frequency of NMDA-associated ion channels opening, having no influence on current amplitude while NMDA agonists are acting [86, 97]. So, the strychnine insensitive glycine binding site on the NMDA receptor channel, discovered by Johnson and Ascher in 1987 [86], represents an interesting target for the development of neuroprotective compounds, such as its selective antagonists, for the treatment of stroke [28, 98, 203].

The glycine site antagonists currently identified can be divided into five main categories depending on their chemical structure: indoles, tetrahydroquinolines, benzoazepines, quinoxalinediones and pyridazinoquinolines [203]. It was shown in different experimental models that the glycine site antagonists can decrease NMDA-induced excitotoxic damage [21, 56, 205, 218]. Neuroprotective properties of various drugs of this group (HA 966, GV-150526, L-687414, ZD-9379, ACPC (1-aminocyclopropane carboxylic acid), and this acid derivate ACEA-1021) were demonstrated in experimental stroke [133, 134, 196, 197].

Mennini *et al.* [127] showed that HA 966 appears to be a partial agonist, while 7-Cl-kynurenic acid and GV 150526A are competitive antagonists at the strychnine-insensitive glycine sites. The nanomolar potency of GV 150526A in reducing NMDA receptor function suggested that GV 150526A could be effective *in vivo* to reduce NMDA receptor over-stimulation, like in brain ischemia. Post-ischemia administration of GV150526 (3-[2-(phenylaminocarbonyl)ethenyl]-4,6-dichloroindole-2-carboxylic acid sodium salt) (3 mg/kg intravenously) up to 6 h after middle cerebral artery occlusion in rats resulted in a significant reduction of the infarction volume measured histologically 24 h later. The neuronal protection by GV150526 was accompanied by functionally significant protection determined by somatosensory evoked potential responses recorded from the primary somatosensory cortex of rats under urethane anesthesia [15]. The neuroprotective activity of GV150526 has been also evaluated by magnetic resonance imaging in a rat model of middle cerebral artery occlusion [166]. The drug was administered at a dose of 3 mg/kg intravenously both before and after (6 h) middle cerebral artery occlusion. Substantial neuroprotection was demonstrated at 6, 24, and 144 h of the stroke when GV150526 was administered before the occlusion. The ischemic volume was reduced by 84% and 72%, compared to control values, when measured from T_2W and DW images, acquired 24 h after middle cerebral artery occlusion. Administration of the same dose of GV150526, 6 h post-ischemia, also resulted in a significant ($p < 0.05$) neuroprotection. The ischemic volume was reduced by 48% from control values when measured from T_2W images and by 45% when measured from DW images. These data confirmed the potential neuroprotective activity of GV150526 when administered either before or up to 6 h after ischemia.

Tatlisumak *et al.* [198] evaluated the effect of the glycine antagonist ZD9379 and demonstrated that ZD9379 initiated 30 min before or 30 min after middle cerebral artery occlusion significantly reduced the infarction volume, as well as number of spreading depressions, that were monitored electrophysiologically for 4.5 h following middle cerebral artery occlusion by continuous recording of cortical direct current potentials and electrocorticogram in a permanent focal ischemia model. It suggested that ZD9379 is neuroprotective and its neuroprotective effect may be related to inhibiting ischemia-related spreading depressions.

Two drugs were given access to clinical trials: *GV-150526A* (*Gavestinel*) and *ACEA-1021* (*Licostinel*). A special study of the safety, tolerability, and pharmacokinetics of loading and maintenance infusions of GV-150526A in patients with acute stroke [37] showed that it is an emerging neuroprotective agent, with no apparent significant central nervous system or hemodynamic effects. Dose-limiting effects appear to be restricted to mild transient and asymptomatic rises in bilirubin and/or transaminases, primarily observed at high maintenance doses, and there were no findings that should preclude further clinical development. The other investigations [3, 111, 218] also confirmed that selective blockers of the glycine site were safer and better tolerated than competitive and non-competitive antagonists of NMDA-receptors. A randomized, double-blind, placebo-controlled trial (GAIN-I) of *Gavestinel* [110] enrolled 1804 patients from 173 centers in 21 countries, has demonstrated neutral results: the drug caused neither harm nor benefit. Analogous data were received in the other randomized, double-blind, placebo-controlled trial (GAIN A), that included 1367 patients from 132 centers in the USA and Canada [168]. Gavestinel administered up to 6 h after an acute ischemic stroke did not improve functional outcome at

3 months. Thus, Gavestinel failed to produce any significant treatment benefits for patients treated within 6 h after ischemic stroke onset [52].

Licostinel (ACEA 1021; 5-nitro-6,7-dichloro-2,3-quinoxalinedione) is a competitive antagonist at the strychnine-insensitive glycine site of NMDA receptors that was an effective neuroprotective agent in animal models of cerebral ischemia. A 5-center trial [3] that enrolled 64 patients who were treated with ascending doses of a short infusion of Licostinel or a placebo within 48 h of an ischemic stroke demonstrated that Licostinel in doses up to 3.0 mg/kg is safe and tolerable in acute stroke patients. At the same time, no difference in the dynamics of neurological deficit, assessed by the National Institutes of Health Stroke Scale, was revealed between the placebo group and the Licostinel-treated patients.

It was proven that AMPA glutamate receptors participate in the development of excitotoxicity and in ischemic damaging action (see Section 4.1.3). This evidence is a reason to regard them as an additional therapeutic target. **AMPA receptor antagonists** decrease sodium intracellular influx preventing membrane depolarization and following intracellular accumulation of calcium and water [93, 101]. Such AMPA receptor antagonists as NBQX, ZK200775, YM90K, and YM872 exhibited prominent neuroprotective properties in experimental focal brain ischemia in rats with transient and permanent middle cerebral artery occlusion, causing the reduction of the infarction volume and increase in the neurological deficits [90–92]. The therapeutic efficacy of AMPA receptor antagonists persisted for several days after middle cerebral artery occlusion [91]. Being combined with the thrombolytic agent rt-PA, they significantly increased thrombolysis efficacy [18, 123]. At the same time, severe adverse effects of these drugs, such as nephrotoxicity for NBQX and pronounced sedation for ZK200775, were revealed. These effects make these drugs unsuitable for stroke patients [208, 211].

Thus, clinical trials of the majority of neuroprotectors most powerful in experiments (glutamate receptor antagonists) were stopped due to a wide range of serious adverse effects. Application of less toxic low affinity or selective NMDA antagonists occurred to be less effective. Taking into consideration the wide distribution of glutamate receptors in the CNS and their importance for many physiological functions including intellectual, memory-associated, psychic, and locomotor, it is hard to speculate about the safety of their "switching off". It is a conformity that the efficacy of neuroprotection that is produced via glutamate receptors blockade directly correlates with the number and severity of adverse side effects, which casts doubt on the expediency of such a therapeutic strategy.

14.3.   Inhibitors of synthesis and pre-synaptic release of glutamate

High toxicity of glutamate receptor antagonists motivated research of other effective means to decrease glutamate excitotoxicity. The inhibition of synthesis and pre-synaptic release of glutamate was proposed as a novel neuroprotective strategy for decreasing post-synaptic excitotoxicity. The group of inhibitors of glutamate release includes BW-619C89, propentofylline, phenytoin, fos-phenytoin, and lubeluzole.

BW-619C89 [4-amino-2-(4-methyl-1-piperazinyl)-5-(2,3,5-trichlorophenyl) pyrimidine] is a derivate of BW-1003C87 compound, chemically related to lamotrigine (lamictal), widely used in epileptology. It was shown that BW-619C89 influences pre-synaptic potential-dependent sodium channels, decreasing glutamate release into the synaptic cleft. In experimental models of focal brain ischemia, BW-619C89, when administered parenterally, reduced neurological deficit and infarction volume (by 58%) after middle cerebral artery occlusion in rats [59, 88, 89, 108, 186, 187]. Doses of 10, 20, 30, and 50 mg/kg given intravenously prior to the occlusion significantly ($p < 0.05$–0.01) reduced the infarction volume in a dose-dependent fashion maximal at 30–50 mg/kg compared to saline controls. Treatment with 30-50 mg/kg of BW-619C89 30, 45, and 60 min after the onset of ischemia also was effective in significantly reducing cortical infarction volume ($p < 0.05$), but there was no effect when the drug was given 5 min after reperfusion [88, 108]. So, in experimental focal brain ischemia the neuroprotective effects of BW-619C89 have been confirmed. However, clinical trials of BW-619C89 (*sipatrigine*) [129, 130, 132–134] revealed such adverse neuropsychiatric effects of the drug as reduced consciousness, agitation, confusion, visual perceptual disturbance, and frank hallucinations. Nausea, vomiting, infusion site reactions, and hyponatremia were also commoner in sipatrigine patients. Phase III of clinical trials was cancelled.

Propentofylline is an adenosine uptake and phosphodiesterase inhibitor required for activation of pre-synaptic α_1-receptors connected with sodium and calcium channels [36, 50]. Changes in the state of pre-synaptic ion channels result in inhibition of neurotransmitter release. In the acute global brain ischemia model, after acute and long-term permanent bilateral common carotid artery occlusion in rats, propentofylline (25 mg/kg per day) showed a neuroprotective effect on cerebral energy state (increase in energy-rich phosphates) and pro-inflammatory cytokines (decrease in concentrations of TNF-α and IL-1β 12- and 19-fold) in parietotemporal cortex and hippocampus [161]. In the gerbil permanent focal cerebral ischemia model, propentofylline at suitable dose improved regional cerebral blood flow in ischemic brain areas [207]. Continuous intravenous infusions of propentofylline (0.01, 0.05, or 0.1 mg/kg per min) initiated 15 min following permanent middle cerebral artery occlusion in rats, produced significant reductions in infarction volumes, the highest dose being the most effective (reduction by 39%; $p < 0.005$) and improved the neurological symptoms when compared with an untreated control group. In contrast to other anti-ischemic agents, such as glutamate receptor antagonists, the drug induced no behavioral disturbances. This study indicated that propentofylline may provide neuroprotective effect against ischemic brain damage following stroke without negative behavioral side effects [72, 153]. A pilot clinical trial of propentofylline in 30 patients with acute ischemic stroke failed to detect significant acceleration of functional recovery (by the Barthel Index); only a trend towards improvement of glucose metabolism in the infarction area was registered. Data for neurological impairment and disability were not in a form suitable for analysis, and data on quality of life, stroke recurrence, thromboembolism, and bleeding were not reported [11]. Such side effects revealed as arterial hypotension and myocardial contractile depression did not allow the trial to be continued.

Anticonvulsant *phenytoin* and its derivate *fos-phenytoin* are also antagonists of pre-synaptic sodium channels with neuroprotective properties [47]. In an experimental study performed in a rat corticostriatal slice preparation, it was shown that phenytoin produced a significant neuroprotection by reducing glutamatergic transmission [20]. Clinical trials of fos-phenytoin in ischemic stroke failed to reveal its significant influence on the outcome, but they showed its relative safety [201]. Adverse effects were hypotonia and somnolence.

Lubeluzole is a benzothiazole derivate that blocks sodium channels. The drug prevents pre-synaptic glutamate release, decreases the content of neurotransmitter in the extracellular space of the peri-infarction area [172, 180], and also inhibits glutamate-induced neurotoxicity of nitric oxide by decreasing the activity of NO-synthase [109, 113]. Experimental works have revealed a prominent neuroprotective effect of lubeluzole in acute focal brain ischemia [6, 26]. At Phase II of a clinical trial in 232 patients with acute ischemic stroke, good safety and tolerability of the drug were demonstrated; being administered from the first 6 h for 5 days in daily dose of 10 mg, the drug was noted to decrease mortality by 10% more than in the placebo group and led to outcome improvement [35]. The drug was well tolerated. Transient elongation of QT interval on ECG was an adverse effect.

Considering the initially revealed influence of lubeluzole on the QT ECG interval, the special cardiovascular safety of lubeluzole was evaluated in patients with ischemic stroke in a double-blind, placebo-controlled trial [73]. Forty-six patients were randomized to receive a continuous daily infusion of lubeluzole 5 mg (loading dose 3.75 mg over 1 h), lubeluzole 10 mg (loading dose 7.5 mg over 1 h), or placebo for 5 days within 24 h of stroke onset. Neither dose of lubeluzole had any statistically or clinically relevant effects on the QTc. Neither was there any significant differences among the three treatment groups in the area under the curve for heart rate, QT interval, QT dispersion, or QTlc. Lubeluzole did not increase the frequency of ECG abnormalities. No ventricular fibrillation, ventricular tachycardia, or torsades de pointes were observed in any of the treatment groups. The authors concluded that in doses to 10 mg/day, lubeluzole had a favorable cardiovascular safety profile and it was well tolerated by patients with ischemic stroke.

At present, several the phase III trials have been finished. The US and Canadian Lubeluzole Ischemic Stroke Study (Lub Int 9) [64, 145], a randomized, double-blind, placebo-controlled study that included 721 patients from 83 centers in the USA and Canada, was conducted to assess the efficacy and safety of lubeluzole in the treatment of ischemic stroke. Patients with clinical symptoms of acute ischemic stroke were randomized to receive either lubeluzole (7.5 mg over 1 h, followed by a continuous daily infusion of 10 mg for up to 5 days) or placebo. Treatment was initiated within 6 h of symptom onset. The overall mortality rate at 12 weeks for lubeluzole-treated patients was 20.7 compared to 25.2% for placebo-treated patients ($p > 0.05$). Controlling for relevant covariates, the degree of neurological recovery (by the National Institutes of Health Stroke Scale) at week 12 significantly favored lubeluzole over placebo ($p = 0.033$). Lubeluzole treatment similarly resulted in significantly greater improvements in functional status (by the Barthel Index) ($p = 0.038$) and overall disability (by the Rankin Scale) ($p = 0.034$) after 12 weeks. A global test statistic confirmed that lubeluzole-treated patients had a more favorable clinical outcome at 12 weeks ($p = 0.041$). The safety profile of lubeluzole resembled that of placebo. So, it was demonstrated that treatment with lubeluzole within 6 h of the onset of ischemic stroke had a non-significant effect on mortality but resulted in improved clinical outcome compared with placebo, with no safety concerns.

The other randomized, parallel-group, double-blind, placebo-controlled study (Lub Int 5) [32] enrolled 725 patients from 107 centers in Europe and Australia. The drug administration regimen was the same as in the Lub Int 9 Study. It was shown that in the total ischemic stroke population, the overall mortality rate at 3 months was similar for lubeluzole (21.0%) and placebo (21.4%). Treatment benefit was related to stroke severity, as determined by the Clinical Global Impression rating, which was a pronounced clinically significant reduction in mortality was noted in the lubeluzole-treated patients for whom stroke severity was mild to moderate, but not in those for whom it was severe. Lubeluzole did not increase morbidity among stroke survivors, as measured by the European Stroke Scale, Barthel Index, and Rankin Scale. No safety concerns were seen with lubeluzole treatment. So, the European trial failed to confirm the significant efficacy of the treatment with intravenous lubeluzole within 6 h of the onset of ischemic stroke in relation to mortality rate which was taken to be the primary end-point. Only among patients with mild to moderate ischemic stroke, lubeluzole decreased mortality without increasing morbidity. The study also failed to confirm lubeluzole influence on clinical stroke outcome.

Meta-analysis that enrolled 1355 cases of ischemic stroke demonstrated positive neuroprotective effect of lubeluzole in patients with mild and moderately severe stroke, whereas no effect was found in severe disease [75].

In recent years a double-blind, placebo-controlled, phase III trial has been held with an 8-h inclusion window to assess the efficacy and safety of an intravenous loading dose of 7.5 mg followed by a daily intravenous dose of 10 mg lubeluzole for 5 days in acute ischemic stroke patients [33, 34]. A total of 1786 patients were randomized: 901 to lubeluzole and 885 to placebo. This study failed to show an efficacy of the drug in the treatment of acute stroke. On the other hand, it was shown that lubeluzole treatment by the current dose schedule was not associated with a significant safety problem.

Special analysis was performed by means of the Cochrane Stroke Group trials register to assess the effectiveness and safety of lubeluzole given in the acute phase of acute ischemic stroke [51]. Five trials involving a total of 3510 patients were included. The quality of the trials did not vary considerably. Sensitivity/subgroups analysis was not performed completely because of lack of data. Lubeluzole given at the doses of 5, 10, and 20 mg/day for 5 days was tested against a placebo-control group. There was no evidence that lubeluzole given at any dose either reduced the odds of death from all causes (OR = 0.93, 95% CI 0.79–1.09) or reduced the odds of death or dependence at the end of follow-up (OR = 1.04, 95% CI 0.91–1.19). On the other hand, given at any dose, lubeluzole was associated with a significant excess of heart-conduction disorders (QT prolonged > 450 msec) at the end of follow-up (OR = 1.43, 95% CI 1.09–1.87). It was concluded that lubeluzole, given in the acute phase of ischemic stroke, is not associated with a significant reduction in death or dependency at the end of scheduled follow-up period but seems to be associated with a significant increase of heart-conduction disorders (QT prolonged >450 msec). Thus, the drug is no longer in development for ischemic stroke.

14.4.   GABA agonists

Revealed importance of imbalance between the excitatory and inhibitory neurotransmitter systems and insufficiency of natural protective mechanisms in the pathogenesis of ischemic stroke (see Section 4.1.1) determined the choice of a new therapeutic strategy aimed at decreasing glutamate excitotoxicity. It is an attempt to remove neurotransmitter imbalance via activation of inhibitory systems.

GABA (gamma-aminobutyric acid) is a basic inhibitory neurotransmitter in the CNS. Its receptors are widely distributed throughout brain structures, practically in all neuronal groups [42]. Presenting a natural mechanism of protection, GABA imposes limitations on an excitatory stimulus on the pre-synaptic level via CABA$_B$ receptors (which are functionally linked to potential-dependent calcium channels of pre-synaptic membranes), as well as on the post-synaptic level via GABA$_A$ receptors (GABA–barbiturate–benzodiazepine receptor complex, functionally linked to potential-dependent chlorine channels) [41, 164, 165, 174, 175].

Radioimmunological study [44] confirmed especially wide distribution of GABA$_A$ receptors in brain structures. These receptors are represented most densely in frontal and temporal cortex, hippocampus, amygdala and hypothalamic nuclei, substantia nigra, peri-aqueduct gray matter, and cerebellar nuclei. To somewhat less extent these receptors are represented in caudate nucleus, putamen, thalamus, occipital cortex, and epiphysis. Activation of post-synaptic GABA$_A$ receptors leads to hyperpolarization of cellular membranes and inhibition of excitatory stimulus provoked by depolarization [27, 94]. All three subunits of a GABA$_A$ receptor (α, β, and γ) bind GABA, but the highest affinity is related to the α-subunit. Barbiturates interact with α and β subunits; benzodiazepines interact only with the γ subunit [10, 212]. It was demonstrated that the affinity of each of the ligands increases when other ligands coherently interact with the receptor.

The basic physiological role of GABA is to modulate glutamate activity, which is the main excitatory neurotransmitter, and also to create steady equilibrium between the excitatory and inhibitory systems [125]. Certain proportions exist between representation of the GABA and glutamate receptors in different brain areas. Neurons most sensitive to ischemia, along with a powerful glutamate-aspartate excitatory "entrance" located on the dendritic structure, possess many GABA-ergic terminals on neuronal somata [125, 178]. In the first seconds of experimental brain ischemia immediate release of glutamate and GABA occurs out of pre-synaptic terminals [61]. However, insufficiency of GABA-ergic inhibitory neurotransmission exists within the first 4 h after induced ischemia [67, 69, 70]. Stimulation of GABA$_A$-receptors in experiment allows regulation of the activity of the glutamate–calcium cascade [61]. Experimental studies on models of focal and global brain ischemia showed that the increase of GABA level due to inhibition of its degradation via such GABA transaminase inhibitor as γ-vinyl-GABA causes significant neuroprotective action, keeping neurons alive and elevating the level of their energy metabolism [1, 176]. The application of GABA$_A$ agonists such as *muscimol* and *chlomethiazole* leads to hyperpolarization of neuronal post-synaptic membranes, stabilizes resting potential of the membrane, and, hence, blocks peri-infarction depolarization and spreading depression waves that increase the area of ischemic brain damage [83]. Besides, it was shown that chlormethiazole had anti-inflammatory properties because it potently and selectively inhibited p38 mitogen-activated protein (MAP) kinase in primary cortical glial cultures. The inhibition of p38 MAP kinase resulted in the attenuation of the induction of c-fos and c-jun mRNA and AP-1 DNA [181]. In experimental ischemic stroke, *muscimol* and *chlomethiazole* significantly decreased the size of infarction [117, 188, 194]. The clinical trial of chlomethiazole given within the first 24 h after stroke onset established its safety for clinical use. The phase III randomized, double-blind, placebo-controlled trial (CLASS), conducted in 1360 patients from 92 centers in Canada and 7 countries in Europe, tested the efficacy and safety of chlomethiazole in patients with acute hemispheric stroke [216]. Patients were treated beginning from the first 12 h with the drug in dose of 75 mg/kg infused over 24 h. The study showed that chlomethiazole had no adverse or beneficial effect on long-term outcome for all patients [213, 214, 216]. However, in patients with total anterior infarction, who comprised 40% of all studied patients, significant improvement of functional recovery (assessed by the Barthel Index) was registered. This result, along with moderately expressed adverse effects of the drug (rhinitis and mildly pronounced sedation) allowed continued clinical study of chlomethiazole in patients with large strokes.

14.5. Glycine

The inhibitory amino acid glycine also deserves attention as an important participant of the complex biochemical cascade in focal cerebral ischemia.

Glycine was traditionally regarded as a neurotransmitter which exhibits its properties on spinal, medullary, and pontine levels being released basically from segmental interneurons and propriospinal systems and inhibiting motor neurons via axo-axonal and axo-dendritic contacts [27, 74, 81, 148]. Later glycine was shown to be an inhibitory neurotransmitter of almost all CNS structures. High density of glycine receptors in brain was revealed not only in stem structures, but also in hemispheric cortex, striatum, hypothalamic nuclei, pathways connecting frontal cortex with hypothalamus, and in cerebellum. Fagg and Foster [42] and Mayor *et al.* [120] concluded that GABA and glycine are equal neurotransmitters providing protective inhibition in the CNS, the role of which gains more importance when the release of glutamate is excessive. Glycine manifests its inhibitory properties via interaction not only with glycine receptors, but also with GABA receptors [42, 85, 120].

Along with this, Johnson and Ascher [86] were the first to prove in experiments that glycine in submicromolecular concentrations is necessary for normal functioning of glutamate NMDA receptors. Activation of NMDA receptors is possible only when glycine binds their specific strychnine-insensitive glycine site, i.e. glycine appears to be their co-agonist [79, 86, 95, 97, 151]. In the normal state *in vivo* usual concentrations of endogenous glycine bind to the glycine site of NMDA receptors completely [86, 95]. The facilitating action of glycine appears at concentrations below 0.1 μM and the binding site of NMDA receptor is saturated between 10 and 100 μM [22]. Addition of a larger concentration of glycine (100 μM, 1 mM) after oxygen deprivation is not responsible for the long term modulation of NMDA receptor activity in the rat hippocampus [87], and increased extracellular concentrations of glycine, such as those observed in ischemia (10–100 μM), do not potentiate NMDA-evoked depolarization *in vivo* and, thereby, excitotoxicity [139, 140].

It is interesting to note that high doses of glycine and of some of its agonists (1-amino-1-carboxycyclopropane which is its full agonist and D-cycloserine which possesses 40–60% of glycine effect) have anticonvulsant activity in experimental animals [105, 156, 157, 173, 189, 202, 204] and potentiate the anticonvulsant effect of clinically established anti-epileptic drugs [155]. It is difficult to explain the revealed effects if activation of excitatory glutamate systems is supposed to be the main mechanism of action of glycine.

Glycine also has general metabolic effects and plays an important role as a conjugator of low-molecular-weight toxic substances generated during ischemic processes [107, 124, 221].

The possibility of the penetration of glycine into brain tissue through the receptor zones markedly expanded its therapeutic facilities. It was shown that 10 min after application of tritium-labeled glycine (^{3}H-glycine) onto eye conjunctiva, intranasally, or on buccal mucous membrane in rats, scintillation spectrometry registered its appearance in various brain structures and cerebrospinal fluid. Within the first 30 min after application of ^{3}H-glycine on eye conjunctiva, its concentration elevated up to peak values in the *chiasma opticum*, and after 2 h in the optical cortex [165].

The pharmaceutical drug *Glycine* was developed in the Laboratory of Metabolic Regulators of the Medical Scientific and Production Complex "Biotics" (Moscow, Russia). It comprises the inhibitory amino acid glycine and a pharmaceutical carrier 0.5–2.0 mass % methyl cellulose (for sublingual or in-the-cheek application). Experiments with Glycine confirmed good penetration of the drug through the receptor zones [165]. Administration of 20 mg/kg of Glycine in rats with frontal lobe focal ischemia resulted in a significant elevation of glycine concentration and an increase in the velocity of the GABA cycle in different brain structures including also the area of the ischemic focus [9]. Besides, a significant reduction of the concentrations of products of oxidative stress in the brain infarction zone and a significant normalization of latency periods of conditioned (behavior) reflexes were shown in rats under Glycine treatment [29, 167].

Safety trials with Glycine in normal subjects (including rats and healthy volunteers) were performed in 1992 and demonstrated good tolerability and completely safe profile of Glycine treatment [102]. Being a natural brain metabolite, Glycine was not toxic even given in concentrations exceeding 10 g/day. The only side effect was a slight sedative state. In early studies, the drug Glycine in doses of 300–600 mg/day was shown to possess anti-stress, stress-protective, and nootropic effects [102].

Beneficial mediator and metabolic properties of the inhibitory amino acid glycine obtained in experiments and clinics as well as the absence of toxicity, the safety profile, and favorable effects of the pharmaceutical drug Glycine demonstrated in animal models with focal brain ischemia provided good evidence of possible neuroprotective properties of the drug in patients with acute ischemic stroke.

14.5.1. Randomized, double-blind, placebo-controlled study of the safety and efficacy of Glycine in carotid ischemic stroke and of its neuroprotective effects

We performed the first clinical trial of the pharmaceutical drug Glycine in patients with acute focal brain ischemia in the Neurology Clinic of the Russian State Medical University [66, 67, 70, 182–184] (Russian patent No. 282398, 1992/1997).

The purpose of the study was to investigate the safety and the efficacy of Glycine in patients with acute ischemic stroke and the ability of this drug to ameliorate the outcome of those patients, as well as to elucidate certain mechanisms of its neuroprotective effects.

The trial was performed after the approval of the Ethics Committee of the study center was obtained, and the patients or their legal representative gave written or witnessed informed consent to participate in the trial.

Patients with acute carotid ischemic stroke were eligible for inclusion in the trial if they (1) were admitted to the Intensive Stroke Unit at the Department of Neurology of the Russian State Medical University within the first 6 h after the onset of stroke, (2) were within the age range 45-75 years, (3) were conscious or mildly obtund (baseline Orgogozo score more than 15). Patients who had experienced a previous stroke with residual neurological impairment, suffered from any other disorder interfering with neurological or functional assessment, or who had a life-threatening concurrent illness were excluded from participation in the trial. Other exclusion criteria were congestive heart failure, acute myocardial infarction (within the previous 6 weeks), ECG findings of ventricular arrhythmia, second- or third-degree atrioventricular block.

The trial initially included 212 patients (out of 775 screened patients before randomization) who were eligible for inclusion (Scheme 14.1). Twelve patients with either intracerebral hemorrhage or ischemic stroke in the vertebrobasilar territory were excluded from the primary efficacy analysis after randomization (the study protocol allowed a CT scan to be performed within the 24 h after the start of trial medication). The target population consisted of 200 patients with carotid ischemic stroke (male/female 110/90; mean age 63.7 ± 10.1 years); 115 patients suffered from left hemispheric stroke, 85 from right hemispheric stroke. There were no significant differences between the treatment groups in the target population with regard to demographic and baseline characteristics (Table 14.2).

Within 6 h after the onset of stroke, individual random and blind assignment was performed on patients to receive sublingual treatment with placebo or one of three doses of Glycine: 0.5, 1.0, or 2.0 g/day for 5 days (see Scheme 14.1). Tablets of placebo and Glycine were similar in appearance and taste. The personnel at the trial site, outcome assessors, the personnel of the Safety Monitoring Committee involved in conducting or monitoring the trial, and also data analysis were blind to the trial drug codes. Concomitant treatment with calcium channel blockers, piracetam (Nootropil), drugs with neurotrophic and neuromodulatory properties (such as gangliosides, low-molecular-weight peptides) and other experimental stroke drugs was prohibited throughout the trial. Routine therapy included hemodilution and aspirin (in all patients) and glycerol (in all cases of severe strokes). The same background therapy allowed us to compare stroke outcomes in the placebo and Glycine groups.

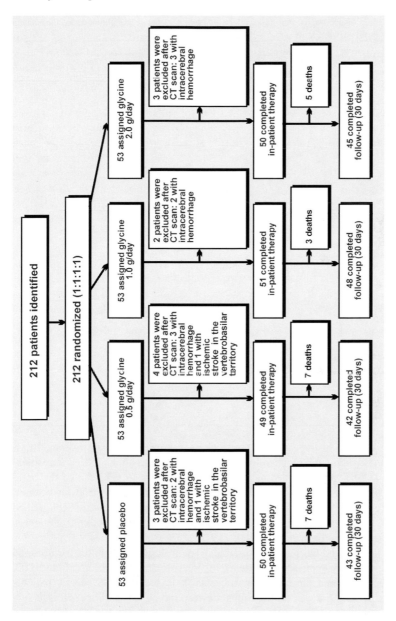

Scheme 14.1. Trial profile.

Table 14.2. Demographic and baseline characteristics of the target population (200 patients with carotid ischemic stroke)

Characteristic	Placebo (n = 50)	Glycine			p
		0.5 g/day (n = 49)	1.0 g/day (n = 51)	2.0 g/day (n = 50)	
Sex (M/F)	28/22	25/24	27/24	30/20	> 0.05
Mean age, years	65.1 ± 9.5	64.5 ± 10.5	67.3 ± 8.7	66.7 ± 9.7	> 0.05
Left/right hemisphere localization	32/18	28/21	28/23	27/23	> 0.05
Mean inclusion time ± SEM, h	5.5 ± 0.2	5.3 ± 0.2	5.4 ± 0.2	5.2 ± 0.3	> 0.05
Mean OSS score at start ± SEM	41.2 ± 2.6	41.1 ± 2.4	38.9 ± 1.9	40.9 ± 2.1	> 0.05
Patients with baseline OSS score ≤ 40 (severe stroke)	19 (38%)	20 (40.8%)	21 (41.2%)	20 (40%)	> 0.05
Patients with baseline OSS score 41–64 (moderate stroke)	23 (46%)	22 (45%)	23 (45.1%)	24 (48%)	> 0.05
Patients with baseline OSS score ≥ 65 (mild stroke)	8 (16%)	7 (14.2%)	7 (13.7%)	6 (12.0%)	> 0.05
Mean SSS score at start ± SEM	27.0 ± 1.4	27.1 ± 1.3	26.5 ± 1.3	26.9 ± 1.2	> 0.05

A medical history, general physical and neurological examinations, ECG, hematological and biochemical tests, and investigations of cerebrospinal fluid (CSF) were included as baseline assessments. A CT scan of the brain was performed within the first 24 h after the beginning of trial medication. Neurological status was assessed by the Scandinavian Stroke Scale (SSS) [170] and the Orgogozo Stroke Score (OSS) [147], and functional status was assessed by the Barthel Index [60, 119].

The SSS and OSS are both graded 10 neurological items; for the SSS and OSS, the maximum score (absence of neurological deficit) is 66 and 100, respectively, and the minimum score is 0. Based on OSS scores, patients were divided into several groups corresponding to severity of ischemic stroke; a score ≥65 corresponded to mild stroke, from 64 to 41 to moderate stroke, from 40 to 26 to severe stroke and ≤25 to extremely severe stroke. The initial division was maintained throughout the trial; the SSS score assessment and other measures were considered within the initial OSS score groups. The Barthel Index evaluates 10 activities of daily living with a maximal value of 100 and a minimal value of 0; a score from 0 to 45 corresponds to severe disability, one from 50 to 70 to moderate disability, one from 75 to 95 to mild disability, and a score of 100 equals no disability [60].

Blood pressure, heart rate, and ECG were repeated within 6–12 h after the initiation of therapy and on days 3, 5, 7, 14, 21, and 30. Neurological assessments were made at the beginning on admission within the first 6 h, at the end of the Glycine treatment (on day 6), and at the end of the trial (on day 30). The Barthel Index was estimated on day 30.

The following methods of laboratory monitoring were selected taking into consideration beneficial mediator and metabolic properties of Glycine observed in experiments and clinics. Laboratory investigations of blood (the detection of autoantibodies (AB) to phencyclidine-binding protein of glutamate NMDA-receptors (see Section 4.1.3)) were carried out on admission to hospital and 6, 9, 12, and 24 h after the onset of stroke (monitoring within the first 24 h), and also on days 3 and 5 (within the early acute period). Investigations of CSF (determination of neurotransmitter amino acids (glutamate, aspartate, glycine, and GABA) levels (see Section 4.1.1) and concentration of thiobarbituric-acid-reactive substances (TBARS) (see Section 4.3.1)) were carried out within the first 6 h before the Glycine treatment and were repeated on day 3. The immune and biochemical investigations were carried out together with colleagues from the Institute of Pharmacology of the Russian Academy of Medical Sciences (K. S. Raevsky, V. S. Kudrin, V. G. Bashkatova) and the Institute of the Human Brain of the Russian Academy of Sciences (S. A. Dambinova and colleagues).

No complications related to lumbar puncture were observed. Mortality and adverse events were followed over the entire trial period of 30 days. Safety was followed by the Safety Monitoring Committee, which reviewed all reports of death and adverse events.

Protocol-specified study end points were the safety, neurological outcome according to the OSS and SSS, and functional outcome according to the Barthel Index and mortality on day 30.

Statistical analysis. Demographic and baseline disease characteristics were compared with the use of the Cohran–Mantel–Haenszel test for general association for nominal categorical variables (e.g., sex) and a one-way ANOVA for continuous variables (e.g., age). One-sample *t* tests were performed with the use of descriptive statistics for each heart rate and blood pressure to evaluate changes versus baseline values.

The primary efficacy analysis was performed on a target population basis as pre-specified in the protocol, i.e. on all 200 patients with carotid ischemic stroke who were correctly included in the trial according to the inclusion and exclusion criteria and completed the follow-up within 30 days.

Descriptive statistics and frequency distributions were generated for the study and point data. The Wilcoxon matched-pairs signed-rank test was used to analyze the statistical significance for the changes in measured parameters and the Mann–Whitney U test was used for pairwise and group comparisons. Dynamics of biochemical parameters was assessed using ANOVA. Mortality rates and disability levels on the Barthel Index among the treatment groups were compared with the use of Fisher's exact test.

Clinical assessment of the Glycine groups showed that there was no significant difference between the 1.0 and 2.0 g/day Glycine groups. A post hoc subgroup analysis regarding biochemical parameters was performed on the 0.5 and 1.0–2.0 g/day Glycine groups.

All statistical tests were interpreted at the 5% two-tailed significance level.

The trial confirmed the safety profile of the Glycine treatment. Slight sedation as a side effect was observed in 9 patients (4.5%). Other marked side effects or adverse events were absent. The Glycine treatment had no statistically significant effects on ECG and hemorheological parameters.

Compared to placebo, there was a lower mortality in the 2.0 g/day Glycine group and a significantly lower mortality in the 1.0 g/day group ($p < 0.05$) (Table 14.3).

A quantitative time-related analysis of the dynamics of neurological deficit (as measured by the mean shift from baseline on the Orgogozo and the Scandinavian Stroke Scales) performed by intention-to-treat and on-treatment methods revealed more rapid improvement of neurological deficits up to days 6 and 30 in patients of the 1.0 and 2.0 g/day Glycine groups versus placebo (Table 14.4), more pronounced in the 1.0 g/day group. In the 0.5 g/day Glycine group the tendency towards acceleration of improvement was noted in only patients with mild to moderate stroke (OSS score >40) up to day 30 (Table 14.4).

When outcome was evaluated with the Barthel Index, the 1.0 and 2.0 g/day Glycine groups were found to have a higher proportion of patients with good recovery (no or mild disability, i.e. a Barthel score >70) than the 0.5 g/day Glycine group and the placebo group. A significant decrease was revealed in the number of patients with severe disability in the 1.0 g/day Glycine group ($p < 0.01$ versus placebo; $p < 0.05$ versus 0.5 g/day Glycine) and the 2.0 g/day Glycine group ($p < 0.05$ versus placebo; $p < 0.05$ versus 0.5 g/day Glycine) (Table 14.5).

After the neurological and functional assessment showed that there was no significant difference between the 1.0 and the 2.0 g/day Glycine groups, we used the 0.5 and 1.0–2.0 g/day Glycine groups for the analysis regarding biochemical parameters.

Table 14.3. Mortality and causes of death according to treatment group

Characteristic	Placebo (*n* = 50)	Glycine		
		0.5 g/day (*n* = 49)	1.0 g/day (*n* = 51)	2.0 g/day (*n* = 50)
Mortality				
All patients	7/50 (14.0%)	7/49 (14.3%)	3/51 (5.9%)*	5/50 (10.0%)
Patients with baseline OSS score ≥ 40 (mild and moderate stroke)	2/31 (6.5%)	2/29 (6.9%)	0/30 (0%)	1/30 (3.3%)
Patients with baseline OSS score 25 - 40 (severe stroke)	2/11 (18.2%)	2/12 (16.7%)	1/11 (9.1%)	2/13 (15.4%)
Patients with baseline OSS score ≤ 25 (extremely severe stroke)	3/8 (37.5%)	3/8 (37.5%)	2/10 (20.0%)	2/7 (28.6%)
Causes of death, No. of patients				
Brain edema	5	3	1	1
PA thromboembolism	1	2	1	2
Cardiac failure	1	1	1	2
Secondary stroke		1		

Note. * $p < 0.05$; compared with the placebo group. PA = Pulmonary artery.

Table 14.4. Change in neurological outcome as measured by the mean shift from baseline on the OSS and SSS

a. Intention-to-treat analysis of 212 identified patients

Stroke scale	Placebo (*n* = 53)	Glycine		
		0.5 g/day (*n* = 53)	1.0 g/day (*n* = 53)	2.0 g/day (*n* = 53)
		At day 6		
OSS				
Patients with mild to moderate stroke (OSS > 40)	+14.05±1.2 (*n* = 32)	+14.43±1.5 (*n* = 30)	+22.05±1.3** (*n* = 31)	+16.93±1.3 (*n* = 31)
Patients with severe stroke (OSS ≤ 40)	+5.5±1.0 (*n* = 21)	+5.89±0.9 (*n* = 23)	+20.17±1.6** (*n* = 22)	+16.11±1.6** (*n* = 22)
SSS				
Patients with mild to moderate stroke (OSS > 40)	+5.11±0.8 (*n* = 32)	+5.24±0.7 (*n* = 30)	+7.25±0.6* (*n* = 31)	+6.85±0.7 (*n* = 31)

(*continued*)

Table 14.4 (continued)

Stroke scale	Placebo (n = 53)	Glycine		
		0.5 g/day (n = 53)	1.0 g/day (n = 53)	2.0 g/day (n = 53)
Patients with severe stroke (OSS ≤ 40)	+3.57±0.9 (n = 21)	+3.22±0.9 (n = 23)	+12.26±1.2** (n = 22)	+10.10±1.1** (n = 22)

	At day 30			
OSS				
Patients with mild to moderate stroke (OSS > 40)	+18.64±2.0 (n = 32)	+24.54±2.8* (n = 30)	+33.52±1.9** (n = 31)	+27.20±2.2** (n = 31)
Patients with severe stroke (OSS ≤ 40)	+13.85±3.1 (n = 21)	+16.23±3.9 (n = 23)	+34.04±2.6** (n = 22)	+31.95±2.7** (n = 22)
SSS				
Patients with mild to moderate stroke (OSS > 40)	+9.52±1.0 (n = 32)	+15.80±1.2** (n = 30)	+21.09±0.7** (n = 31)	+19.94±1.2** (n = 31)
Patients with severe stroke (OSS ≤ 40)	+10.27±2.4 (n = 21)	+11.12±2.6 (n = 23)	+23.19±1.7** (n = 22)	+22.25±1.9** (n = 22)

b. On-treatment analysis of 200 included patients

Stroke scale	Placebo (n = 50)	Glycine		
		0.5 g/day (n = 49)	1.0 g/day (n = 51)	2.0 g/day (n = 50)
	At day 6			
OSS				
Patients with mild to moderate stroke (OSS > 40)	+14.28±1.1 (n = 31)	+14.58±1.4 (n = 29)	+22.28±1.3** (n = 30)	+17.09±1.2 (n = 30)
Patients with severe stroke (OSS ≤ 40)	+5.77±0.9 (n = 19)	+6.32±0.9 (n = 20)	+20.42±1.5** (n = 21)	+16.72±1.5** (n = 20)
SSS				
Patients with mild to moderate stroke (OSS > 40)	+5.15±0.7 (n = 31)	+5.28±0.6 (n = 29)	+7.31±0.5* (n = 30)	+6.91±0.7 (n = 30)
Patients with severe stroke (OSS ≤ 40)	+3.74±0.8 (n = 19)	+3.40±0.7 (n = 20)	+12.61±1.1** (n = 21)	+10.87±1.0** (n = 20) (continued)

Table 14.4 (*continued*)

Stroke scale	Placebo (n = 50)	Glycine		
		0.5 g/day (n = 49)	1.0 g/day (n = 51)	2.0 g/day (n = 50)
		At day 30		
OSS				
Patients with mild to moderate stroke (OSS > 40)	+18.92±1.9 (n = 31)	+24.77±2.8* (n = 29)	+33.97±1.7** (n = 30)	+27.61±2.1** (n = 30)
Patients with severe stroke (OSS ≤ 40)	+14.78±2.9 (n = 19)	+17.47±3.7 (n = 20)	+34.71±2.5** (n = 21)	+33.35±2.6** (n = 20)
SSS				
Patients with mild to moderate stroke (OSS > 40)	+9.70±0.9 (n = 31)	+16.00±1.1** (n = 29)	+21.29±0.6** (n = 30)	+20.10±1.1** (n = 30)
Patients with severe stroke (OSS ≤ 40)	+10.83±2.3 (n = 19)	+12.04±2.5 (n = 20)	+23.82±1.6** (n = 21)	+23.48±1.7** (n = 20)

Note: Patients who died during the 30-day study period were assigned the worst score of each of the neurological scales after their death. Positive values indicate improvement. * $p < 0.05$; ** $p < 0.01$; compared with placebo.

Table 14.5. Functional outcome on day 30
a. Intention-to-treat analysis of 212 identified patients

Functional outcome	Placebo (n = 53)	Glycine		
		0.5 g/day (n = 53)	1.0 g/day (n = 53)	2.0 g/day (n = 53)
All patients				
– death	8/53 (15.1%)	8/53 (15.1%)	3/53 (5.7%)	6/53 (11.3%)
– severe disability	10/53 (18.9%)	8/53 (15.1%)	1/53 (1.8%)**§	3/53 (5.7%)*§
– moderate disability	21/53 (39.6%)	20/53 (37.7%)	8/53 (15.1%)*§	11/53 (20.7%)
– mild or no disability	14/53 (26.4%)	17/53 (32.1%)	41/53 (77.4%)**§	33/53 (62.3%)
Patients with mild to moderate stroke (OSS > 40)				
– death	2/32 (6.3%)	2/30 (6.7%)	0/31 (0%)	1/31 (3.2%)
– severe disability	1/32 (3.1%)	1/30 (3.3%)	0/31 (0%)	0/31 (0%)
– moderate disability	17/32 (53.1%)	14/30 (46.7%)	3/31 (9.7%)**	5/31 (16.1%)*
– mild or no disability	12/32 (37.5%)	13/30 (43.3%)	28/31 (90.3%)*	25/31 (80.7%)
Patients with severe stroke (OSS ≤ 40)				
– death	6/21 (28.6%)	6/23 (26.1%)	3/22 (13.7%)	5/22 (22.7%)
– severe disability	9/21 (42.9%)	7/23 (30.4%)	1/22 (4.5%)**	3/22 (13.7%) (*continued*)

Table 14.5 (continued)

Functional outcome	Placebo (n = 53)	Glycine		
		0.5 g/day (n = 53)	1.0 g/day (n = 53)	2.0 g/day (n = 53)
– moderate disability	4/21 (19.0%)	6/23 (26.1%)	5/22 (22.7%)	6/22 (27.3%)
– mild or no disability	2/21 (9.5%)	4/23 (17.4%)	13/22 (59.1%)*	8/22 (36.3%)

b. On-treatment analysis of 200 included patients

Functional outcome	Placebo (n = 50)	Glycine		
		0.5 g/day (n = 49)	1.0 g/day (n = 51)	2.0 g/day (n = 50)
All patients				
– death	7/50 (14.0%)	7/49 (14.3%)	3/51 (5.9%)	5/50 (10.0%)
– severe disability	9/50 (18.0%)	6/49 (12.2%)	0/51 (0%)**§	2/50(4.0%)*§
– moderate disability	20/50 (40.0%)	19/49 (38.8%)	8/51 (15.7%)*§	10/50 (20.0%)
– mild or no disability	14/50 (28.0%)	17/49 (34.7%)	40/51 (78.4%)**§	33/50 (66.0%)
Patients with mild to moderate stroke (OSS > 40)				
– death	2/31 (6.5%)	2/29 (6.9%)	0/30 (0%)	1/30 (3.3%)
– severe disability	1/31 (3.2%)	1/29 (3.5%)	0/30 (0%)	0/30 (0%)
– moderate disability	16/31 (51.6%)	13/29 (44.8%)	3/30 (10.0%)**	4/30 (13.3%)*
– mild or no disability	12/31 (38.7%)	13/29 (44.8%)	27/30 (90.0%)*	25/30 (83.3%)
Patients with severe stroke (OSS ≤ 40)				
– death	5/19 (26.3%)	5/20 (25.0%)	3/21 (14.3%)	4/20 (20.0%)
– severe disability	8/19 (42.1%)	5/20 (25.0%)	0/21 (0%)**	2/20 (10.0%)
– moderate disability	4/19 (21.1%)	6/20 (30.0%)	5/21 (23.8%)	6/20 (30.0%)
– mild or no disability	2/19 (10.5%)	4/20 (20.0%)	13/21 (61.9%)*	8/20 (40.0%)

Note: Severe disability – Barthel score 0–45; moderate disability – Barthel score 50–70; mild or no disability – Barthel score 75–100.
* $p < 0.05$; ** $p < 0.01$, compared with the placebo group; § $p < 0.05$, compared with 0.5 g/day Glycine group.

Carried out within the first 6 h before starting the Glycine treatment, the immunoferent analysis of blood serum AB to glutamate NMDA receptors revealed increased levels of AB to NMDA-BP in all groups of patients with acute ischemic stroke in comparison with a control group ($p < 0.01$) (see Section 4.1.3). Initial levels of AB to glutamate NMDA receptors were significantly higher in patients with severe stroke. The predominance of increased levels in severe patients remained for 12 h after the stroke onset as compared to the group with mild to moderate stroke in the placebo and 0.5 g/day Glycine groups ($p = 0.01$ and $p = 0.01$, respectively). At the same time, first application of 1.0–2.0 g/day of Glycine induced an early normalization of AB titers to NMDA-BP in those patients ($p < 0.01$ versus placebo and $p < 0.01$ versus 0.5 g/day Glycine). Levels of AB titer in the 1.0–2.0 g/day Glycine group remained significantly lower during the period from 6 to 12 h ($p < 0.01$) in comparison with both the placebo and 0.5 g/day Glycine groups (Table 14.6; Fig. 14.1). In patients with mild to moderate stroke a tendency towards normalization of AB titers to NMDA-receptors was demonstrated in all groups of the study by 9–12 h after the stroke onset (see Table 14.6; Fig. 14.1).

Neurotransmitter amino acid analysis which was performed within the first 6 h after the stroke onset before starting the Glycine treatment demonstrated a significant increase in levels of excitatory amino acids (aspartate, glutamate) versus control levels ($p < 0.001$), this corresponding to previous investigations [67, 70, 71, 184, 185]. A tendency was revealed towards elevation of glycine levels versus control ($p > 0.05$). GABA concentrations in CSF of patients with mild to moderate ischemic stroke

Table 14.6. Dynamics of levels of serum AB to NMDA-BP (ng/ml) in patients with carotid ischemic stroke

Groups of patients	Placebo	Glycine	
		0.5 g/day	1.0–2.0 g/day
Mild to moderate stroke, n	31	29	60
on admission[1]	4.86±0.8[§]	3.51±0.9[§]	3.55±0.4[§]
6 h	3.95±0.7	3.45±0.7	2.30±0.8
9 h	2.54±0.3	2.74±0.8	1.85±0.7[†]
12 h	2.72±0.3	2.78±0.3	1.84±0.4
24 h	2.45±0.3	2.05±0.3	1.89±0.2
72 h	2.80±0.5	2.10±0.4	1.42±0.6
120 h	1.63±0.1	1.59±0.1	1.55±0.4
Severe stroke, n	19	20	41
on admission[1]	5.04±0.9[§*]	4.75±0.7[§*]	5.10±0.6[§*]
6 h	5.10±0.7*	5.05±0.8*	2.40±0.4[††#]
9 h	7.90±1.2*	6.41±1.1*	2.23±0.4[††#]
12 h	7.30±1.5*	4.95±0.6*	2.90±0.5[††#]
24 h	3.20±0.6	2.70±0.3	1.95±0.5
72 h	3.50±0.5	2.55±0.3	1.70±0.1
120 h	1.67±3	1.30±0.4	1.66±0.1

[1] Before the Glycine treatment.
[§] $p < 0.01$, versus control; * $p < 0.01$, ANOVA, severe stroke versus mild to moderate stroke; [†] $p < 0.05$, [††] $p < 0.01$, ANOVA, 1.0–2.0 g/day Glycine versus placebo; [#] $p < 0.01$, ANOVA, 1.0–2.0 g/day Glycine versus 0.5 g/day Glycine.

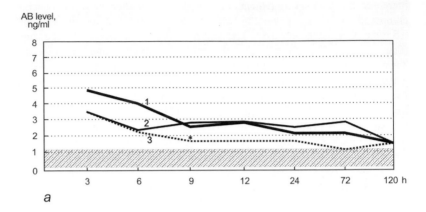

a

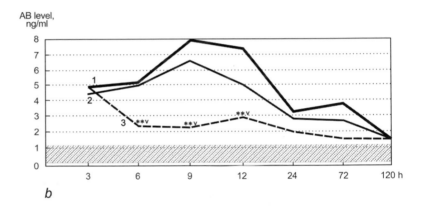

b

Figure 14.1. Dynamics of titers of AB to NMDA-BP in acute ischemic stroke: a) moderately severe stroke (OSS > 40); b) severe stroke (OSS < 39). 1 – placebo; 2 – Glycine 0.5 g/day; 3 – Glycine 1.0–2.0 g/day. Shaded area corresponds to normal range. Significance of the differences versus placebo group: * $p < 0.05$, ** $p < 0.01$. Significance of the differences versus patients treated with Glycine 0.5 g/day: $^v p < 0.01$.

were shown to be unchanged in comparison with the control group, while in patients with severe stroke GABA levels were significantly lower than control values ($p < 0.01$). There were no significant differences in amino acid levels within the first 6 h between all groups of patients with acute ischemic stroke (Table 14.7).

Table 14.7. Dynamics of CSF concentrations of amino acids (µM) in patients with carotid ischemic stroke

Group of patients	Placebo	Glycine	
		0.5 g/day	1.0–2.0 g/day
Mild to moderate stroke, *n*	31	29	60
Glutamate			
within the first 6 h	26.30±3.5**	25.50±4.35**	21.96±3.2**
day 3	16.66±1.84	17.88±1.7	14.89±1.16
Aspartate			
within the first 6 h	79.88±7.13**	82.88±7.73**	90.38±7.28**
day 3	98.33±1.73	80.63±8.03	66.90±5.03#§§
Glycine			
within the first 6 h	24.21±4.4	28.60±5.72	35.91±2.79
day 3	29.93±2.8	32.59±4.92	34.16±3.33
GABA			
within the first 6 h	0.87±0.49	0.97±0.87	0.68±0.29
day 3	1.75±0.87	1.65±0.78	1.94±0.68
Severe stroke, *n*	19	20	41
Glutamate			
within the first 6 h	33.86±5.17**	32.30±4.15**	35.84±3.47**
day 3	38.49±6.46	36.37±5.92	23.32±1.97#§§†
Apartate			
within the first 6 h	70.05±6.38**	65.25±5.78**	59.78±2.48**
day 3	54.98±4.88	56.48±5.03	41.18±3.08#§†
Glycine			
within the first 6 h	27.53±3.59	30.99±3.86	40.03±2.53
day 3	36.44±4.12	28.86±4.39	28.99±1.99§
GABA			
within the first 6 h	0.097±0.097*	0.097±0.097*	0.19±0.097*
day 3	0.097±0.097	0.29±0.097	1.94±0.49#§††

Note: Concentrations are given as means ± SD.
* $p < 0.01$, ** $p < 0.001$, Mann–Whitney *U* test, versus control;
# $p < 0.05$, ## $p < 0.01$, Mann–Whitney *U* test, 1.0–2.0 g/day Glycine versus placebo;
§ $p < 0.05$, §§ $p < 0.01$, Wilcoxon matched-pairs signed-rank test, day 3 versus within the first 6 h;
† $p < 0.05$, †† $p < 0.01$, Mann–Whitney *U* test, 1.0–2.0 g/day Glycine versus 0.5 g/day Glycine.

On day 3, no significant differences were demonstrated between the placebo and the 0.5 g/day Glycine groups with regard to excitatory and inhibitory amino acid levels in CSF. Application of Glycine in doses of 1.0–2.0 g/day was accompanied by statistically significant changes in amino acid concentrations (versus the placebo and 0.5 g/day Glycine groups) that were more pronounced in patients with severe stroke. There was a significant reduction in glutamate levels by 35% ($p < 0.05$ versus placebo, $p < 0.01$

versus day 1 and $p < 0.05$ versus 0.5 g/day of Glycine) and in aspartate levels by 31% ($p < 0.05$ versus placebo, $p < 0.05$ versus day 1 and $p < 0.05$ versus 0.5 g/day of Glycine) in CSF of patients with severe stroke, while GABA concentrations significantly increased by day 3 ($p < 0.01$ versus placebo, $p < 0.05$ versus day 1 and $p < 0.01$ versus 0.5 g/day of Glycine). In CSF of patients with mild to moderate stroke, application of Glycine in doses of 1.0–2.0 g/day was accompanied by analogical tendencies in changes in amino acid concentrations, but significance was obtained only for dynamics of aspartate levels ($p < 0.01$ versus placebo and $p < 0.01$ versus day 1; Table 14.7).

Determination of lipid peroxidation products revealed a slight tendency towards elevation of concentrations of TBARS ($p > 0.05$ versus control group) within the first 6 h in all groups of patients with acute ischemic stroke before starting the Glycine treatment, without a significant difference between the groups (see Section 4.3.1). In patients with mild to moderate stroke the tendency to continuous increase of TBARS by day 3 was demonstrated in the placebo group, while in both Glycine groups a reduction in concentrations of lipid peroxidation products was found that was significant in the 1.0–2.0 g/day Glycine group ($p < 0.05$ versus placebo). In patients with severe stroke the tendency to an increase in TBARS by day 3 was demonstrated in all groups of the study. However, in both Glycine groups the rate of this elevation was shown to be significantly lower as compared to the placebo group: by 19.58% in the 1.0–2.0 g/day Glycine group, by 45% in the 0.5 g/day Glycine group, and by 68.3% in the placebo group ($p < 0.05$, 1.0–2.0 g/day versus the placebo group) (Table 14.8).

Thus, the randomized, double-blind, placebo-controlled study confirmed the absence of side effects and adverse events of Glycine, which corresponded to results of previous investigations [102] and could be connected with properties of glycine as a natural product of brain metabolism.

Table 14.8. Dynamics of CSF level of lipid peroxidation products (TBARS, µM) in patients with carotid ischemic stroke

Groups of patients	Placebo	Glycine	
		0.5 g/day	1.0–2.0 g/day
Mild to moderate stroke, *n*	31	29	60
within the first 6 h	3.03±0.7	3.02±0.55	2.82±0.8
day 3	3.95±0.94	2.48±1.2	1.48±0.54[#]
Severe stroke, *n*	19	20	41
within the first 6 h	3.29±1.4	3.1±1.3	2.86±1.4
day 3	5.54±1.8	4.5±0.5	3.42±1.8[#]

[#] $p < 0.05$, Mann–Whitney U test, 1.0–2.0 g/day Glycine versus placebo.

The intention-to-treat and on-treatment analysis demonstrated a tendency to improvement of neurological recovery and functional outcome in patients with acute ischemic stroke treated with 1.0–2.0 g/day Glycine for 5 days after the event. A tendency to a decrease in mortality rate was also shown. The positive effects of Glycine were more pronounced in severely affected patients.

The early normalization of AB titer to NMDA-BP in the 1.0–2.0 g/day Glycine group also prevailed in severely affected patients and corresponded to an accelerated restoration of altered neurological functions, which possibly reflected the improvement of the functional state of glutamate NMDA-receptors under the influence of Glycine (see Section 4.1.3) [66–68].

Neurotransmitter amino acid analysis confirmed the effects of Glycine on glutamate and aspartate excitotoxicity. The Glycine treatment in doses 1.0–2.0 g/day induced not only a statistically significant reduction in glutamate and aspartate concentrations in CSF up to day 3 (versus placebo and versus day 1), but also a significant increase in GABA levels (versus placebo and versus day 1); thus, it provided evidence of a reduction of the imbalance between excitatory and inhibitory neurotransmitter systems. It is interesting that application of 1.0–2.0 g/day of Glycine in severely affected patients was accompanied by a statistically significant reduction of CSF concentration of glycine up to day 3 after the stroke onset. Perhaps the marked decrease of CSF glycine level under the influence of the drug Glycine reflected a tendency to normalization of amino acid metabolism in brain tissue (as a result of improved participation of glycine in anabolic processes of the cells).

As revealed in the 1.0–2.0 g/day Glycine group, the significant reduction in TBARS levels in patients with mild to moderate stroke and the deceleration of increase in concentrations of lipid peroxidation products in severely affected patients probably reflected not only neurotransmitter effects of the drug Glycine, but also its general metabolic properties.

In conclusion, the trial suggests that sublingual application of 1.0–2.0 g/day Glycine started within 6 h after the onset of acute carotid ischemic stroke is safe and can exert favorable clinical effects. These results will be tested in further trials with a larger number of patients.

14.5.2. Glycine effects on functional state of brain in patients with carotid ischemic stroke (neurophysiological monitoring)

The influence of Glycine on the functional state of supra-segmental motor, sensory, and nonspecific brain systems and the segmental peripheral neuromotor system was studied in 80 patients (45 males and 35 females;

average age 63.5 ± 9.7 years) in the acute period of ischemic stroke in the carotid artery territory (left in 41 cases and right in 39 cases). We performed neurophysiological monitoring of: EEG with mapping, somatosensory evoked potentials (SSEP) with mapping, transcranial electrical stimulation (TES), and electromyography (EMG). Clinical and baseline assessments corresponded to those given in Section 14.5.1. On admission, 50 (62%) patients had moderately severe disease and 30 (38%) had severe disease. Within the first 6 h after the stroke onset, 40 patients received Glycine and the other 40 received placebo. Taking into consideration the results of the study of dose-dependent efficacy of Glycine (see Section 14.5.1), the drug was administered at the optimal dose of 1 g/day. Patients were examined on admission (on the background of the first administration of the drug (Glycine or placebo) and during 4 h that followed) and then repeatedly on day 6 (at the end of the Glycine treatment).

EEG and mapping EEG monitoring were performed on the universal neurophysiological analyzer "Brain Surveyor" manufactured by "Saico" (Italy) with autonomic block of enhancers. The patient lay supine with eyes closed in an isolated darkened room. Baseline EEG signal was acquired according to the standard technique [224]. Background EEG was registered for 4–5 min before mapping, then during mapping and for 1.5 min after its termination. The bipolar and monopolar derivations with 16 electrodes placed on the scalp according to the International system ("10–20" Jasper, 1958) were used during EEG recording. Auricular homolateral electrodes were applied as reference. Mapping EEG was performed for 3–5 min with analysis epoch of 1 min long. Control values of mapping EEG indexes were obtained from 30 healthy volunteers.

SSEP were performed using "Basis" neurophysiological analyzers manufactured by "O.T.E. Biomedica" (Italy) and "Brain Surveyor" manufactured by "Saico" (Italy). Short latency SSEP were recorded by disc collodium and single-use superficial electrodes according to specialized program from standard software "Library 49" (manufactured by "O.T.E. Biomedica"). The active electrode of the first channel was placed contralaterally to the stimulation point, 2 cm behind cortical C3 point (by the International system "10–20"); the second active electrode was located above C7 *processus spinosus*; the third electrode was placed on the Erb's point, ipsilaterally to the stimulation point. Reference electrodes of the first and the second channels were located in Fz point (by the International system "10–20"), that of the third channel was placed on a shoulder ipsilateral to stimulation in the innervation area of C4-C5 spinal segments. The ground electrode was located in the upper third of the forearm contralateral to stimulation. The median nerve was stimulated with standard bipolar forked electrode placed 2 cm proximally to transversal ligament of wrist between tendons of *m. flexor carpi radialis* and *m. flexor digitorum superficialis.* For SSEP recording, 200–250 responses were averaged. Thus, on the first channel we registered thalamocortical complex (N20), on the second the nucleus complex of spinal posterior horn (N13), on the third the complex of median nerve and brachial plexus strands (N10) [40]. The difference between latent periods of N13 and N20 peaks (in msec) corresponded to the sensory central conduction time (sCCT) from segmental sensory structures of cervical enlargement of a spinal cord to contralateral thalamocortical complex. Average values of SSEP acquired in 30 healthy volunteers served as control.

Mapping SSEP was performed on a "Brain Surveyor" neurophysiological analyzer manufactured by "Saico" (Italy). The distributions of SSEP on the scalp diagram were studied in 16-channels recordings. The specialized program "Evoked potentials" from the standard software "Mapping Library" was used. Along with detection of latency values (in msec), peak amplitudes (in μV), inter-peak intervals (in msec), mapping SSEP allowed the study of peculiarities of distribution of amplitude values on scalp diagrams. Twenty healthy individuals were used as controls.

Transcutaneous transcranial electrical stimulation (TES) was performed for detection of functional state of fast conduction phasic strands of the corticospinal pathway. The study was carried out using a "Basis" neurophysiological analyzer manufactured by "O.T.E. Biomedica" (Italy) with special unit for TES. The most physiological monopolar TES technique was used. The active electrode (anode) was located corresponding to the cortical motor projection of a hand (7 cm laterally to Cz point and 1 cm in front of the line that joins Cz and *polus auriculus*). The reference cathode was placed pericranially. The extent of painful sensations in patients was decreased by creating low sub-electrode resistance (lower than 3 kΩ) and minimal duration of electrical stimuli (0.05 msec). Stimulation was performed by electrical rectangular wave current with 70–86 μA intensity. Central motor-evoked potential (CMP) was recorded by cutaneous bipolar electrode placed on *thenar* contralateral to stimulation, following perpendicular to projection of *m. abductor pollicis brevis* strands. CMP parameters were analyzed after averaging of 8–10 potentials. We took into account the minimal latency value and the maximal amplitude value of CMP. The comparison of latencies of CMP, M-response and F-wave obtained after stimulation of *n. medianus* in the same muscle (from the same recording electrode) allowed detection of motor central conduction time (mCCT), the time of conduction from pyramidal cells of motor cortex to peripheral motoneurons of the contralateral half of the corresponding spinal segment:

$$mCCT = CMP - (M + F - 1) / 2,$$

where CMP, M, and F are latency volumes of CMP, M-response, and F-wave; 1 msec is the time of central synaptic delay. Control volumes of all the parameters were obtained from 22 healthy volunteers.

Complex **EMG** study was performed on a "Basis" neurophysiological analyzer manufactured by "O.T.E. Biomedica" (Italy) by the standard technique [8]. We registered shape, latency, duration, and maximal amplitude of M-responses in hands and feet muscles after orthodromic stimulation of the corresponding peripheral nerve in a distal point. The action potential of a peripheral nerve was recorded by ring finger electrodes during antidromic stimulation of the same nerve trunk. We studied its amplitude, shape, and latency. Motor (mCV) and sensory (sCV) conduction velocities were measured for median, ulnar, tibial, and peroneal nerves of both sides by Hodes [80] and Mavor and Shiazawa [8] methods.

Glycine effects on brain spontaneous bioelectrical activity evidently manifested already 3–5 min after its first administration as an increase in power of α and β EEG frequency ranges in both hemispheres. Along with this, a trend towards decreasing of the initial (on admission to the clinic) inter-hemispheric asymmetry and to acceleration of frequency characteristics of main EEG rhythms were marked. Ten to thirty minutes after Glycine administration significant elevation of α-activity power in both occipital lobes, especially pronounced in the affected hemisphere, as well as moderate diffuse increase in power of slow activity (predominantly in θ-range) were registered. By the end of the first hour of monitoring, 36/40 (90%) patients exhibited pronounced normalization of EEG pattern: the energy level of slow activity decreased, α-index significantly increased not only in healthy, but also in the affected hemisphere, and representation of inter-hemispheric asymmetry dwindled in occipital lobes. In the placebo group EEG changes were of casual and varied character, no significant tendencies being registered during 4 h of monitoring.

In patients treated with Glycine statistical comparison of EEG maps on day 1 (before the drug administration) and on day 6 (at the end of the Glycine treatment) revealed significant decrease in power of slow frequency ranges ($p < 0.01$ for δ-activity, $p < 0.0001$ for θ-activity) practically even in all zones of both hemispheres (Plate 3). It corresponded to positive clinical dynamics such as a consciousness improvement, general activation of patients, and regress of signs of edema. It is noteworthy that in the left hemispheric lesion, these positive changes were more pronounced than in the right one, and their manifestation was more even throughout both hemispheres. In patients with right hemispheric infarction, decrease of power of slow frequency ranges predominated in the left ("intact") hemisphere.

The comparison of EEG maps between patients from the Glycine and placebo groups with identical stroke severity showed a significant difference in power of slow ranges on day 6 after the stroke onset, while their EEG maps were comparable on admission (Plate 4). In patients treated with Glycine power of δ- and θ-activity was less pronounced in both hemispheres, especially on the ischemized side ($p < 0.001$ and $p < 0.001$, respectively).

It was important that in the Glycine group 68% of patients (17/25) with moderately severe stroke and 33% patients (5/15) with severe stroke did not show the formation of a focus of slow δ- and θ-activity in the projection of ischemic lesion. The presence of such a focus on days 5–7 after the stroke onset appears to be the main EEG feature for focal morphological brain defect and steadfast functional deficit. That was typical for all patients with severe stroke and in 60% of those with moderately severe stroke from the placebo group.

Statistical analysis of the dynamics of α-activity on the background of Glycine treatment (by day 6 after the event) in patients with moderately severe stroke confirmed a tendency towards normalization of its zonal distribution and to decrease in the extent of inter-hemispheric asymmetry. In severe strokes, significant increase in power of α-range in the occipital lobes was registered and prevailed in the ischemized hemisphere ($p < 0.05$ for contralateral and $p < 0.01$ for ipsilateral hemispheres; Plate 5). The comparison of quantitative characteristics of α-range on day 6 after the stroke onset between the Glycine and placebo groups of comparable severity revealed significantly higher values of α-index ($p < 0.05$) in the affected hemisphere in patients treated with Glycine (Plate 6).

The study of SSEP and TES allowed the influence of Glycine on functional state of motor and sensory central conduction pathways to be objectified (Table 14.9). Already after the first Glycine administration, a tendency towards speeding up of motor central conduction (reduction of mCCT) was revealed on both sides: of ischemized and intact hemispheres. This acceleration was significant ($p < 0.001$ versus the initial level, before Glycine administration) in patients with severe stroke with higher values of mCCT at baseline (retardation of motor central conduction). By day 6, full normalization of mCCT was registered on the side of ischemia in patients treated with Glycine ($p < 0.05$ versus the initial level, before Glycine administration; $p < 0.01–0.001$ versus the placebo group) that allowed, taking into account our previous data [103, 182], to prognosticate good recovery of altered motor functions with the most complete improvement of paresis. At the same time, in the placebo group, mCCT asymmetry persisted and even worsened in patients with moderately severe illness due to retardation of conduction ($p < 0.01$) on the side of ischemized hemisphere. That gave evidence concerning steadfast morphological changes of fast conducting phasic strands of the corticospinal pathway [12, 103, 182].

The TES data have been confirmed by the clinical quantitative analysis of the dynamics of motor deficit in all studied patients. The motor status was assessed by the Original Stroke Scale [65]. In patients treated with Glycine more rapid improvement of motor function was found on day 30 ($p < 0.04$ versus placebo), the difference being more pronounced in patients with severe stroke ($p < 0.001$ versus placebo). Significant reverse correlation was revealed between the increase in the clinical motor score (that reflected improvement of motor function) by days 6 and 30 and the prolongation of mCCT by day 6 ($r = -0.47$; $p = 0.03$ and $r = -0.73$; $p = 0.007$, respectively).

Table 14.9. The dynamics of mCCT and sCCT (msec) in patients with carotid ischemic stroke

Characteristic	Stroke severity	Group of patients	Day 1				Day 6	
			on admission		after the first dose			
			affected side	"intact" side	affected side	"intact" side	affected side	"intact" side
mCCT, msec	Moderate	Glycine (n = 26)	5.73±0.57	5.30±0.38	5.53±0.24	5.16±0.48	5.25±0.49$^\#$	4.82±0.35
		Placebo (n = 24)	5.14±0.49	5.31±0.58	5.23±0.51	5.37±0.81	6.63±0.62*	4.16±0.41
	Severe	Glycine (n = 14)	8.49±0.87	4.24±0.53	5.76±0.91$^{\#***}$	3.36±0.36	5.04±0.39$^{\#\#*}$	4.29±0.62
		Placebo (n = 16)	7.38±1.16	3.80±0.27	7.60±0.97	3.75±0.41	8.76±1.03	3.45±0.70
sCCT, msec	Moderate	Glycine (n = 26)	5.72±0.59	5.51±0.50	5.52±0.40	6.65±0.38	5.47±0.26$^\#$	5.96±0.30
		Placebo (n = 24)	4.87=0.39	5.80±0.51	4.90±0.45	5.73±0.70	6.73±0.30**	5.84±0.61
	Severe	Glycine (n = 14)	5.36±0.10	5.46±0.54	5.83±0.57	5.89±0.82	6.37±0.28*	5.24±0.18
		Placebo (n = 16)	5.49±0.76	5.82±0.62	5.71±0.05	6.03±0.75	6.45±0.28*	5.89±0.50

$^\#$ $p < 0.01$, $^{\#\#}$ $p < 0.001$, compared with the placebo group;
* $p < 0.01$, ** $p < 0.001$, compared with day 1 (before treatment).

The influence of Glycine on the sensory central conduction was shown only in patients with moderately severe stroke and manifested as a prevention of "afferent retardation" phase (elongation of sCCT) on the side of ischemia. In the placebo group, "afferent retardation" phase developed on days 5–7 after the stroke onset ($p < 0.01$–0.001 versus the initial level, before Glycine administration; see Table 14.9). Positive dynamics of sCCT preceded the most complete clinical recovery of sensory functions [103, 182]. Along with this, in patients with severe stroke treated with Glycine, no differences in sCCT dynamics versus patients from the placebo group were revealed.

The EMG study of functional state of segmental peripheral neuromotor system showed no significant differences in values of mCV and sCV for median, ulnar, tibial, and peroneal nerves of both sides between patients from the Glycine and placebo groups. At the same time, in patients with moderately severe stroke, more complete normalization of conduction via motor ($p = 0.04$) and sensory ($p = 0.05$) peripheral strands was registered in the Glycine group versus the placebo group up to day 30 after the event. In patients with severe stroke, 1 h after the first Glycine administration statistically significant increase in mCV by 17.5% ($p = 0.05$) was registered for all studied peripheral nerves; however, the following dynamics of mCV and sCV did not differ from those in the comparable in severity placebo-treated group ($p > 0.05$).

Thus, the neurophysiological investigation objectified effects of the pharmaceutical drug Glycine on the functional state of brain in patients with carotid ischemic stroke. Analyzing the dynamics of spontaneous brain bioelectrical activity, we have revealed significantly more complete and accelerated normalization of EEG pattern by day 6 after the stroke onset in patients treated with Glycine. It manifested by the significant decrease in power of general and focal slow activity of δ- and θ-ranges. In some cases the complete averting of the development of δ- and θ-foci in the projection of ischemized brain area was demonstrated under Glycine treatment. Along with this, the increase in power of α-activity with normalization of its zonal distribution and vanishing of inter-hemispheric asymmetry was found in the Glycine group.

The study of TES showed the significant beneficial effect of Glycine on the motor central conduction that closely correlated with clinical regress of motor deficit and prognosticated good motor recovery. An action of Glycine on sensory central conduction (according to SSEP data) was not evident.

So, the normalization of neurophysiological findings, which have favorable prognostic significance [45, 182], confirms the positive influence of the pharmaceutical drug Glycine on functional recovery in patients with ischemic stroke.

The first positive results of application of Glycine as a pharmaceutical drug in acute ischemic stroke demands that large trials be arranged. This attempt to reduce neurotransmitter imbalance via activation of inhibitory systems appears to be one of the safest and potentially beneficial among therapeutic strategies aimed at decreasing glutamate excitotoxicity.

REFERENCES

1. Abel, M. S., McCandless, D. W., 1992, *J Neurochem.* **58**: 740–744.
2. Albers, G. W., Atkinson, R. P., Kelley, R. E., Rosenbaum, D. M., 1995, *Stroke.* **76**: 254–258.
3. Albers, G. W., Clark, W. M., Atkinson, R. P., *et al.*, 1999, *Stroke.* **30** (3): 508–513.
4. Albers, G. W., Saenz, R. E., Moses, J. A., Choi, D. W., 1991, *Stroke.* **22**: 1075–1077.
5. Allen, G. S., Ahn, H. S., Preziosi, T. J., *et al.*, 1983, *New Engl J Med.* **308**: 619–624.
6. Aronowski, J., Strong, R., Grotta, J. C., 1996, *Neuropharmacology.* **35**: 689–693.
7. Azcona, A., Lataste, X., 1990, *Drugs.* **40** (2): 52–57.
8. Badalyan, L. O., Skvortsov, I. A., 1986, *Clinical Electroneuromyography.* Meditsina, Moscow, 386 (in Russian).
9. Badikov, V. I., Gitel, E. P., Ivanova, N. Ya., *et al.*, 1990, *Bull Exp Biol Med.* **110** (9): 236–237 (in Russian).
10. Barnard, E. A., Darlison, M. G., Seebury, P., 1987, *Trends Neurosci.* **10**: 502–509.
11. Bath, P. M., Bath, F. J., Asplund, K., 2000, *Cochrane Database Syst Rev.* (2): CD000162.
12. Berardelli, A., Inghilleri, M., Cruccu, G., *et al.*, 1991, *Electroenceph Clin Neurophysiol.* **81** (5): 389–396.
13. Berridge, M. J., 1985, *Triangi.* **3** (4): 79–90.
14. Boast, C. A., Gergard, S. C., Pastor, G., *et al.*, 1988, *Brain Res.* **442**: 395–398.
15. Bordi, F., Pietra, C., Ziviani, L., Reggiani, A., 1997, *Exp Neurol.* **145**: 425–433.
16. Bridgers, S. L., Koch, G., Munera, C., *et al.*, 1991, *Stroke.* **22**: 153.
17. Buchan, A. M., 1990, *Cerebrovasc Brain Metab Rev.* **2**: 1–26.
18. Buchan, A. M., Xue, D., Slivka, A., *et al.*, 1989, *Neurosci.* **15**: 804.
19. Busa, W. B., Nuccitelli, R., 1984, *Am J Physiol.* **246**: 409–438.
20. Calabresi, P., Picconi, B., Saulle, E., *et al.*, 2000, *Stroke.* **31** (3): 766–772.
21. Chen, J., Graham, S. H., Moroni, F., Simon, R. P., 1993, *J Pharmacol Exp Ther.* **267**: 937–941.
22. Chizhmakov, I. V., Kishin, N. I., Krishtal, O. A., *et al.*, 1989, *Neurosci Lett.* **99** (1–2): 131–136.
23. Choi, D. W., 1990, *Cerebrovasc Brain Metab Rev.* **2**: 105–137.
24. Klein, R. C., Castellino, F. J., 2001, *Curr Drug Targets.* **2** (3): 323–329.
25. Coull, B. M., 1994, Randomized trial of CGS 19755, a glutamate antagonist, in acute ischaemic stroke treatment. In *Abst Am Acad Neurol Ann Meet.*
26. Culmsee, C., Junker, V., Wolz, P., *et al.*, 1998, *Eur J Pharmacol.* **342** (1): 93–201.
27. Curtis, D. R., Hosli, L., Johnston, G. A. R., Johnston, I. H., 1968, *Exp Brain Res.* **6**: 1–18.
28. Dannhardn, G., Kohl, B. K., 1998, *Curr Med Chem.* **5** (4): 253–263.
29. Davalos, A., Naveiro, J., Noya, M., 1996, *Stroke.* **27**: 1060–1065.
30. Davis, S. M., Lees, K. R., Albers, G. W., *et al.*, 2000, *Stroke.* **31**: 347–354.

31. Desbourdes, J. M., Ades, P. E., Giggiari, M., 1989, *Agressologie.* **30**: 438–440.
32. Diener, H. C., 1998, *Cerebrovasc Dis.* **8**: 172–181.
33. Diener, H. C., 1999, *Stroke.* **30**: 234.
34. Diener, H. C., Cortens, M., Ford, G., 2000, *Stroke.* **31** (11): 2543–2551.
35. Diener, H. C., Hacke, W., Hennerichi, M., *et al.*, 1996, *Stroke.* **27**: 76–81.
36. Dunwiddie, T. V., 1985, *Int Rev Neurobiol.* **27**: 63–139.
37. Dyker, A. G., Lees, K. R., 1999, *Stroke.* **30** (5): 986–992.
38. Dyker, A. G., Lees, K. R., 1999, *Stroke.* **30** (9): 1796–1801.
39. Edwards, K., and the CNS 1102-008 Study Group, 1996, *Neurology.* **46** (1): 424.
40. Elsen, A., Steward, J., Nudelman, K., *et al.*, 1979, *Neurology.* **29**: 827.
41. Erecinska, M., Nelson, D., Wilson, D. R., *et al.*, 1984, *Brain Res.* **304**: 19–23.
42. Fagg, G. E., Foster, A. C., 1983, *Neurosci.* **96**: 701–719.
43. Farber, L., Chien, K., Mittnacht, S., 1981, *Ann J Pathol.* **102**: 271–281.
44. Ferrarese, C., Appollonio, F., Frigo, M., *et al.*, 1989, *Neurology.* **39** (3): 443–445.
45. Fidler, S. M., 1993, *Clinical and Neurophysiological Study of Functional State of Brain in Acute Period of Hemispheric Ischemic Stroke.* Candidate's dissertation, Moscow, 263 (in Russian).
46. Fisher, M., 1994, *Cerebrovasc Dis.* **4**: 245.
47. Fisher, M., 1995, *Eur Neurol.* **35**: 3–7.
48. Fisher, M., 1997, *Stroke.* **28**: 866–872.
49. Fisher, M., 1999, *The Neuroscientist.* **6**: 392–401.
50. Fredholm, B. B., Dunwiddie, T. V., 1988, *Trends Pharmacol Sci.* **9**: 130–134.
51. Gandolfo, C., Sandercock, P., Conti, M., 2002, *Cochrane Database Syst Rev.* (1): CD001924.
52. Gavestinel produces no benefit for stroke patients, study finds, 2001. *Clin Res Manag.* **2** (11): 173–175.
53. Gelmers, H. J., 1984, *Acta Neurol Scand.* **69**: 232–239.
54. Gelmers, H. J., Gorter, K., De Weerdt, C. J., Wiezer, H. J. A., 1988, *New Engl J Med.* **318**: 203–207.
55. George, C. P., Goldberg, M. P., Choi, D. W., Steinberg, J. K., 1988, *Brain Res.* **440**: 375–379.
56. Gill, R., Hagreaves, R. J., Kemp, J. A., 1995, *J Cerebr Blood Flow Metab.* **15**: 197–204.
57. Gotoh, O., Mohamed, A., McCulloch, J., *et al.*, 1986, *J Cerebr Blood Flow Metab.* **6**: 321–331.
58. Gotti, B., Benavides, J., McKenzie, E. T., Scatton, B., 1990, *Brain Res.* **522**: 290–307.
59. Graham, S. H., Chen, J., Lan, J., *et al.*, 1994, *J Pharmacol Exp Ther.* **269**: 854–859.
60. Grander, C. V., Hamilton, B. R., Gresham, G. E., *et al.*, 1989, *Arch Phys Med Rehabil.* **70**: 100–103.
61. Green, A. R., Cross, A. J., Snape, M. F., De Souza, R. J., 1992, *Neurosci Lett.* **138**: 141–144.
62. Greenberg, J. H., Uematsu, D., Araki, N., *et al.*, 1990, *Stroke.* **21** (4): 72–77.
63. Grotta, J., 1994, *Stroke.* **25**: 255.
64. Grotta, J., 1997, *Stroke.* **28**: 2338–2346.
65. Gusev, E. I., Skvortsova, V. I., 1990, *Case Record Form for the Examination and Treatment of Patients with Ischemic Stroke*, Moscow, pp. 1–44 (in Russian).
66. Gusev, E. I., Skvortsova, V. I., Dambinova, S. A., *et al.*, 1996, *Cerebrovasc Dis.* **6** (2): 44.
67. Gusev, E. I., Skvortsova, V. I., Dambinova, S. A., *et al.*, 2000, *Cerebrovasc Dis.* **10** (1): 49–60.

68. Gusev, E. I., Skvortsova, V. I., Izykenova, G. A., *et al.*, 1996, *Korsakoff J Neurol Psychiatr.* **5**: 68–72 (in Russian).
69. Gusev, E. I., Skvortsova, V. I., Kovalenko, A. V., Sokolov, M. A., 1999, *Korsakoff J Neurol Psychiatr.* **2**: 65–70 (in Russian).
70. Gusev, E. I., Skvortsova, V. I., Raevsky, K. S., *et al.*, 1995, Detection of neurotransmitter amino acid content in CSF of patients with acute ischemic stroke. In *Functional Studies as a Basis for Drug Elaboration*, Moscow, pp. 133–134 (in Russian).
71. Gusev, E. I., Skvortsova, V. I., Raevsky, K. S., *et al.*, 1997, *Eur J Neurol.* **4** (1): 78.
72. Haag, P., Schneider, T., Schabitz, W., Hacke, W., 2000, *J Neurol Sci.* **175** (1): 52–56.
73. Hacke, W., Lees, K. R., Timmerhuis, T., *et al.*, 1998, *Cerebrovasc Dis.* **8** (5): 247–254.
74. Hammerstad, J. F., Murray, J. E., Culter, R. W. P., 1971, *Brain Res.* **35**: 357–367.
75. Hantson, L., Wessel, T., 1998, *Stroke.* **29**: 287.
76. Harrison, N. L., Simmonds, M. A., 1985, *Br J Pharmacol.* **84**: 381–391.
77. Hasegawa, Y., Fisher, M., Baron, B. M., Metcalf, G., 1994, *Stroke.* **25**: 1241–1246.
78. Heiss, W. D., Holthoff, V., Pawlik, G., Neveling, M., 1990, *J Cerebr Blood Flow Metab.* **10**: 127–132.
79. Henderson, G., Johnson, J., Ascher, P., 1990, *J Physiol.* **430**: 189–212.
80. Hodes, R., Larrabee, M. G., German, W., 1948, *Arch Neurol Psychiatr.* **60**: 340–365.
81. Hopkin, J., Neal, M. J., 1971, *Br J Pharmacol.* **42**: 215–223.
82. Horn, J., Haas, R., Vermuelen, M., Limburg, M., 1999, *Stroke.* **30**: 242.
83. Hossman, K. A., 1996, *Cerebrovasc Brain Metab Rev.* **8**: 195–208.
84. Izumi, Y., Roussel, S., Pinard, E., Seylaz, J., 1991, *J Cerebr Blood Flow Metab.* **11**: 1025–1030.
85. James, T. A., Starr, M. S., 1979, *Eur J Pharmacol.* **57**: 115–125.
86. Johnson, J. W., Ascher, P., 1987, *Nature.* **325**: 529–531.
87. Jones, M. G., Szatkowski, M. S., 1995, *Neurosci Lett.* **201** (3): 227–230.
88. Kawaguchi, K., Graham, S. H., 1997, *Brain Res.* **749** (1): 131–134.
89. Kawaguchi, K., Henshall, D. C., Simon, R. P., 1999, *Eur J Pharmacol.* **364** (2-3): 99–105.
90. Kawasaki-Yatsugi, S., Ichiki, C., Yatsugi, S., 1998, *Naunyn Schmiedebergs Arch Pharmacol.* **358** (5): 586–591.
91. Kawasaki-Yatsugi, S., Ichiki, C., Yatsugi, S., 2000, *Neuropharmacology.* **39** (2): 211–217.
92. Kawasaki-Yatsugi, S., Shimizu-Sasamata, M., Yatsugi, S., Yamaguchi, T., 1998, *J Pharm Pharmacol.* **50** (8): 891–898.
93. Keinanen, K., Wisden, W., Sommer, B., *et al.*, 1990, *Science.* **299**: 556–560.
94. Kelly, J. S., Krnjevic, K., Morris, M. E., Yim, G. K. W., 1969, *Exp Brain Res.* **7**: 11–31.
95. Kemp, J. A., Leeson, P. D., 1993, *Trends Pharmacol Sci.* **14**: 20–25.
96. Kinoshita, Y., Ueyama, T., Senba, E., *et al.*, 2001, *J Neurotrauma.* **18** (4): 435–445.
97. Kleckner, N. W., Dingledine, R., 1988, *Science.* **24**: 835.
98. Klein, R. C., Castellino, F. J., 2001, *Curr Drug Targets.* **2** (3): 323–329.
99. Kochhar, A., Zivin, J. A., Lyden, P. D., Mazarella, V., 1988, *Arch Neurol.* **45**: 148–153.
100. Koek, W., Woods, J. H., Winger, G. D., 1988, *J Pharmacol Exp Ther.* **245**: 969–974.
101. Koh, J., Golgberg, M. P., Hartley, D. M., Choi, D. W., 1990, *J Neurosci.* **10**: 693–705.
102. Komissarova, I. A., Gudkova, J. A., Soldatenkova, T. A., *et al.*, 1992, Medical drug with anti-stress, stress-protecting and nootropic properties. Patent of Russian Federation. *Invent Discov*, 2025124 (in Russian).

103. Kovalenko, A. V., 1992, *Functional State of Efferent and Afferent Brain Pathways in Acute Ischemic Stroke on the Background of Neuroprotective Therapy.* Candidate's dissertation, Moscow, 205 (in Russian).
104. Lampl, Y., Gilad, R., Geva, D., *et al.*, 2001, *Clin Neuropharmacol.* **24** (1): 11–15.
105. Lapin, I. P., 1981, *Eur J Pharmacol.* **71**: 495–498.
106. Lataste, X., Maurer, W., Whitehead, J., *et al.*, 1992, Application of sequential methods of clinical trial in stroke: The Asclepios Study. *2nd World Congress of Stroke*, 16.
107. Lavretskaya, Ye. F., 1985, *Pharmacological Regulation of Mental Processes*, Moscow (in Russian).
108. Leach, M. J., Swan, J. H., Eisenthal, D., *et al.*, 1993, *Stroke.* **24**: 1063–1067.
109. Lees, K. R., 1998, *Lancet.* **351**: 1447–1448.
110. Lees, K. R., Asplund, K., Carolei, A., Davis, S. M., *et al.*, 2000, *Lancet.* **355** (9219): 1949–1954.
111. Lees, K. R., Lavelle, J. F., Hobbinger, S. F., 1998, *Cerebrovasc Dis.* **8**: 20.
112. Lekieffre, D., Benavides, J., Scatton, B., Nowicki, J. P., 1997, *Brain Res.* **776** (1–2): 88–95.
113. Lesage, A. S., de Loore, K. L., Osikowska-Evers, B., *et al.*, 1994, *Soc Neurosci Abst.* **20**: 185.
114. Leung, L. W. S., Desborough, K. A., 1988, *Brain Res.* **463**: 148–152.
115. Limburg, M., 1996, *Stroke.* **27**: 172.
116. Limburg, M., Hijdra, A., 1990, *Eur Neurol.* **30**: 121–122.
117. Lyden, P. D., 1997, GABA and neuroprotection. In *Neuroprotective Agents and Cerebral Ischemia* (Green, A. R., Cross, A. J., eds.) Academic Press, London, pp. 233–258.
118. Mabe, H., Nagai, H., Takagi, T., *et al.*, 1986, *Stroke.* **17**: 501–505.
119. Mahoney, F. I., Barthel, D. W., 1965, *Md. State Med J.* **14**: 61–65.
120. Mayor, F., Valdivieso, F., Ugar, M., 1991, *J Amino Acid Neurotransmit Adv Biochem Psychopharmacol.* **29**: 551–560.
121. McDonald, J. W., Silverstein, F. S., Johnston, M. V., 1990, *Neurosci Lett.* **109**: 234–238.
122. Meadows, M. E., Fisher, M., Minematsu, K., 1994, *Cerebrovasc Dis.* **4**: 26–31.
123. Meden, P., Overgaard, K., Sereghy, T., Boysen, G., 1993, *J Neurosci.* **119**: 209–216.
124. Meister, A., 1957, *Biochemistry of the Amino Acids.* New York.
125. Meldrum, B. S., 1989, Excitotoxicity in ischemia: an overview. In *Cerebrovasc Dis – 16th Res Conf*, Raven Press, New York, pp. 47–60.
126. Minematsu, K., Fisher, M., Li, L., *et al.*, 1993, *Neurology.* **43**: 397–403.
127. Mennini, T., Mancini, L., Reggiani, A., Trist, D., 1997, *Eur J Pharmacol.* **336** (2–3): 275–281.
128. Mohr, J. P., Orgogozo, J. M., Harrison, M. J. G., *et al.*, 1994, *Cerebrovasc Dis.* **4**: 197–203.
129. Muir, K. W., Hamilton, S. J., Lunnon, M. W., *et al.*, 1998, *Cerebrovasc Dis.* **8** (1): 31–37.
130. Muir, K. W., Holzapfel, L., Lees K. R., 2000, *Cerebrovasc Dis.* **10** (6): 431–436.
131. Muir, K. W., Lees, K. R., 1994, Initial experience with remacemide hydrochloride in patients with acute ischemic stroke. *2nd Int Conf Neuroprotect Agents.* New York.
132. Muir, K. W., Lees, K. R., 1995, *Ann NY Acad Sci.* **765**: 315–316.
133. Muir, K. W., Lees, K. R., 1995, *Stroke.* **26**: 503–513.
134. Muir, K. W., Lees, K. R., 1995, *Stroke.* **26**: 1183–1188.
135. Muir, K. W., Lees, K. R., 1996, *Cerebrovasc Dis.* **6**: 75–383.

136. Muir, K. W., Lees, K. R., 1998, *Stroke.* **29** (5): 918–923.
137. Norris, J. W., LeBrun, L. H., Anderson, B. A., 1994, *Cerebrovasc Dis.* **4**: 194–196.
138. Nowak, L., Bregestovski, P., Ascher, P., 1984, *Nature.* **307**: 462–465.
139. Obrenovitch, T. P., Hardy, A. M., Urenjak, J., 1997, *Brain Res.* **746** (1–2): 190–194.
140. Obrenovitch, T. P., Urenjak, J., Zilkha, E., 1994, *Br J Pharmacol.* **113** (4): 1295–1302.
141. Oczkowski, W. J., Hachinski, V., Bogousslavsky, J., *et al.*, 1989, *Stroke.* **20**: 604–608.
142. Ohman, J., Heiskanen, O., 1988, *J Neurosurg.* **69**: 683–688.
143. Olney, J. W., Labruyere, J., Price, M. T., 1989, *Science.* **244**: 1360–1362.
144. Onal, M. Z., Fisher, M., 1996, *Drugs Today.* **32** (7): 573–592.
145. Onal, M. Z., Fisher, M., 1997, *Eur Neurol.* **38**: 141–154.
146. O'Neill, M., Allain, H., Bentue-Ferrer, D., *et al.*, 1995, *Eur Neurol.* **35** (1): 28–36.
147. Orgogozo, J. M., Dartigues, J. F., 1986, In *Acute Brain Ischemia. Medical and Surgical Therapy* (Battistini, N., ed.) Raven Press, New York, pp. 282–289.
148. Osborne, R. H., Bradford, H. F., Jones, D. G., 1973, *J Neurochir.* **21**: 407–419.
149. Ozyurt, E., Graham, D. I., Woodruff, G. N., McColloch, J., 1988, *J Cerebr Blood Flow Metab.* **8**: 138–143.
150. Park, C. K., Nehls, D. G., Graham, D. I., *et al.*, 1988, *Ann Neurol.* **24**: 543–551.
151. Park, C. K., Nehls, D. G., Graham, D. I., *et al.*, 1988, *J Cerebr Blood Flow Metab.* **8**: 757–762.
152. Park, C. K., Nehls, D. G., Teasdale, G. M., McCulloch, J., 1989, *J Cerebr Blood Flow Metab.* **9**: 617–622.
153. Park, C. K., Rudolphi, K. A., 1994, *Neurosci Lett.* **178** (2): 235–238.
154. Patel, J., Zinkand, W. C., Thompson, C., *et al.*, 1990, *J Neurochem.* **54**: 849–854.
155. Peterson, S. L. 1991, *Eur J Pharmacol.* **199**: 341–348.
156. Peterson, S. L., Boehnke, L. E., 1989, *Exp Neurol.* **104**: 113–117.
157. Peterson, S. L., Schwade, N. D., 1993, *Epilepsy Res.* **15**: 141–148.
158. Petruk, K. C., West, M., Mohr, G., *et al.*, 1988, *J Neurosurg.* **68**: 505–517.
159. Philippon, J., Grob, R., Dagreou, F., *et al.*, 1986, *Acta Neurochir (Wien).* **82**: 110–114.
160 Pickard, J. D., Murray, G. D., Illingworth, R., *et al.*, 1989, *Br Med J.* **298**: 636–642.
161. Plaschke, K., Grant, M., Weigand, M. A., *et al.*, 2001, *Br J Pharmacol.* **133** (1): 107–116.
162. Poignet, H., Nowicky, J. P., Scatton, B., 1992, *Brain Res.* **596**: 320–324.
163. Prange, H., Hartung, J., Hertel, G., *et al.*, 1991, Treatment of acute stroke with flunarizine i.v. *Int Conf on Stroke*, Geneva, Switzerland, 39.
164. Rayevsky, K. S., Georgiev, V. P., 1986, *Mediator Amino Acids: Neuropharmacological and Neurochemical Aspects.* Moscow, 240 (in Russian).
165. Rayevsky, K. S., Romanova, G. A., Kudrin, V. S., *et al.*, 1997, *Bull Exp Biol Med.* **123** (4): 370–373 (in Russian).
166. Reggiani, A., Pietra, C., Arban, R., *et al.*, 2001, *Eur J Pharmacol.* **419** (2–3): 147–153.
167. Romanova, G. A., Kudrin, V. S., Malikova, L. A., 1995, *Pharmacol Res.* **31**: 128.
168. Sacco, R. L., DeRosa, J. T., Haley, E. C., 2001, *JAMA.* **285** (13): 1719–1728.
169. Sauter, A., Wiederhold, K. H., Rudin, M., 1990, *Stroke.* **21** (1): 1–158.
170. *Scandinavian Stroke Study Group Stroke*, 1985. **16**: 885–890.
171. Schanne, F. A., Young, E. E., Farber, J. L., *et al.*, 1979, *Science.* **206**: 700–702.
172. Scheller, D., Kolb, J., Szathmary, S., *et al.*, 1995, *J Cerebr Blood Flow Metab.* **15** (1): 379.
173. Scolnick, P., Marvizon, J., Jackson, B., *et al.*, 1989, *Life Sci.* **45**: 1647–1655.
174. Sergeyev, P. V., Shimanovsky, N. D., 1987, *Receptors to Metabolic Active Compounds.* Meditsina, Moscow, 560 (in Russian).

175. Sergeyev, P. V., Shimanovsky, N. D., Petrov, V. I., 1999, *Receptors*. Sem' Vetrov, Moscow-Volgograd, 640 (in Russian).
176. Shuaib, A., Hasan, S., Kalra, J., 1992, *Brain Res.* **590**: 13–17.
177. Siesjo, B. K., 1981, *J Cerebr Blood Flow Metab.* **1**: 155–185.
178. Siesjo, B. K., 1986, *Eur Neurol.* **25** (1): 45–56.
179. Sills, M. A., Loo, P. S., 1989, *Mol Pharmacol.* **36**: 160–165.
180. Silver, B., Weber, J., Fisher, M., 1995, *Clin Neuropharmacol.* **19** (2): 101–128.
181. Simi, A., Ingelman-Sundberg, M., Tindberg, N., 2000, *J Cerebr Blood Flow Metab.* **20** (7): 1077–1088.
182. Skvortsova, V. I., 1993, *Clinical and Neurophysiological Monitoring and Neuroprotective Therapy in Acute Ischemic Stroke*. Doctoral dissertation. Moscow, 379 (in Russian).
183. Skvortsova, V. I., Gusev, E. I., Komissarova, I. A., *et al.*, 1995, *Korsakoff J Neurol Psychiatr.* **1**: 11–19 (in Russian).
184. Skvortsova, V. I., Rayevsky, K. S., Kovalenko, A. V., *et al.*, 1999, *Korsakoff J Neurol Psychiatr.* **2**: 34–39 (in Russian).
185. Skvortsova, V. I., Raevskii, K. S., Kovalenko, A. V., *et al.*, 2000, *Neurosci Behav Physiol.* **30** (5): 491–495.
186. Smith, S. E., Hodges, H., Sowinski, P., *et al.*, 1997, *Neuroscience.* **77** (4): 1123–1135.
187. Smith, S. E., Lekieffre, D., Sowinski, P., Meldrum, B. S., 1993, *Neuroreport.* **4**: 1339–1342.
188. Snape, M. F., Baldwin, H. A., Cross, A. J., Green, A. R., 1993, *Neuroscience.* **53**: 837–844.
189. Stark, L., Peterson, S. L., Albertson, T. E., 1990, In *Kindling 4* (Wada, J. A., ed.) Raven Press, New York, pp. 267–281.
190. Steinberg, G. K., George C. P., DeLaPaz, R., *et al.*, 1988, *Stroke.* **19**: 1112–1118.
191. Steinberg, G. K., Saleh, J., Kunis, D., 1988, *Neurosci Lett.* **89**: 193–197.
192. Steinberg, G. K., Saleh, J., Kunis, D., *et al.*, 1989, *Stroke.* **20**: 1247–1252.
193. Strand, T., Wester, P. O., *et al.*, 1993, A double blind randomized pilot trial of magnesium therapy in acute cerebral infarction. *7th Nordic Meet on Cerebrovasc Dis.* 37.
194. Sydserff, S. G., Cross, A. J., Green, A. R., 1995, *Neurodegeneration.* **4**: 323–328.
195. Synthelabo Recherche, 1996, Press Release, Paris, February 6.
196. Takano, K., Tatlisumak, T., Formato, J., *et al.*, 1997, *Stroke.* **28**: 1255–1262.
197. Tatlisumak, T., Takano, K., Meiler, M. R., Fisher, M., 1998, *Stroke.* **29**: 190–195.
198. Tatlisumak, T., Takano, K., Meiler, M. R., Fisher, M., 2000, *Acta Neurochir.* **76**: 331–333.
199. The American Nimodipine Study Group, 1992, *Stroke.* **23**: 3–8.
200. The Trust Study Group, 1990, *Lancet.* **336**: 1205–1209.
201. Tietjen, G. E., Dombi, T., Pulsinelli, W. A., *et al.*, 1996, *Neurology.* **46** (1): 424.
202. Toth, E., Lajtha, A., Sarhan, S., *et al.*, 1983, *Neurochem Res.* **8**: 291–302.
203. Tranquillini, M. E., Reggiani, A., 1999, *Expert Opin Investig Drugs.* **8** (11): 1837–1848.
204. Tricklebank, M. D., Saywell, K., 1990, *Soc Neurosci Abst.* **16**: 462.
205. Tridgett, R., Foster, A. C., 1988, *Br J Pharmacol.* **85**: 890.
206. Tsuda, T., Kogure, K., Nishioka, K., Watanabe, T., 1991, *Neuroscience.* **44**: 335–341.
207. Turcani, P., Tureani, M., 2001, *J Neurol Sci.* **183** (1): 57–60.
208. Turski, L., Huth, A., Sheardown, M., *et al.*, 1998, *Proc Natl Acad Sci USA.* **95**: 10960–10965.

209. Uematsu, D., Greenberg, J. H., Hickey, W. F., Reivich, M., 1989, *Stroke.* **20**: 1531–1537.
210. Van den Kerckhoff, W., Rewers, L. R., 1985, *J Cerebr Blood Flow Metab.* 459–460.
211. Wahlgren, N. G., 1995, Cytoprotective therapy for acute ischaemic stroke. In *Stroke Therapy* (Fisher, M., ed.) Butterworth & Heinemann, Boston, pp. 315–350.
212. Wahlgren, N. G., 1997, A review of earlier clinical studies on neuroprotective agents and current approaches. In *Neuroprotective Agents and Cerebral Ischemia* (Green, A. R., Cross, A. J., eds.) Academic Press, pp. 337–363.
213. Wahlgren, N. G., 1997, Clomethiazole Acute Stroke Study Collaborative Group. *Cerebrovasc Dis.* **7** (4): 19.
214. Wahlgren, N. G., 1998, Clomethiazole Acute Stroke Study Collaborative Group. *Cerebrovasc Dis.* **8**: 20.
215. Wahlgren, N. G., MacMahon, D., De Keyser, J., *et al.*, 1994, *Cerebrovasc Dis.* **4**: 204–210.
216. Wahlgren, N. G., Rawasinha, K. W., Rosolacci, T., *et al.*, 1999, *Stroke.* **30**: 21–28.
217. Warkins, J. C., Olverman, H. J., 1987, *Trends Neurosci.* **10**: 265–272.
218. Warner, D. S., Martin, H., Ludwig, P., *et al.*, 1995, *J Cerebr Blood Flow Metab.* **15**: 188–196.
219. Wiernsperger, N., Gygax, P., Hofmann, A., 1984, *Stroke.* **15**: 679–685.
220. Williams, K., Romano, C., Dichter, M. A., Molinoff, P. B., 1991, *Life Sci.* **48**: 469–498.
221. Williams, R., 1963, In *Biogenesis of Natural Compounds* (Dernfeld, R., ed.) New York, pp. 368–404.
222. Wong, M. L., Bongiorno, P. B., Giold, P. W., Licinio, J., 1995, *Neuroimmunomodulation.* **2**: 141–148.
223. Yang, Y., Li, Q., Ahmad, F., Shuaib, A., 2000, *Neurosci Lett.* **285** (2): 119–122.
224. Zenkov, L. R., Ronkin, M. A., 1982, *Functional Diagnostics of Nervous Diseases.* Meditsina, Moscow, 432 (in Russian).

Chapter 15

Secondary Neuroprotection

Taking into consideration the modern concept of delayed morphological completion of infarction and the possibility of a certain part of the penumbral brain tissue to survive for at least 48–72 h from the stroke onset, great importance is attached to elaboration of the secondary neuroprotection approaches aimed at averting delayed neuronal death connected with oxidant stress reactions, local inflammation, trophic dysfunction, and apoptosis (Table 15.1). These processes take part not only in the "up-formation" of brain infarction, but also in long-term reconstitution of the neuro–immune–endocrine system, and promote both progression of atherogenesis and chronic brain tissue damage for months after the stroke (see Chapters 6–10). Secondary neuroprotection can be started relatively later, 6–12 h after the vascular event, and should be most intensive within the first 7 days of the illness. It is important to mention not only its therapeutic, but also preventive value. Correction of delayed consequences of ischemia can lead to retardation of vascular encephalopathy development in the post-stroke period.

Table 15.1. Main directions in secondary neuroprotection

Direction	Basic drug groups	Representatives	Present state of study
Antioxidants	Free radical scavengers	Tirilazad mesylate (U-74006F) Phenyl-*t*-butyl nitrone (PBN)	Efficacy is not proven, **X** Preclinical studies
	NO-synthase blockers	7-Nitroindazole 1-(2-Fluoromethyl-phenyl)-imidazole Aminoguanidines	Preclinical studies Preclinical studies Preclinical studies *(continued)*

Table 15.1 (continued)

Direction	Basic drug groups	Representatives	Present state of study
	Selen-organic compound of complex antioxidant action	Ebselen	Studies are in progress
Inhibitors of local inflammation	Antibodies to intercellular adhesion molecules (anti-ICAM)	Enlimomab Human anibodies to leukocyte integrins, CD11/CD18	Severe adverse effects, **X** Preclinical studies, start of clinical studies
	Pro-inflammatory cytokine inhibitors	Endogenous antagonists of TNF and IL-1β receptors Zinc protoporphyrin (ZnPP)	Preclinical studies Preclinical studies
	Endogenous anti-inflammatory cytokines	TGF-β_1 IL-10	Preclinical studies Preclinical studies
Statins	Statins	3-Hydroxy-3-methylglutaryl coenzyme A (HMG-CoA) reductase inhibitors	Preclinical studies, start of clinical studies
Estrogens	Estrogens	Estrogens	Preclinical studies, start of clinical studies
Trophic factors	Neurotrophic factors	Basic fibroblastic growth factor (bFGF) Brain-derived neurotrophic factor (BDNF) Insulin-dependent growth factor (IGF) Osteogenic protein-1 (OP-1)	Studies are in progress Preclinical studies Preclinical studies Preclinical studies
Neuromodulators	Neuropeptides	Semax (ACTH 4–10) Cerebrolysin NAP (NAPVSIPQ)	Studies are in progress Studies are in progress Preclinical studies, start of clinical studies
Regulators of receptor structures	Gangliosides	GM1	Studies are in progress

Note: **X** – trials are stopped.

15.1.   Antioxidants

Antioxidant therapy is an important direction of secondary neuroprotection. Already in the 1960s a wide range of antioxidants, different in chemical structure and mechanism of action, was known; the necessity of their differentiated administration was proven. In the 80s it was shown that in the earliest period of focal brain ischemia it is expedient to use "scavengers" for free radicals and drugs that destroy hydroperoxides (with sulfide and thiol groups): 2,3-dimercaptopropane-sulfone (Unithiol, BAL, Antoxol, Dimercaprol, Dicaptol, Dithioglycerin), sodium thiosulfate, etc. For a later period administration of tocopherols and carotenoids was recommended, as these drugs bind catalysts and inactivate singlet oxygen [57]. However, attempts to use 2,3-dimercaptopropane-sulfone, sodium thiosulfate, and tocopherols in the complex therapy of acute ischemic stroke showed that the "contribution" of these drugs to the final result of treatment is negligible [88, 89, 95, 224, 225].

It has now been shown that such endogenous brain antioxidants as superoxide dismutase (SOD), glutathione peroxidase, catalase, lasaroids, and iron chelates are important determinants in the defense mechanisms against lesion formation after ischemia [132, 175, 185, 265]. Neuroprotective properties of the drugs that bind free radicals were demonstrated in rat models of focal brain ischemia caused by permanent occlusion of the middle cerebral artery [132, 272]. These antioxidants significantly decreased the infarction volume, which substantiated the need to try them in clinic.

Tirilazad mesylate (U-74006F) is a 21-aminosteroid, acting as a free radical scavenger and an inhibitor of lipid peroxidation. The drug causes statistically significant reduction of the infarction size in rats and rabbits with permanent and transient focal brain ischemia [184, 215, 272]. The neurological outcome after transient and permanent middle cerebral artery occlusion in rats and rabbits treated with tirilazad was also better, and the ratio of ischemic neurons to total neurons in the cortex and the subcortex was lower than in control [184, 215]. The differences versus control were especially pronounced in cases of combination of tirilazad with magnesium or mannitol, which provided better neuroprotection from the effects of focal ischemia than did therapy with tirilazad or magnesium alone.

Randomized trial of tirilazad (the STIPAS study) administered in three daily doses (0.6, 2, or 6 mg/kg) in 111 patients with ischemic stroke showed its good safety profile. The drug was well tolerated at all used doses including 6 mg/kg [239]. In the pilot study of tirilazad efficacy in stroke [129], mortality decrease was registered as well as increase in the quota of patients with good functional recovery. At the same time, the results of the randomized, double-blind, placebo-controlled trial of tirilazad, that enrolled

556 patients from 27 centers in the USA, failed to confirm a significant neuroprotective effect of the drug in stroke patients [195, 208, 239]. In this trial tirilazad was administered within the first 6 h of the stroke as 150 mg bolus, then 1.5 mg/kg 4 times daily for 3 days. The number of favorable outcomes (evaluated by the Glazgo Outcome scale) comprised 61.9 versus 68.4% in the placebo group ($p > 0.05$). Good functional recovery (assessed by the Barthel Index) was registered in 62.8 and 70.1%, respectively; $p > 0.05$). A special investigation studied the influence of tirilazad on infarction volume [259]. Only in the subgroups of male patients and of those with a cortical infarct, tirilazad significantly reduced infarction volume. These effects were reduced to non-significant trends after adjustment for imbalances in baseline characteristics. The conclusion was made that tirilazad had no significant action on infarction volume. Attempts to increase tirilazad dose were unsuccessful due to side effects that appeared [97, 149, 186].

At present, six clinical trials assessing the safety and efficacy of tirilazad in 1757 patients with acute ischemic stroke have been completed [50]. A systematic review of all these randomized, controlled trials showed that tirilazad did not alter early case fatality (odds ratio (OR) 1.11, 95% confidence interval (CI) 0.79 to 1.56) or end-of-trial case fatality (OR 1.12, 95% CI 0.88 to 1.44). A just-significant increase in death and disability, assessed as either the expanded Barthel Index (OR 1.23, 95% CI 1.01 to 1.51) or Glasgow Outcome Scale (OR 1.23, 95% CI 1.01 to 1.50) was observed. Tirilazad significantly increased the rate of infusion site phlebitis (OR 2.81, 95% CI 2.14 to 3.69). Functional outcome (expanded Barthel Index) was significantly worse in pre-specified subgroups of patients: females (OR 1.46, 95% CI 1.08 to 1.98) and subjects receiving low-dose tirilazad (OR 1.31, 95% CI 1.03 to 1.67); a non-significantly worse outcome was also seen in patients with mild to moderate stroke (OR 1.40, 95% CI 0.99 to 1.98) [248]. Thus, it was demonstrated that tirilazad mesylate increased death and disability by about one fifth when given to patients with acute ischemic stroke. Although further trials of tirilazad are now unwarranted, analysis of individual patient data from the trials may help elucidate why tirilazad appears to worsen outcome in acute ischemic stroke.

The mechanisms of the neuroprotective activity of the *alpha-phenyl-N-tert-butyl nitrone* (*PBN*) have been investigated extensively. PBN has been shown to be a free radical scavenger forming stable nitroxides after interaction with radicals [67]. It was found that PBN inhibits inducible NO-synthase (iNOS) gene induction and prevents the formation of its neurotoxic products in rat stroke models and significantly suppresses protein p38 activation induced by IL-1β and by H_2O_2 in rat primary astrocytes. It was also revealed that PBN decreases the amount of reactive oxygen species produced in mitochondrial respiration [67]. It is important that PBN protects brain tissue in experimental stroke even if administered several hours after the event, decreasing cytotoxic edema and extent of neuronal damage in rat brain [31, 218, 279]. Chronic low-level PBN administration to old experimental animals reversed their age-enhanced susceptibility to stroke even several days after the last dose.

Nitric oxide and peroxynitrite release can be inhibited by neuronal (n) or inducible (i) NOS blockers [66]. The *selective nNOS blockers* 7-nitroindazole and 1-(2-trifluoromethylphenyl)-imidazole significantly reduced the infarction volume after focal and global cerebral ischemia in animals [61, 180]. Blockade of iNOS can relatively selectively be achieved by *aminoguanidines* [39, 66]. According to Iadecola *et al.* [110], aminoguanidines possess protective properties even if administered 24 h after the stroke onset, clearly presenting interest in viewing the possibility of their clinical application.

Ebselen, 2-phenyl-1,2-benzisoselenazol-3(2H)-one (PZ 51, DR3305), is a seleno-organic compound which has glutathione peroxidase-like activity and also reacts with peroxynitrite and can inhibit enzymes such as lipoxygenases, NO synthases, NADPH oxidase, protein kinase C, and H^+/K^+-ATPase. Numerous *in vitro* experiments using isolated liposomes and microsomes, as well as isolated cells and organs have established that ebselen protects against oxidative stress and inflammatory reactions [191]. Its antioxidant and anti-inflammatory properties have been proven in a variety of *in vivo* models [191, 200]. Ebselen decreased lipid peroxidation of membrane-associated phospholipids and lipoxygenase in the arachidonate cascade, blocked the production of superoxide anion by activated leukocytes, inhibited iNO-synthase, and protected against effects of peroxynitrite [66]. Experimental studies in rats and dogs have revealed that ebselen is neuroprotective after transient and permanent focal brain ischemia, able to inhibit both vasospasm and tissue damage, which correlates with its inhibitory effects on oxidative processes [48, 191, 245]. Two hours after transient focal ischemia induced in rats, intravenous ebselen significantly reduced the volume of gray matter damage in the cerebral hemisphere (by 53.6% compared with vehicle, $p < 0.02$), axonal damage by 46.8% ($p < 0.002$) and oligodendrocyte pathology by 60.9% ($p < 0.005$). At the same time, the neurological deficit score was decreased by 40.7% ($p < 0.05$) [112].

Ebselen was also shown to diminish cytochrome *c* release at 12 and 24 h after induction of transient middle cerebral artery occlusion in mice. It was accompanied by the decrease of both DNA fragmentation (determined by terminal deoxynucleotidyl transferase-mediated DNA nick-end labeling – TUNEL) and brain damage volume at 3 days after ischemia. Furthermore, ebselen increased the number of neuronal nuclei of immunopositive cells at

21 days after ischemia. All these facts demonstrated that ebselen attenuated neuronal apoptosis by inhibiting of the mitochondrial pathway of its induction connected with cytochrome *c* release [179].

Herin *et al.* [103] revealed that ebselen readily reversed dithiothreitol potentiation of NMDA-mediated currents in cultured neurons and in Chinese hamster ovary cells expressing wild-type NMDA NR1/NR2B receptors; therefore, the drug could act as a modulator of the NMDA receptor redox modulatory site.

Unlike many inorganic and aliphatic selenium compounds, ebselen has low toxicity as metabolism of the compound does not liberate the selenium moiety, which remains within the ring structure. Subsequent metabolism involves methylation, glucuronidation, and hydroxylation [191].

A clinical pilot study that enrolled 300 patients showed that ebselen was safe and well tolerated and significantly improved functional outcome even when given up to 48 h of the stroke [191, 274]. Oral administration of the drug (150 mg) or placebo was started immediately after admission and was continued for 2 weeks. Ebselen treatment achieved a significantly better outcome than placebo at 1 month ($p = 0.023$, Wilcoxon rank sum test) but not at 3 months ($p = 0.056$, Wilcoxon rank sum test). The improvement was significant in patients who started ebselen treatment within 24 h of stroke but not in those who started treatment after 24 h. There was a corresponding improvement in the modified Mathew Scale and modified Barthel Index scores. It was concluded that early treatment with ebselen improved stroke outcome. An additional randomized, double-blind, placebo-controlled trial of ebselen was conducted in patients with complete occlusion of the middle cerebral artery [182]. Ebselen or placebo granules suspended in water (150 mg) were orally administered within 12 h of onset and continued for 2 weeks. Although the intent-to-treat analysis of 99 patients (43 given ebselen and 56 given placebo) did not reach statistical significance in reduction of the infarction volume ($p = 0.099$), the protocol-compatible analysis of 83 patients with complete occlusion of the middle cerebral artery (34 given ebselen and 49 given placebo) showed a significant reduction using ebselen treatment ($p = 0.034$). A good outcome was seen in approximately 15% more patients from the ebselen group, but the difference between the two groups was not significant. There was a corresponding significant reduction in the volume of the brain infarction and an improvement in the outcome of patients who started treatment within 6 h of onset. These findings suggest that ebselen can protect the brain from ischemic damage in the acute stage. Thus, ebselen appears to be a promising neuroprotective agent. Further trials are now in progress.

15.2. Anti-intercellular adhesion molecule (anti-ICAM) antibodies

Another strategy of the secondary neuroprotection is directed to inhibition of inflammatory processes in the ischemized zone. Taking into consideration the role of leukocytes in the development of local inflammation and microvascular–cellular cascade reactions underlying the reperfusion damage of brain tissue [65, 186] (see Section 8.2), it is interesting to study efficacy of antibodies to intercellular adhesion molecules (anti-ICAM). Expression of ICAM-1 on vascular endothelium enhances adhesion of leukocytes to it [54, 229, 238]. Adhered neutrophils are can activate free radicals, proteases, and toxic pro-oxidant metabolites, causing additional damage to brain tissue [104]. Mechanical obstruction of small arteries may cause the phenomenon of "no-reflow" [151]. Experimental studies [27, 282, 283] showed that antibodies to ICAM-1 prevent activation of leukocytes and microcirculation disturbances, decrease the infarction size, and improve outcome of transient, but not long-term focal brain ischemia.

The randomized, placebo-controlled clinical trial of the pharmaceutical drug *Enlimomab* (murine monoclonal antibodies to ICAM-1) was performed in 625 patients with ischemic stroke [59, 221, 247]. The treatment started from the first 6 h after the stroke onset and was given intravenously in doses 160 mg on day 1 and 40 mg on the next 4 days. At day 90, the Modified Rankin Scale score was worse in patients treated with Enlimomab than with placebo ($p = 0.004$). Fewer patients had symptom-free recovery on Enlimomab than placebo ($p = 0.004$), and more died (22.2 versus 16.2%). The negative effect of the drug was apparent on days 5, 30, and 90 of treatment ($p = 0.005$). There were significantly more adverse events with Enlimomab treatment than placebo, primarily infections, pneumonias, and fever. Probably, such adverse effects were connected with complement-mediated reactions induced by the drug. Thus, it was shown that anti-ICAM therapy with Enlimomab is not an effective treatment for ischemic stroke and may significantly worsen stroke outcome.

Despite the negative results of the Enlimomab Acute Stroke Trial, much attention is paid to studies of therapeutic effects of humanized antibodies directed against the leukocyte integrins, CD11/CD18 [65]. These antibodies were shown to decrease the infarction volume during a transient, but not a permanent focal brain ischemia, like Enlimomab [34, 38, 69]. Clinical trials of the drug have been started [74].

15.3.   Pro-inflammatory cytokine inhibitors; anti-inflammatory factors

As mentioned in Chapter 8, pro-inflammatory cytokine IL-1β is a basic trigger of local inflammatory reaction in the ischemic area. Besides, IL-1β mediates excitotoxicity via glutamate NMDA receptors, promotes NO-synthase generation, and activates superficial adhesion molecules [118, 271, 275]. In experimental models of short-term focal brain ischemia, excessive expression of endogenous antagonists of IL-1β receptors or administration of IL-1β inhibitor, zinc protoporphyrin (ZnPP), led to a reduction of both infarction area and cytotoxic edema zone [19, 120]. Identical results were obtained in experimental inhibition of other pro-inflammatory cytokines [47, 83, 190, 203]. The suppression of TNF and its receptor, TNF-R1, in mice 30 min after induction of ischemic stroke blocked stroke-related damage at two levels, the primary ischemic and the secondary inflammatory injury [167], which was accompanied by a marked decrease in both infarction volume and mortality. At present the elaboration of techniques of pro-inflammatory cytokine blockade is going on the pre-clinical level, and studies of dose- and time-dependent effects of their antagonists are being carried out on different animal models of focal brain ischemia [65].

At the same time, experimental studies have demonstrated that the progression of ischemic brain damage might occur not only due to elevation of pro-inflammatory cytokine level, but also due to insufficiency of anti-inflammatory and neurotrophic factors [14, 28, 33, 70, 115, 173, 178]. It was found that preliminarily administered anti-inflammatory cytokines, such as IL-10 and transforming growth factor β₁ (TGF-β₁), significantly inhibited production of IL-1β, IL-6, IL-8, and TNF-α [33], adhesion of leukocytes to endothelium [89], as well as the synthesis of superoxide anion and production of intermediate reactive products of oxygen metabolism and NO metabolites in macrophages [53, 102, 253]. IL-10, administered to rats after middle cerebral artery occlusion, significantly ($p < 0.01$) reduced infarction size by 20.7-40.3% compared to vehicle [235]. Neuroprotective effects of TGF-β₁ are connected not only with its anti-inflammatory action, but also with stabilization of calcium homeostasis [203] and an increase in *bcl-2* gene expression, which leads to significant inhibition of free radical damage and to suspension of necrosis and apoptosis [107, 122]. In recent experiments [285], it was shown that   TGF-β₁ suppressed expression of pro-apoptotic protein Bad under ischemic conditions induced by transient middle cerebral artery occlusion, increased Bad phosphorylation, and activated the mitogen-activated protein kinase (MAPK)/Erk pathway (see Chapter 10), which may contribute to its neuroprotective activity. The activation of MAPK was demonstrated to be necessary for the antiapoptotic effect of TGF-β₁ [285].

The accordance of *in vivo* and *in vitro* data suggests a potential neuroprotective role of IL-10 and TGF-β1 against brain ischemia in a human when administered exogenously or made available from endogenous sources [79].

15.4. Statins and estrogens

In studies of recent years neuroprotective properties of statins (3-hydroxy-3-methylglutaryl coenzyme A (HMG-CoA) reductase inhibitors) and estrogens widely used in clinical practice were revealed [58, 252]. These drugs were shown to decrease brain tissue sensibility to focal ischemia and frequency of ischemic strokes in patients with ischemic heart disease.

Several large clinical trials have demonstrated that *statins* reduce serum cholesterol levels and the incidence of cardiovascular diseases. However, overlap and meta-analyses of these clinical trials suggest that the beneficial effects of statins may extend beyond their effects on serum cholesterol levels [52, 56]. Because statins also inhibit the synthesis of isoprenoid intermediates in the cholesterol biosynthetic pathway, they may have pleiotropic effects on the vascular wall. Their ability to decrease the incidence of ischemic stroke highlights some of their non-cholesterol effects since serum cholesterol levels are poorly correlated with the risk for ischemic stroke [52, 152, 261]. Statins both upregulate endothelial nitric oxide synthase (eNOS) and inhibit inducible nitric oxide synthase (iNOS). The application of statins in experimental animals with induced focal brain ischemia improved cerebral blood flow via upregulation of eNOS and reduced the infarction size [58, 260]. According to Pahan *et al.* [189], statins also inhibited the cytokine-mediated (IL-1β, TNF-α) upregulation of iNOS and NO production in rat astrocytes and macrophages. Along with that, statins suppressed oxidation of lipoproteins and decreased the activity of free radical processes in brain tissue [37, 108].

Laufs *et al.* [144] found that the novel HMG-CoA reductase inhibitor rosuvastatin dose-dependently up-regulated eNOS expression and activity and protected from focal cerebral ischemia in mice. After 2-h middle cerebral artery occlusion, this statin caused reduction of the infarction volume by 27, 56, and 50% (for 0.2, 2, and 20 mg/kg, respectively). At the same time, serum cholesterol and triglyceride levels were not significantly lowered by the treatment. So, it was confirmed that the effects of the statin were independent of changes in cholesterol levels.

The precise molecular mechanisms underlying the neuroprotective effect of *estrogens* remain obscure. The ways through which estrogens increase both metabolism and blood flow in certain regions of the brain became only recently better understood. It has been shown that these hormones increase secretion of neuromediators, stimulate formation of new synapses, and can activate certain genes responsible for production of anti-apoptotic proteins and growth factors [210]. Estrogens dilate cerebral vessels, acting through increased synthesis of nitric oxide and by stimulating such compounds as prostacyclin and a potent vasodilator—epoxyeikosatrienoic acid. So, during brain ischemia the physiological stimulation by estrogen of both brain metabolism and cerebral blood flow becomes biased towards increased release of vasodilating substances. Other brain protective properties of estrogens are related to attenuation of the excitotoxic effects of glutamate and to activation of enzymes scavenging free oxygen radicals. Moreover, estrogens can diminish free radicals synthesis and act as free radicals scavengers themselves. There are experimental data about estrogen activation of Bcl-2 protein synthesis that reflects inhibition of apoptotic mechanisms in the ischemic region [210].

Several studies have provided evidence to suggest that estrogens result in a significant reduction (approximately 50%) in the size of ischemic zone in the middle cerebral artery occlusion model of stroke in rats and mice [43, 212, 251]. It was found that pre- or post-stroke estrogen administration prevented or reversed acute stroke-induced autonomic dysfunction in rats and that endogenous estrogen levels in males could contribute to this neuroprotection [212]. The volume of brain infarction in the female rats after the occlusion of both the common carotid arteries and one of middle cerebral arteries (1.5 h) followed by reperfusion (24 h) was inversely correlated with circulating estrogen levels. More severe post-ischemic changes (increased neutrophil accumulation, elevated antioxidant enzyme, and lactate dehydrogenase activities) and injury accompanied by the decline in circulating estrogen levels in normal cycling female rats, indicating that estrogen is probably the major hormonal player in female resistance to ischemia [153].

Because of these results, research is now aimed at determining the optimal doses of statins and estrogens to be applied in acute ischemic stroke as well as the time window and treatment regimens providing their maximal efficacy.

15.5. Neurotrophic factors

Great importance in the development of ischemic brain damage is trophic support insufficiency (see Chapters 9 and 12), the level of which determines the choice between apoptotic or anti-apoptotic defense programs "switching on" as well as influencing necrosis and reparation processes [13, 96, 141, 266].

The synthesis of trophic factors and their receptors is a natural protective reaction of brain which occurs from the first minutes of ischemia. When the expression of genes encoding growth factors is active and fast, brain ischemia may not cause infarction formation for a long time [246]. In a case when the ischemic damage has developed, high level of trophic factors may provide neurological improvement, even if a permanent morphological lesion that has caused this deficit remains [256]. Thus, the other important direction of the secondary neuroprotection is the clinical application of drugs which can elevate the trophic potential of the brain.

Growth factors, especially neurotrophins such as NGF and BDNF (see Chapter 9), are potential therapeutic agents that could be used in the treatment of acute stroke because they possess neuroprotective, reparative, and proliferative properties, should these proteins be made transportable through the blood–brain barrier (BBB) *in vivo*. In the 1970s the protective influence of NGF on neuronal survival, energy metabolism, and protein synthesis in brain tissue was shown under ischemic conditions [126]. However, large size of NGF polypeptide molecule does not allow its permeation through the BBB, this imposing limitations on its therapeutic application. Later, neuroprotective properties of other trophic factors were also revealed. Being injected into hippocampal structures a few minutes before glutamate application, basic fibroblastic growth factor (bFGF) multiply decreased the number of dead neurons and inhibited calcium influx into cells [158]. Intravenous application of BDNF [214, 276, 277], bFGF [64, 241], TGF-β_1 [83], insulin-dependent growth factor (IGF) [158], and osteogenic protein-1 (OP-1) [155] showed protective effects in animal (rat, mouse, and cat) models of focal brain ischemia with significant reduction in the infarction size. A special study was devoted to disclosing potential mechanisms of infarction volume reduction by bFGF. It was shown that this trophic factor decreases DNA fragmentation and prevents reduction of immunoreactivity of Bcl-2 in the ischemic hemisphere [13]. So, a possible mechanism of bFGF neuroprotective action appears to be the inhibition of apoptotic process in the ischemic brain.

Along with neuroprotective action, bFGF and OP-1 exhibited prominent regeneration and proliferative properties [12, 130, 131]. Being administered intravenously 24 h after the stroke onset, they did not reduce the infarction volume but significantly improved the functional outcome (recovery of sensorimotor function of the impaired limbs), presumably due to enhancement of new neuronal sprouting and synapse formation in the intact brain tissue.

One approach to the BBB problem is to attach the non-transportable peptide to a brain targeting vector, which is a peptide or peptidomimetic monoclonal antibody that is transported into brain from blood via an endogenous BBB transport system [234, 284]. Conjugates of BDNF and the OX26 monoclonal antibody to the transferrin receptor, as well as conjugates of bFGF and the OX26-SA, were used in rats subjected to 24 h of permanent middle cerebral artery occlusion. A single intravenous injection of the conjugate BDNF/OX26, equivalent to a dose of 50 μg//kg of the BDNF, decreased the infarction volume by 65%. A single intravenous injection of the conjugate bio-bFGF/OX26-SA, equivalent to a dose of 25 μg/kg bFGF, produced an 80% reduction in the infarction volume in parallel with a significant improvement of neurological deficit. The neuroprotection was time-dependent: there was lower (but still significant) reduction in the infarction volume if the conjugates are administered at 60 min after arterial occlusion, whereas no significant reduction in the infarction volume was observed if treatment was delayed 2 h. So, this study demonstrates that significant positive effect can be achieved with the noninvasive intravenous administration of trophic factors, such as BDNF and bFGF, providing the peptides are conjugated to a BBB drug targeting system [234, 284].

The other therapeutic perspective is connected with creation of encapsulated neurotrophic factor-secreting cell grafting for ischemic injury [68]. Fujiwara *et al.* [68] established bFGF-secreting cell line by genetic manipulation, enveloped these cells into polymer capsules, which consist of a semipermeable membrane, and implanted them into the striatum of rats. At 6 days after implantation, these rats received homolateral to implantation middle cerebral artery occlusion using interluminal suture technique. At 24 h after occlusion, approximately 30% reduction in the infarction volume was revealed in the encapsulated bFGF-secreting cell grafting groups versus the encapsulated naive BHK cell grafting group or the without implantation group. It is important that the retrieved capsules continued to secrete bFGF. There was no significant difference in bFGF secretion between the capsules before and after transplantation. A large number of viable BHK-bFGF cells were observed within the full length of the retrieved capsule. These results indicate that encapsulated bFGF-secreting cell grafting exerts a protective effect on ischemic injury can be considered as a potential therapeutic approach.

The Phase II clinical trial of bFGF has demonstrated a good tolerability and safety profile of the drug in humans; the neuroprotective and reparative effects were also confirmed [62].

15.6.   Neuropeptides

Tight interaction between all delayed consequences of ischemia as well as between their trigger mechanisms allows them to be influenced not only separately, but also to use common regulators controlling and modulating expression of second messengers, cytokines, and other signaling molecules as well as triggering genetic programs of apoptosis, anti-apoptotic defense and neurotrophic support enhancement. Such modulatory effects alleviate general disintegration in the complex and often multi-directed molecular biochemical mechanisms, recovering their normal balance. Neuropeptides play an important role in these events as endogenous regulators of CNS functions. Their molecules being short amino acid chains are "chopped" out of larger protein precursors by proteolytic enzymes (so-called processing) only "in suitable place and time" in relation to demands of the human body [51]. Neuropeptides exist for only seconds, but the duration of their effects may be measured in hours [10]. Endogenous formation of a neuropeptide in response to any environmental change leads to release of various other neuropeptides induced by the primary one. If their action were similarly directed, their effect would be summarized and long-lasting. The release of a peptide may be regulated by several regulatory peptides earlier in the cascade. Thus, the effector sequence of peptide totality forms a so-called peptide regulatory continuum, each of the regulatory peptides of which is able to induce or inhibit the release of a number of other peptides. As a result, primary effects of one or another peptide may develop in time as chain reactions and cascade-like processes [10].

Neuropeptide structures are peculiar in that they have many ligand-binding groups determined to different cellular receptors. This is a "molecular explanation" of their polyfunctional mode of action. Physiological activity of neuropeptides many times exceeds effects of non-peptide compounds. In relation to place of their release, neuropeptides may embody mediator function (signal transduction from one cell to another), modulate reactivity of certain neuronal pools, stimulate or inhibit release of hormones, regulate tissue metabolism or fulfill functions of physiologically active effectors (vasomotor, natriuretic, and other modes of regulation). It is known that neuropeptides can regulate activity of pro- and anti-inflammatory cytokines via modulation of their receptor activity. And the recovery of cytokine imbalance in this case is conducted more effectively than those produced when influences are directed onto separate cytokine systems. As usual, effects of neuropeptides on cytokine systems are accompanied by their influence on nitric oxide generation and other oxidative events. Many neuropeptides exhibit important neurotrophic and growth properties. Taking into consideration that neuropeptides easily permeate through the BBB (in contrast to polypeptide chains of growth factors), it is hard to overestimate their potential therapeutic importance.

In recent years much attention has been given to study of neuropeptides that are structurally connected with adrenocorticotropic hormone (ACTH). It was shown that ACTH-like peptides influence synaptic membrane viscosity [106], modulate receptor functions, influence protein phosphorylation, inhibit microglial activation and excessive synthesis of neurotoxic cytokines and ligands to NMDA receptors, and possess intrinsic neurotrophic and regenerative effects [105, 109, 168].

Semax is the synthetic analog of ACTH 4–10, consisting of the amino acid sequence Met-Glu-His-Phe-Pro-Gly-Pro (MEHFPGP), which is identical to the amino acid fragment of α-melanocyte-stimulating hormone (α-MSH). It deserves special attention. It was synthesized under the guidance of Prof. I. P. Ashmarin in the Institute of Molecular Genetics of the Russian Academy of Sciences and Lomonosov Moscow State University [10, 198]. In experiments conducted on brain ischemia models as well as under clinical conditions, neuroprotective effects of Semax manifesting in its influence on practically all mechanisms of delayed neuronal death (such as microglial activation and cytokine imbalance, inflammation, NO synthesis, oxidant stress, trophic dysfunction) were revealed [8, 93, 177, 227, 286].

15.6.1.   Semax (ACTH 4–10)

Semax is the first Russian non-depleting nootropic drug that belongs to the neuropeptide group. It has several noteworthy advantages over its known analogs (Table 15.2): total absence of hormonal activity, toxic and side effects, 24-fold longer action than of its natural analog, and penetration through BBB 4 min after intranasal administration [201, 202]. The half-life of Semax in the human body is only several minutes long, but its therapeutic action after single administration lasts for 20–24 h [202]. The prolonged action of Semax is connected with its consecutive degradation. Its fragments EHFPGP (Glu-His-Phe-Pro-Gly-Pro) and HFPGP (His-Phe-Pro-Gly-Pro) are also stable neuropeptides that independently modulate cholinergic neurotransmission and NO generation and thus impart the majority of the effects of Semax [8, 9] (Fig. 15.1).

Table 15.2. Semax and its nearest analogs

Semax (ACTH 4–10)	– Met-Glu-His-Phe-Pro-Gly-Pro
ACTH 4–7	– Met-Glu-His-Phe
ORG 2766	– (O_2)Met-Glu-His-Phe-dKPhe
Ebiratid	– (O_2)Met-Glu-His-Phe- dKPhe -NH-$(CH_2)_8$-NH_2

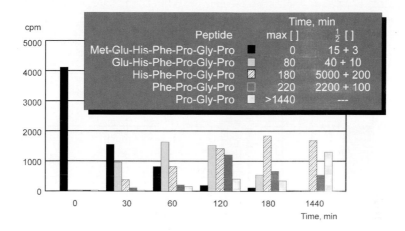

Figure 15.1. Sequential degradation of Semax [9].

Recent investigations have shown that Semax specifically binds plasma membranes of subcortex nuclei in rats with the dissociation constant $(2.41 \pm 1.02) \times 10^{-9}$ M and $B_{max} = (33.5 \pm 7.9) \times 10^{-15}$ M/mg protein [80]. At present, it is thought that Semax acts through an effector mechanism and induces intracellular processes via binding some receptors located on plasma membranes of brain cells. The preliminary experimental data show that Semax receptors appear to be similar to some subtypes of melanocortin receptors such as MC_3, MC_4, and MC_5. Taking into consideration that α-MSH is a pro-opiomelanocortin derivate which shares the first 13 amino acid sequence with ACTH [111], such possibility does not seem unlikely.

Strong trophic action of Semax has been shown on cholinergic neurons both in complete and glucose- or oxygen-lacking medium. Survival of neurons became 2-fold higher ($p < 0.05$) when Semax was added in doses of 100 nM and 10 μM, this being comparable with the same effect of NGF [54, 81] (Fig. 15.2). The specificity of Semax action on cholinergic neurons of subcortical nuclei was confirmed by the experiment where the absence of positive effect of the drug on GABAergic neurons and granular cerebellum cell survival was demonstrated [2]. This direct effect on cholinergic neurons was accompanied by significant increase in acetylcholinesterase activity in specific brain structures. This usually correlates with memory and learning improvement [2]. Semax has been shown to influence acetylcholinesterase activity due to transcription of a gene encoding one of the isoforms of the enzyme [6].

Surviving
neurons, %

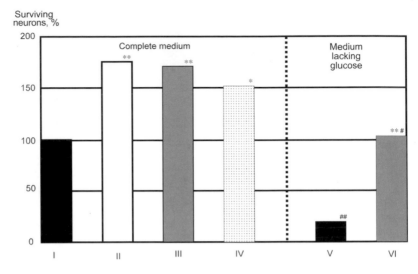

Figure 15.2. Influence of Semax on survival of cholinergic neurons in rat brain [81]. In complete medium: I – control, II – NGF, III – Semax (100 ng/ml), IV – Semax (10 µg/ml). In medium lacking glucose: V – control, VI – Semax (100 ng/ml). Significance of the differences versus controls: * $p < 0.05$, ** $p < 0.01$. Significance of the differences versus complete medium: [#] $p < 0.05$, [##] $p < 0.01$.

A study conducted with microelectrode techniques demonstrated that Semax influences the conductivity of an electrotonic synapse between two identified neurons [121]. Two-fold increase in synaptic transmission coefficient was registered 14–30 min after administration of the neuropeptide in a dose of 200 nM and lasted for 5–6 h.

Even very low doses of Semax (3–30 µg/kg) demonstrated prominent nootropic effect, increased brain adaptation abilities, enhanced its resistance to stress as well as to hypoxia and ischemia in experimental animal models. Higher doses of the drug (150–300 µg/kg), which are not toxic and retain nootropic properties of the small doses, showed additionally pronounced antioxidative, antihypoxic, angioprotective, and neurotrophic effects [9, 124, 125, 264].

Together with our colleagues from Lomonosov Moscow State University (I. P. Ashmarin, V. B. Koshelev), we performed a randomized, blind trial of Semax on 30 male rats after bilateral permanent occlusion of common carotid arteries [273]. It was shown that intraperitoneal Semax application in a daily dose of 300 µg/kg (subdivided into 4 doses injected in 15, 60, 120 min and 5 h after the occlusion) significantly decreased neurological deficit (assessed by the McGrow [171] and the Rudolphi [46] scales) within the first 1.5–6.5 h after the induced ischemia, compared to the control groups of the placebo-treated and pseudo-operated rats ($p = 0.03$, ANOVA) (Fig. 15.3). Increased animal survival and prolongation of the period before the first mortal outcome were also found in the group of rats treated with Semax (Fig. 15.4). In the control groups, 50% of mortalities were registered within the first 7 h after the occlusion, while in rats treated with Semax no mortalities were found during this period, and 50% of mortalities were registered only at 10 h after the ischemia induction.

Significant decrease in frequency of hemorrhage complications ($p < 0.01$ versus placebo) and improvement of neurological deficit ($p < 0.05$ versus placebo) were demonstrated in rats with focal brain ischemia after middle cerebral artery occlusion under Semax treatment [123–125].

Safety trials with Semax in healthy volunteers confirmed safety profile and absence of toxicity and hormone activity of the drug [10, 125, 165, 198, 199, 264].

The experimental data about potential neuroprotective properties of Semax, introduction of its convenient intranasally administered preparation to daily medical practice, and absence of toxicity, hormone activity, and marked side effects of the drug posed the need for carrying out a trial of Semax in patients with acute ischemic stroke.

The purpose of the present trial was to investigate the safety and efficacy of Semax in patients with acute carotid ischemic stroke.

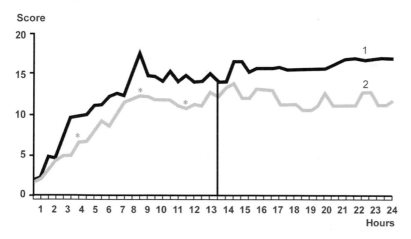

Figure 15.3. Dynamics of total score assessed by the McGrow Scale in rats with global brain ischemia. 1 – control; 2 – Semax. Significance of the differences versus controls: * $p < 0.05$.

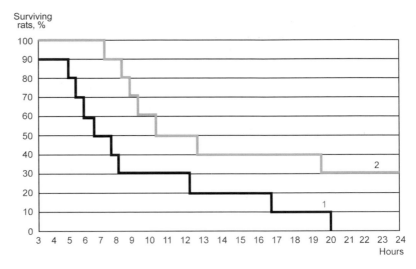

Figure 15.4. Percentage of rats surviving for 24 h after the induction of global brain ischemia.
1 – control; 2 – Semax.

15.6.1.1. Randomized, double-blind, placebo-controlled study of the safety and efficacy of Semax in carotid ischemic stroke and of its neuroprotective effects

We have performed the first clinical trial of the pharmaceutical drug Semax in patients with acute focal brain ischemia in the Neurology Clinic of the Russian State Medical University [91, 93, 177, 227, 286] (Russian patent No. 2124365, 1997/1999).

The purpose of the study was to investigate the safety and the efficacy of the Semax in patients with acute ischemic stroke and the ability of this drug to ameliorate the outcome of those patients, as well as to make more precise certain mechanisms of its neuroprotective effects.

The trial was performed after the approval of the Ethics Committee of the study center was obtained, and the patients or their legal representative gave written or witnessed informed consent to participate in the trial.

Patients with acute carotid ischemic stroke were eligible for inclusion in the trial if they (1) were admitted to the Intensive Stroke Unit at the Department of Neurology of the Russian State Medical University within the first 6 h after the onset of stroke, (2) were within the age range 45–75 years, (3) were conscious or mildly obtunded (baseline Orgogozo score more

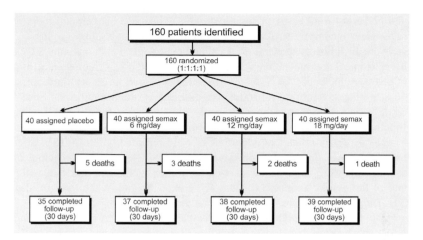

Scheme 15.1. Trial profile.

than 15). Patients who had experienced a previous stroke with residual neurological impairment, suffered from any other disorder interfering with neurological or functional assessment, or who had a life-threatening concurrent illness were excluded from participation in the trial. Other exclusion criteria were congestive heart failure, acute myocardial infarction (within the previous 6 weeks), ECG findings of ventricular arrhythmia, second- or third-degree atrioventricular block.

The target population consisted of 160 patients with carotid ischemic stroke (85 male and 75 female; mean age 61.3 ± 9.6 years); 97 patients suffered from left hemispheric stroke, 63 from right hemispheric stroke. 68 patients were admitted to hospital within the first 6 h after the stroke onset, 92 patients between 6 and 12 h. There were no significant differences between the treatment groups in the target population with regard to demographic and baseline characteristics (Table 14.2).

Screened patients randomly and blindly received intranasal treatment with placebo or one of three doses of Semax: 6, 12, 18 mg/day for 5 days (Scheme 15.1). The personnel at the trial site, outcome assessors, the personnel of the Safety Monitoring Committee involved in conducting or monitoring the trial and also data analysis were blind to the trial drug codes. Concomitant treatment with calcium channel blockers, Glycine, piracetam (Nootropil), and other drugs with neurotrophic and neuromodulatory properties (such as gangliosides, Cerebrolysin) was prohibited throughout the trial. Routine therapy included hemodilution and aspirin (in all patients) and glycerol (in all cases of severe strokes). The same background therapy allowed us to compare stroke outcomes in the placebo and Semax groups.

A medical history, general physical and neurological examinations, ECG, hematological and biochemical tests and investigation of cerebrospinal fluid (CSF) were included. A CT scan of the brain was performed within the first 12 h of illness. Neurological status was assessed by the Scandinavian Stroke Scale (SSS) [213] and the Orgogozo Stroke Score (OSS) [187]. The functional status was assessed by the Barthel Index (BI) [78, 161].

Blood pressure, heart rate, and ECG were repeated within 6–12 h after the initiation of therapy and on days 3, 5, 7, 14, 21, and 30. Neurological assessments were made at the start on admission within the first 6 h, at the end of the Semax treatment (on day 6), and at the end of the trial (on day 30). The Barthel Index was estimated on day 30.

The following methods of laboratory monitoring were selected taking into consideration beneficial metabolic properties of Semax observed in experiments and clinics. Laboratory investigations of blood (the levels of autoantibodies (AB) to phencyclidine-binding protein of glutamate NMDA-receptors (see Section 4.1.3), to S100β protein and MBP (see Sections 9.2 and 9.3); the activity of leukocytic elastase (LE) and its inhibitor α-1-antitrypsin (α-1-A) (see Section 9.3); and concentration of anti-apoptotic protein Bcl-2) were carried out on admission to hospital and 6, 9, 12, and 24 h after the onset of stroke (monitoring within the first 24 h), and also on days 3 and 5 (within the early acute period). Investigations of CSF (neurotransmitter amino acids (glutamate, aspartate, glycine, and GABA) levels (see Section 4.1.1), concentrations of thiobarbituric-acid-reactive substances (TBARS), cGMP, N-acetylneuraminic acid (see Section 4.3.1), levels of Bcl-2 and neuroptrophins NGF and BDNF (see Section 9.1)) were carried out within the first 6 h before the Semax treatment and were repeated on day 3. The immune and biochemical investigations were carried out together with colleagues from the Institute of Pharmacology of the Russian Academy of Medical Sciences (K. S. Raevsky, V. S. Kudrin, V. G. Bashkatova), the Institute of the Human Brain of the Russian Academy of Sciences (S. A. Dambinova and colleagues), the Mental Health Research Center (T. P. Klushnik *et al.*) and Anokhin Institute of Normal Physiology (V. V. Sherstnyov, M. A. Gruden, *et al.*) of the Russian Academy of Medical Sciences.

No complications related to lumbar puncture were observed. Mortality and adverse events were followed over the entire trial period of 30 days. Safety was followed by the Safety Monitoring Committee, which reviewed all reports of death and adverse events.

Protocol-specified study end points were the safety, neurological outcome according to the OSS and SSS, functional outcome according to the Barthel Index, and mortality on day 30.

Statistical analysis. Demographic and baseline disease characteristics were compared with the use of the Cohran–Mantel–Haenszel test for general association for nominal categorical variables (e.g., sex) and a one-way ANOVA for continuous variables (e.g., age). One-sample *t* tests were performed with the use of descriptive statistics for each heart rate and blood pressure to evaluate changes versus baseline values.

The primary efficacy analysis was performed on a target population basis as pre-specified in the protocol, i.e. on all 200 patients with carotid ischemic stroke who were correctly included in the trial according to the inclusion and exclusion criteria and completed the follow-up within 30 days.

Descriptive statistics and frequency distributions were generated for the study and point data. The Wilcoxon matched-pairs signed-rank test was used to analyze the statistical significance for the changes in measured parameters and the Mann-Whitney *U* test was used for pair-wise and group comparisons. Dynamics of biochemical parameters were assessed using ANOVA. Mortality rates and disability levels on the Barthel Index among the treatment groups were compared with the use of Fisher's exact test.

All statistical tests were interpreted at the 5% two-tailed significance level.

There were no significant differences between treated groups and placebo in the target population with regarded to demographic and baseline characteristics, this allowing comparative analysis of clinical and immune biochemical parameters in follow up (Table 15.3).

Table 15.3. Demographic and baseline characteristics of the target population (160 patients with carotid ischemic stroke)

| Characteristic | Placebo (n = 40) | Semax | | | p |
		6 mg/day (n = 40)	12 mg/day (n = 40)	18 mg/day (n = 40)	
Gender (m/f)	26/14	25/15	25/15	23/17	> 0.05
Mean age, years	63.3±8.7	65.5±9.5	64.7±10.1	66.7±10.5	> 0.05
Left/right hemisphere localization	25/15	24/16	23/17	25/15	> 0.05
Mean inclusion time, h	7.5±1.3	8.0±1.5	7.7±1.7	8.3±2.1	> 0.05
Mean *OSS* score on admission day	40.7±3.1	39.88±2.1	40.1±2.5	38.91±2.7	> 0.05
Patients with baseline OSS score ≤ 40 (severe stroke)	19 (47.5%)	20 (50%)	19 (47.5%)	20 (50%)	> 0.05
Patients with baseline OSS score > 40 (moderate stroke)	21 (52.5%)	20 (50%)	21 (52.5%)	20 (50%)	> 0.05
Mean *SSS* score on admission day	26.67±1.5	25.87±1.3	27.0±1.3	26.10±1.5	> 0.05

Table 15.4. Mortality and causes of death according to treatment group

Characteristic	Placebo (n = 40)	Semax		
		6 mg/day (n = 40)	12 mg/day (n = 40)	18 mg/day (n = 40)
Mortality				
All patients	5/40 (12.5%)	3/40 (7.5%)	2/40 (5%)*	1/40 (2.5%)*
Patients with baseline OSS score ≥ 40 (mild and moderate stroke)	1/21 (4.8%)	0/20 (0%)	0/21 (0%)	0/20 (0%)
Patients with baseline OSS score < 40 (severe stroke)	4/19 (21.1%)	3/20 (15.0%)	2/19 (10.5%)*	1/20 (5.0%)*
Causes of death, No. of patients				
Brain edema	3	3	2	1
PA thromboembolism	1	0	0	0
Cardiac failure	1	0	0	0

* $p < 0.05$; compared with the placebo group. PA = pulmonary artery.

Our findings confirmed the safety profile of the Semax treatment. Only paleness of mucous of nasal cavity was observed in 10%, and a slight (within 10% difference from usual parameters) increase in glucose blood level was registered in 7.4% patients suffering from diabetes mellitus, which was quickly corrected. Significant influence on blood pressure level and ECG parameters as well as other adverse events were absent.

Analysis of clinical data revealed significant ($p < 0.05$, Fisher's exact test) reduction of mortality in patients treated with Semax in doses of 12 and 18 mg/day versus patient treated with 6 mg/day or placebo (Table 15.4). The decrease in mortality in the 6 mg/day group was not significant ($p > 0.05$). Mortality among severe patients with progressive course of stroke was 10.5% (2/19) and 5% (1/20) in the 12 mg/day and 18 mg/day groups, respectively, and 15% (3/20) and 21.05% (4/19) in the 6 mg/day and placebo groups, respectively.

The comparison of the follow up clinical findings in the Semax and placebo groups showed statistically significant more rapid improvement of neurological outcome in patients with moderate ($p < 0.01–0.05$) and severe stroke ($p < 0.001–0.05$) treated with Semax in doses of 12 and 18 mg/day (Table 15.5). At the same time, Semax administered in a dose of 6 mg/day did not lead to significant improvement of neurological outcome compared with the placebo group ($p > 0.05$) (see Table 15.5).

Table 15.5. Change in neurological outcome as measured by the mean shift from baseline on the OSS and SSS

Stroke scale	Placebo (n = 40)	Semax		
		6 mg/day (n = 40)	12 mg/day (n = 40)	18 mg/day (n = 40)
At day 6				
OSS				
Patients with mild to moderate stroke (OSS > 40)	+15.9±1.1	+16.5±1.3	+22.0±1.5*	+18.3±1.8*
Patients with severe stroke (OSS ≤ 40)	+5.5±1.5	+5.9±1.5	+19.5±2.7**	+22.7±3.1***
SSS				
Patients with mild to moderate stroke (OSS > 40)	+6.4±1.7	+7.5±1.5	+12.0±1.8*	+8.93±1.7
Patients with severe stroke (OSS ≤ 40)	+2.0±1.7	+3.5±1.7	+10.0±1.2**	+13.4±1.6***^
At day 30				
OSS				
Patients with mild to moderate stroke (OSS >40)	+29.1±2.9	+31.1±2.3	+39.0±2.3**	+35.0±1.7*^
Patients with severe stroke (OSS ≤ 40)	+23.3±2.9	+27.5±2.3	+37.5±2.7***	+41.5±2.9***
SSS				
Patients with mild to moderate stroke (OSS > 40)	+13.2±1.5	+16.1±1.6	+21.4±1.4***	+18.7±0.9**^
Patients with severe stroke (OSS ≤ 40)	+12.1±2.2	+13.5±2.0	+18.7±1.8**	+20.1±2.3***

Note: Patients who died during the 30-day study period were assigned the worst score of each of the neurologic scales after their death. Positive values indicate improvement.
* $p < 0.01$, ** $p < 0.001$, *** $p < 0.0001$, compared with placebo;
^ $p < 0.01$, compared with 6 mg/kg Semax.

In patients with mild to moderate stroke (on admission OSS score > 40) and prevalence of focal neurological symptoms, the efficacy of Semax in doses of 12 and 18 mg/day was observed in the form of increased total activity and emotional improvement from the first day of the drug administration. Statistical analysis among patients with mild to moderate stroke showed significant improvement of neurological outcome on day 6 (at the end of therapy) and on 30 day in patients treated with Semax in doses of 12 and 18 mg/day doses ($p < 0.01$–0.05 versus the placebo and 6 mg/day groups). The most pronounced therapeutic effect was observed in the 12 mg/day group ($p < 0.05$ versus the 18 mg/day group) (see Table 15.5).

In patients with severe stroke (on admission OSS score of 26–40) acceleration of regression of general and autonomic symptoms was revealed in patients from the 12 and 18 mg/day Semax groups. All patients that were obnubilated on the day of admission recovered consciousness on day 2 of Semax treatment. Consciousness disturbances and meningeal symptoms remained on day 5 of stroke only in 6 patients with subtotal hemispheric infarction who died later. Regressive course of stroke was noted in 84.7% (50/59) patients with severe stroke treated with Semax (in the placebo group 63.2%; 12/19). The mortality in severe stroke was 10.2% in patients treated with Semax, whereas it was 21.1% in the placebo group ($p < 0.05$; Fisher's exact test). Statistical analysis among patients with severe stroke showed significant improvement of neurological outcome on days 6 and 30 in the 12 and 18 mg/day Semax groups versus the placebo and 6 mg/day groups. The most pronounced positive changes in neurological status were observed in the 18 mg/day group (see Table 15.5).

The assessment of separate focal symptoms (by the Original Stroke Scale [90]) revealed statistically significant acceleration of motor ($p < 0.04$–0.05) and speech ($p < 0.03$–0.04) recovery in the 12 and 18 mg/day Semax groups versus placebo.

When functional outcome was evaluated with the Barthel Index, the 12 and 18 mg/day Semax groups were found to have a higher proportion of patients with good recovery (no or mild disability, i.e. a Barthel score >70) than the 6 mg/day Semax group and the placebo group (Table 15.6). The best functional recovery was found in patients with moderate stroke who received Semax in a dose of 12 mg/day and in patients with severe stroke treated with Semax in a dose of 18 mg/day (see Table 15.6).

It is important to mention that the efficacy of Semax depended on hour in which therapy was started (within first 2–6 or 6–12 h after the stroke onset): the advantages of early application (within first 6 h) of the drug were established and manifested themselves by the best neurological outcome (Table 15.7).

Cytokine status and CRP analysis, which was performed on admission (within the first 12 h after the stroke onset, before starting the treatment), demonstrated an imbalance between pro- and anti-inflammatory peptide systems with a significant increase in inflammatory triggers (IL-1β, $p < 0.05$–0.01; TNF-α, $p < 0.05$–0.01) and C-reactive protein (CRP, $p < 0.05$) CSF levels in all the studied groups (Table 15.8). This corresponded to our previous investigations [177, 228]. There were no significant differences in cytokines and CRP levels between the Semax and placebo groups ($p > 0.05$, see Table 15.8).

On day 3, significantly stronger decrease in pro-inflammatory cytokines (IL-1β, IL-8) and CRP levels as well as increase in anti-inflammatory

Table 15.6. Functional outcome on day 30

Functional outcome	Placebo ($n = 40$)	Semax		
		6 mg/day ($n = 40$)	12 mg/day ($n = 40$)	18 mg/day ($n = 40$)
All patients				
– death	5/40 (12.5%)	3/40 (7.5%)	2/40 (5%)	1/40 (2.5%)
– severe disability	9/40 (22.5%)	4/40 (10%)	2/40 (5%)[§]	2/40 (5%)[§*]
– moderate disability	16/40 (40%)	18/40 (45%)	6/40 (15%)[§§**]	7/40 (17.5%)[§**]
– mild or no disability	10/40 (25%)	15/40 (37.5%)	30/40 (75%)[§§§**]	30/40 (75%)[§§§**]
Patients with mild to moderate stroke (OSS > 40)				
– death	1/21 (4.8%)	0/20 (0%)	0/21 (0%)	0/20 (0%)
– severe disability	1/21 (4.8%)	0/20 (0%)	0/21 (0%)	0/20 (0%)
– moderate disability	11/21 (52.4%)	10/20 (50%)[§§§]	1/21 (4.8%)***	2/20 (10%)***
– mild or no disability	8/21 (38.1%)	10/20 (50%)	20/21 (95.2%)[§§§***]	18/20 (90%)*
Patients with severe stroke (OSS ≤ 40)				
– death	4/19 (21.1%)	3/20 (15%)	2/19 (10.5%)	1/20 (5%)
– severe disability	8/19 (42.1%)	4/20 (20%)	2/19 (10.5%)[§]	2/20 (10%)[§]
– moderate disability	5/19 (26.3%)	8/20 (40%)	5/19 (26.4%)	5/20 (25%)
– mild or no disability	2/19 (10.5%)	5/20 (25%)	10/19 (52.6%)[§*]	12/20 (60%)[§§*]

Note: Severe disability – Barthel score 0–45; moderate disability – Barthel score 50–70; mild or no disability – Barthel score 75–100.
[§] $p < 0.05$, [§§] $p < 0.01$, [§§§] $p < 0.001$, compared with the placebo group.
* $p < 0.05$; ** $p < 0.01$, *** $p < 0.001$, compared with 6 mg/day Semax group.

Table 15.7. Dependence of the efficacy of Semax on time between stroke onset and beginning of treatment (shifts of total clinical score)

Hours from stroke onset to the first examination	OSS		SSS	
	day 6	day 30	day 6	day 30
2–5 h, $n = 58$	22.5±2.5	38.7±2.2	13.2±1.6	20.3±1.5
6–12 h, $n = 62$	16.5±1.5	23.1±2.3	8.5±1.5	15.4±1.9
p	0.004	<0.001	0.009	0.03

Table 15.8. Dynamics of CSF concentrations of cytokines and CRP in patients with carotid ischemic stroke

Cytokines		Placebo (n = 40)	Semax		
			6 mg/day (n = 40)	12 mg/day (n = 40)	18 mg/day (n = 40)
IL-1β (pg/ml)	within the first 12 h	4.2±0.9^	5.2±0.7^^	4.8±0.8^^	4.3±0.9^^
	day 3	5.5±0.8^^	4.8±0.9^	2.3±0.5**§#^	2.1±0.4***§#^^
TNF-α (pg/ml)	within the first 12 h	24.8±4.2^	23.1±5.7^^	28.2±5.1^	26.2±4.1^
	day 3	20.2±3.5	25.3±5.5	45.1±3.2***§§#	41.3±4.7***§#
IL-8 (pg/ml)	within the first 12 h	185.8±42.1	168.7±41.3	177.3±49.7	193.9±69.2
	day 3	390.4±51.7##^^^	120.5±35.1	65.7±12.3***§§§##	57.8±9.9***§§§#^^
CRP (µg/ml)	within the first 12 h	36.4±9.2^	38.7±9.5^	35.7±7.9^	33.7±6.9^
	day 3	110.7±14.3###^^^	18.9±4.1^	6.8±1.7***§§§##^	4.8±1.2***§§§#^^
IL-10 (pg/ml)	within the first 12 h	22.5±4.3	25.3±5.7	23.7±4.6	21.7±5.1
	day 3	27.0±4.5	34.9±6.1	45.3±6.3*##	42.1±5.5*#
TGF-β₁ (pg/ml)	within the first 12 h	79.3±21.4	85.9±23.7	81.9±21.9	75.7±25.9
	day 3	102.7±39.2	127.5±37.9	131.0±41.3	127.6±33.1

* $p < 0.05$, ** $p < 0.01$, ***$p < 0.001$, Mann–Whitney U test, versus placebo;
§ $p < 0.05$, §§ $p < 0.01$, §§§ $p < 0.001$, Mann–Whitney U test, versus 6 mg/day Semax;
# $p < 0.05$, ## $p < 0.01$, ### $p < 0.001$, Wilcoxon matched-pairs signed-rank test, day 3 versus the first 6 h;
^ $p < 0.05$, ^^ $p < 0.01$, ^^^$p < 0.001$, Mann–Whitney U test, versus control.

(IL-10) and modulatory (TNF-α) cytokines were revealed in patients treated with Semax in doses of 12 and 18 mg/day in comparison with the placebo and the 6 mg/day Semax groups (see Table 15.8, Fig. 15.5). Administration of Semax in a dose of 6 mg/day was not accompanied by statistically significant changes in cytokines and CRP levels versus placebo.

A close correlation was found between the decrease in CRP level by day 3 and the elevation of total clinical score by day 30 (with OSS, $r = 0.7$, $p = 0.04$; with SSS, $r = 0.5$, $p = 0.05$) in all studied patients, confirming the efficacy of the Semax treatment in cases when "inflammatory marker", CRP, was significantly decreased in CSF.

Δ3 Concentration, %

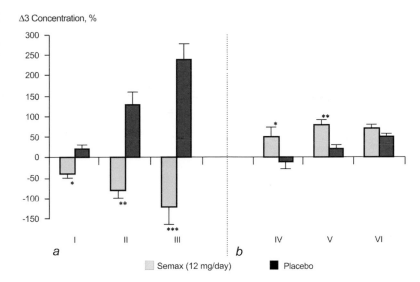

Figure 15.5. Dynamics of cytokines and CRP concentrations in CSF by day 3 after stroke onset: a) pro-inflammatory cytokines and CRP; b) anti-inflammatory cytokines. I – IL-1β; II – IL-8; III – CRP; IV – TNF-α; V – IL-10; VI – TGF-β₁. Significance of the differences versus placebo group: * $p < 0.05$, ** $p < 0.01$, *** $p < 0.001$.

To specify the influence of Semax on intensity of local inflammatory reaction, the dynamic changes in clinical and laboratory parameters was compared in patients with similar CSF concentrations of IL-10 and TNF-α in the first hours after the stroke onset (13 patients in each of two groups, treated with Semax or placebo) (Fig. 15.6). Significant differences in absolute (pg/ml) and relative (%) increase in IL-10 and TNF-α levels, as well as in changes in other cytokines and CRP levels and in shifts of the total clinical score (by OSS and SSS) were shown between these groups in spite of similar initial cytokine concentrations. The Semax treatment caused the more positive (in terms of prognosis) clinical and cytokine dynamics versus placebo: pronounced increase in IL-10 ($p < 0.01$), TNF-α ($p < 0.01$) and TGF-β₁ ($p < 0.05$) levels and marked decrease in IL-8 ($p < 0.01$) and CRP ($p < 0.001$) concentrations by day 3, as well as significant elevation of the total clinical score ($p < 0.001$ by OSS and SSS) by day 6 after the stroke onset (see Fig. 15.6).

Additionally, we studied activity of leukocytic elastase (LE) and its inhibitor α-1-antitrypsin (α-1-A) in blood serum of all treated patients. On

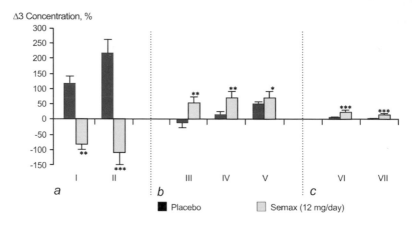

Figure 15.6. Dynamics of cytokines and CRP concentrations in CSF by day 3 and the shifts of the total clinical score by day 6 in stroke patients with initially (on day 1) comparable levels of IL-10 and TNF-α: a) pro-inflammatory cytokines and CRP; b) anti-inflammatory and modulatory cytokines; c) total clinical score; I – IL-8; II – CRP; III – TNF-α; IV – IL-10; V – TGF-$β_1$; VI – Orgogozo Stroke Scale (OSS); VII – Scandinavian Stroke Scale (SSS). Significance of the differences versus placebo group: * $p < 0.05$, ** $p < 0.01$, *** $p < 0.001$.

admission, within the first 12 h, LE activity showed increase in all patients with acute ischemic stroke versus controls (397 ± 33.5 and 258 ± 44 µM/min, respectively, $p < 0.01$) without significant differences between the Semax and placebo groups ($p > 0.05$). At the same time, the α-1-A level showed a tendency to decrease in all patients in comparison with the normal range (19.8 ± 5.7 and 29.9 ± 1.2 IU/ml, respectively, $p < 0.07$ versus control), this corresponding to our previous investigations (see Section 9.3). In the 12 and 18 mg/day Semax groups, significant decrease (38.8%, $p < 0.05$) in LE level was registered by day 21 after the stroke onset, while in the placebo and the 6 mg/day groups LE activity only tended to decrease (by 15.6 and 19.5%, respectively). The α-1-A concentration elevated by day 21 in patients from the 12 and 18 mg/day Semax groups (by 71.2%, $p < 0.05$, and 68.5 %, $p < 0.05$, respectively), while there was practically no change in the placebo and the 6 mg/kg groups (increase by 2.3 and 4.1%, respectively). Such LE and α-1-A dynamics also demonstrated the abatement of inflammatory process and the decrease in degranulation activity of neutrophils in the ischemized area under the Semax treatment in doses of 12 and 18 mg/kg.

The immunological analysis, performed within the first hours of the illness (before starting the treatment), found significant increase in the serum levels of autoantibodies to S100β and MBP in all groups of patients with acute ischemic stroke in comparison with the control group ($p < 0.01$) (see Chapter 9). Repeated study on day 3 failed to reveal any significant dynamics of the level of autoantibodies in the placebo and the 6 mg/kg Semax groups. At the same time, in patients treated with Semax in doses of 12 and 18 mg/day, autoantibody titer significantly decreased ($p < 0.01$ versus placebo and $p < 0.01$ versus 6 mg/day Semax) by day 3.

A correlation was confirmed between the decrease in primary antibodies to S100β and MBP and the reduction of LE activity in blood serum ($r = 0.75$, $p = 0.04$ and $r = 0.67$, $p - 0.05$, respectively), reflecting interrelationship between the abatement of inflammatory response and the reduction of autoimmune reactions. It can be supposed that anti-inflammatory effects of Semax may cause the improvement of BBB condition and restrict the permeation of neuroantigens into blood with prevention of following autoimmunization.

The immunoenzyme analysis showed highly significant dose-independent increase in Bcl-2 concentration in CSF of all the patients treated with Semax: on day 3 after the stroke onset Bcl-2 levels were 15.7 ± 4.5, 17.5 ± 5.1, and 14.5 ± 4.1 U/ml in the 6, 12, and 18 mg/kg Semax groups versus 0.2 ± 0.1 U/ml in the placebo group, $p < 0.0001$. A similar tendency was registered for blood serum levels on day 3 (25.1 ± 4.5, 23.9 ± 4.3, and 31.1 ± 5.1 U/ml, respectively, versus 0.2 ± 0.1 U/ml in the placebo group, $p < 0.05$) and especially on day 7 (238 ± 37, 202 ± 34, and 250 ± 45 U/ml, respectively, versus 0.3 ± 0.1 U/ml in the placebo group, $p < 0.001$). It is important that there were no marked differences between the groups in Bcl-2 concentration in CSF and blood serum within the first hours of the illness (before treatment) (Fig. 15.7). On day 21, serum levels of Bcl-2 were comparable in the Semax and placebo groups (18.3 ± 4.9, 23.1 ± 5.3, and 17.9 ± 3.5 in the 6, 12, and 18 mg/kg Semax groups, respectively, versus 21.5 ± 4.7 U/ml in the placebo group).

Along with this, on day 3 a significant increase in BDNF concentration in CSF was demonstrated in the Semax groups (by 118–120%, $p < 0.01$ versus the findings of day 1), while in the placebo group increase in BDNF level was only by 70.2% (Table 15.9). The comparison of BDNF concentrations on day 3 between the Semax and placebo groups revealed significant differences only in patients treated with high doses (12 and 18 mg/day) of Semax. Interestingly, no influence of Semax on NGF concentration was observed; its increase was comparable between the groups (by 38.5–40% in the Semax groups and by 30.9% in the placebo group, $p > 0.05$) (see Table 15.9). Selective action of Semax to neurotrophins may be connected with the predisposition of the peptide to glial tissue.

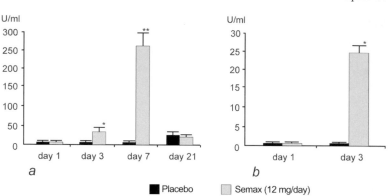

Figure 15.7. Dynamics of Bcl-2 level in blood serum (a) and CSF (b) in patients treated with Semax (12 mg/day) and placebo. Significance of the differences versus placebo: * $p < 0.05$, ** $p < 0.01$.

Findings of dynamic investigation of TBARS demonstrated continued increase in their CSF concentration by day 3 after the stroke onset in patients from the placebo group (see Section 4.3). At the same time, the opposite tendency, to reduction of their levels, was found in all the Semax groups. Differences with placebo were significant in patients treated with Semax in doses of 12 and 18 mg/kg ($p < 0.05$–0.01 versus placebo). Differences with the first hours TBARS level were significant ($p < 0.05$) only in the 18 mg/kg group (see Table 15.9).

Investigation of SOD activity in CSF confirmed earlier found relation between its increase by day 3 of stroke and positive clinical dynamics by day 30 (see Section 4.3). The Semax treatment led to the significant increase in SOD activity in the 12 and 18 mg/kg groups (on day 3, $p < 0.05$ versus day 1, $p < 0.01$–0.001 versus placebo).

The dynamic study of cGMP showed continued increase in its CSF concentration by day 3 in all stroke patients (see Section 4.3). However, the degree of this increase was significantly lower in patients treated with Semax in doses of 12 and 18 mg/kg ($p < 0.05$–0.01 versus placebo).

It is proposed that an important role in regulation of interrelationship between pro-inflammatory cytokine synthesis, reactions of oxidative stress, and cGMP synthesis is played by the *NO* system (Scheme 15.2). Presumably, the decrease in lipid peroxidation processes, the activation of SOD synthesis, as well as the reduction of cGMP level might be caused by depression of NO generation due to normalizing influence of Semax on cytokine balance and the increase in pro-inflammatory factor levels. This hypothesis was confirmed in experimental work, performed together with

Table 15.9. Dynamics of CSF levels of NGF, BDNF, TBARS, SOD, cGMP, and NANA in patients with carotid ischemic stroke

Cytokines		Placebo (n = 40)	Semax		
			6 mg/day (n − 40)	12 mg/day (n = 40)	18 mg/day (n = 40)
BDNF (μM)	within the first 12 h	24.8±4.2^	23.1±4.7^	28.2±3.1^	26.2±4.1^
	day 3	42.2±4.5^	50.8±6.5^^	62.1±5.2**##^^	57.2±5.7*##^^
NGF (μM)	within the first 12 h	4.2±0.9^	5.2±0.7^	4.8±0.8^	4.3±0.9^
	day 3	5.5±0.8^	7.2±0.9#^	6.7±0.8#^	6.0±0.7#^
TBARS (μM)	within the first 12 h	3.1±0.6	2.9±0.6	3.1±0.4	3.3±0.5
	day 3	4.6±0.7^^	2.7±0.5	2.3±0.4**	2.1±0.3**#
SOD (U/liter)	Within the first 12 h	3.3±0.4^^	3.1±0.5^	3.1±0.3^	3.2±0.5^
	day 3	2.0±0.3#	3.6±0.5^^	4.2±0.5***#^^^	4.7±0.6**#^^
cGMP (nM)	within the first 12 h	4.9±0.9^^^	4.5±0.7^^^	3.9±0.6^^^	5.1±1.1^^^
	day 3	12.7±1.3###	10.1±0.8###^	7.4±0.8*§##^^^	8.9±0.9**#^^^
NANA (mg/liter)	within the first 12 h	62.5±12.3^^^	58.1±10.2^^^	57.5±9.5^^^	65.7±11.5^^^
	day 3	121.5±15.1##^^^	53.3±9.9***^^^	30.5±7.7***§#^^^	37.7±9.3***#^^^

* $p < 0.05$, ** $p < 0.01$, ***$p < 0.001$, Mann–Whitney U test, versus placebo;
§ $p < 0.05$, §§ $p < 0.01$, §§§ $p < 0.001$, Mann–Whitney U test, versus 6 mg/day Semax;
# $p < 0.05$, ## $p < 0.01$, ### $p < 0.001$, Wilcoxon matched-pairs signed-rank test, day 3 versus the first 6 h;
^ $p < 0.05$, ^^ $p < 0.01$, ^^^$p < 0.001$, Mann–Whitney U test, versus control.

the Laboratory of Human and Animal Physiology of the Lomonosov Moscow State University in the model of global cerebral ischemia in rats [16]. After 4 h of bilateral permanent occlusion of carotid arteries, NO level in rat brain tissue was detected using the electron paramagnetic resonance method. Animals treated with Semax showed the lowest NO level compared to control group without neuroprotection ($p < 0.001$, Fig. 15.8). Prevention of NO increase was accompanied by significant regression of neurological deficit in rats with global cerebral ischemia ($p < 0.05$).

Taking into consideration the tendency to normalization of cGMP levels in CSF of patients treated by Semax and the experimental data about the influence of cGMP on calcium influx through agonist-dependent channels of

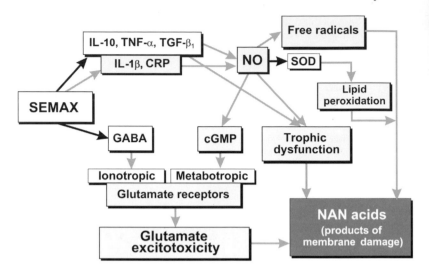

Scheme 15.2. Sequence of metabolic events caused by Semax. Black arrows show activating, grey – inhibitory effects.

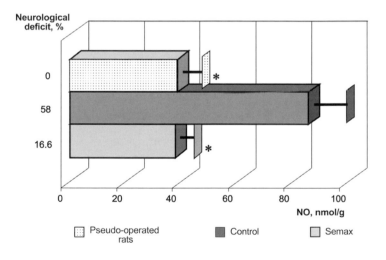

Figure 15.8. Concentration of NO in rat brain tissue 4 h after bilateral carotid occlusion. Significance of the differences versus controls: * $p < 0.001$.

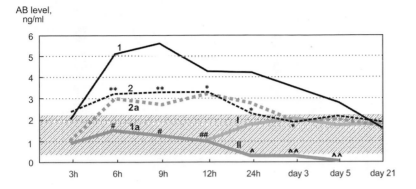

Figure 15.9. Dynamics of serum levels of antibodies to phencyclidine-binding protein of NMDA receptors in patients with carotid ischemic stroke. 1 – placebo; 2 – Semax. 1a and 2a – patients with extremely severe stroke treated with placebo and Semax, accordingly. I – surviving patients; II – deceased patients. Shaded region corresponds to normal level. Significance of the differences versus placebo group: * $p < 0.05$, ** $p < 0.01$. Significance of the differences versus patients with extremity severe stroke treated with Semax: # $p < 0.05$, ## $p < 0.01$. Significance of the differences versus surviving patients: ^ $p < 0.05$, ^^ $p < 0.01$.

NMDA-receptors, it was interesting to analyze changes in titers of antibodies to phencyclidine-binding protein of NMDA receptors in the studied groups (Fig. 15.9). As shown in all screened patients, increased antibodies titers were detected from the first hours of stroke onset (see Section 4.1.3). Significant decrease of antibody titers was revealed in patients receiving Semax in doses of 12 and 18 mg/kg from 6 h to 3 days after the stroke onset ($p < 0.01$ compared with placebo using ANOVA). It can be speculated that the normalization of anti-NMDA antibody levels reflects not only reduction of common autoimmune processes, but also the improvement of functional state of NMDA receptors.

There were no significant differences between CSF glutamate and aspartate concentrations in the Semax and placebo groups ($p > 0.05$), but a significant increase in GABA level was demonstrated in patients treated with Semax in doses of 12 and 18 mg/kg ($p < 0.01$, compared with placebo).

It was very important to assess the "morphological result" of all revealed metabolic effects of Semax. The concentration of N-acetylneuraminic acid (NANA) in CSF was chosen as a criterion of neuron membrane damage (see Section 4.3). Prevention of NANA increase in patients from the Semax groups was demonstrated, while elevation of its level by 94.4% was registered in the placebo group. In patients treated with Semax in doses of 12

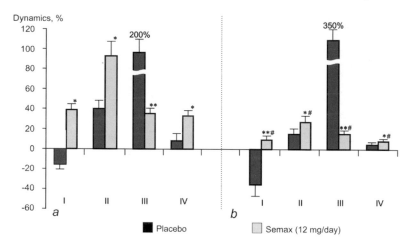

Figure 15.10. Dynamics of cytokines and CRP concentrations by day 3 and the shifts of the total clinical score by day 6 after stroke onset in dependence on time between admission and start of treatment: a) treatment started within 2–5 h; b) treatment started within 6–12 h after stroke onset. I – TNF-α; II – IL-10; III – CRP; IV – total clinical score by OSS. Significance of the differences versus placebo group: * $p < 0.05$, ** $p < 0.01$. Significance of the differences versus 2–5 h: [#] $p < 0.05$.

and 18 mg/kg, we could see significant decrease in NANA level by day 3 ($p < 0.05$ versus day 1, $p < 0.001$ versus placebo). Such NANA dynamics reflects the reduction of intensity of neuronal membrane disruption which accompanies necrotic damage, so confirms the neuroprotective properties of Semax.

The comparative analysis of Semax effects in dependence on time the therapy was started (the first 2–6 or 6–12 h after the stroke onset) revealed that normalization of the majority of laboratory parameters was significantly more pronounced in cases of early drug administration (within first 6 h) and correlated with the increase in the total clinical score assessed by the OSS and SSS, as well as with improvement of functional outcome assessed by the BI (Fig. 15.10).

Thus, the immune-modulating and neurotrophic activity of the drug was demonstrated to shift the balance towards prevalence of anti-inflammatory and trophic factors (IL-10, TNF-α, TGF-β_1, BDNF, NGF) over pro-inflammatory factors (IL-1β, IL-8, CRP, LE), to increase anti-apoptotic defense (elevation of Bcl-2), as well as to reduce peroxidation processes (decrease of TBARS and elevation of SOD activity). Simultaneous anti-inflammatory, antioxidant, neurotrophic, and anti-apoptotic action of Semax

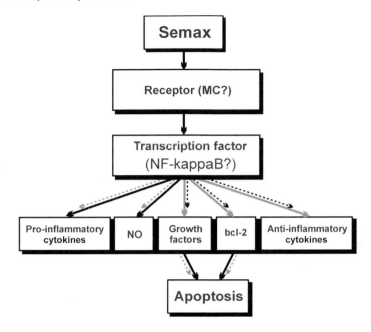

Figure 15.11. Hypothetical effector mechanism of Semax effects. Black arrows show activating and grey arrows show inhibitory effects; dotted line shows the influence of Semax.

can be connected with its influence on molecular trigger mechanisms of delayed post-ischemic events (Fig. 15.11). As the amino acid sequence of Semax is identical to that of α-MSH, and the proposed mechanism of its action is realized via MC-like receptors, we can speculate that, like exogenous α-MSH, Semax may inhibit NF-kappaB activation and IkappaBα degradation [111, 164]. NF-kappaB is a ubiquitous and important transcription factor for genes that encode pro-inflammatory cytokines as those known to be inhibited by Semax and α-MSH [40, 142, 154]; NF-kappaB also influences the expression of adhesion molecules, c-myc, and MnSOD genes [114, 134]. To prove the hypothesis about the connection of neuroprotective effects of Semax with its effector mechanisms of action and its modulation of transcription factors, further experimental investigations are required.

Thus, the first randomized, double-blind, placebo-controlled trial has demonstrated that neuropeptide Semax is safe and well-tolerated in treatment of patients with carotid ischemic stroke and can exert favorable clinical effects. Intranasal administration of Semax in daily doses of 12–18 mg/kg

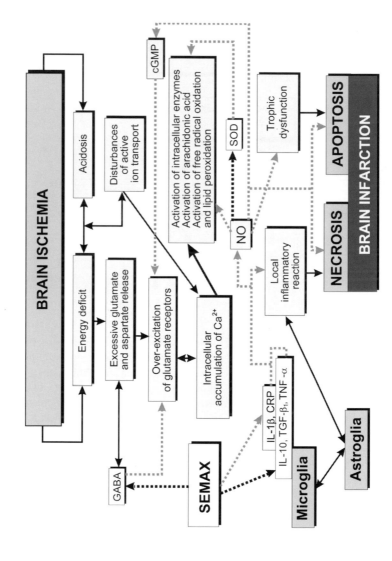

Scheme 15.3. Influence of Semax on the mechanisms of acute focal brain ischemia. Black arrows show activating, grey – inhibitory effects. Dotted lines show Semax influence.

given for 5 days significantly decreased 30-day mortality, improved clinical outcome of stroke and functional recovery in patients with different stroke severity. Investigation of dose-depended efficacy of the drug showed that the optimal daily dose in patients with mild to moderate stroke (OSS > 40) was 12 mg/kg, and a daily dose of 18 mg/kg was optimal in severe stroke patients (OSS < 40). The most pronounced positive effects of Semax were registered in cases of its early administration (within the first 6 h after the stroke onset).

It can be suggested that Semax induces a chain of metabolic transformations: significant decrease in inflammatory factors and increase in anti-inflammatory and neurotrophic factors concentrations can lead to inhibition of local inflammation in the ischemic focus and to improvement of brain trophic maintenance (Scheme 15.3). These reactions can result in the inhibition of NO generation, elevation of SOD activity, and reduction of lipid peroxidation and cGMP synthesis. The decrease in cGMP production can cause changes in glutamate receptor activity. So, the metabolic transformations induced by Semax actually organize a closed circuit of interrelated reactions that is probably one of explanations of long post-action of the drug. Semax positively influences all the main mechanisms of delayed post-ischemic events, emphasizing the possibility of effective neuropeptide brain protection.

These results should be verified in further trials with a larger number of patients.

15.6.1.2. Influence of Semax on functional state of brain in patients with carotid ischemic stroke

To study how Semax influences functional brain activity, neurophysiological monitoring (EEG with coherent and frequency spectrum analysis and mapping, as well as somatosensory evoked potentials (SSEP) with their mapping [87, 94], see Section 14.5.2) was performed in 80 patients (41 males and 39 females; average age 65.7 ± 7.6 years) in the acute period of ischemic stroke in the carotid artery territory (left in 47 cases and right in 33 cases). Clinical and baseline assessments corresponded to those given in Section 15.6.1.1. On admission, 40 (50%) patients had moderately severe disease and 40 (50%) had severe disease. From the first 6 h after the stroke onset, 40 patients received Semax and the other 40 received placebo. Taking into consideration the results of the study of dose-dependent efficacy of Semax (see Section 15.6.1.1), the drug was administered in a dose of 12 mg/day. Patients were examined on admission (on the background of the first administration of the drug (Semax or placebo) and during 4 h that followed) and then repeatedly on day 6 (at the end of the Semax treatment). Methods have been described in Section 14.5.2.

After the first administration of Semax we registered an increase in α-range power more pronounced in the "intact" hemisphere in 87% (20/23) of patients with moderately severe stroke; it was accompanied by a decrease in focal slow rhythm power in several patients in the projection of the ischemic lesion (Plate 7). However, when we compared EEG maps statistically before and after the administration of the first dose, we revealed no significant changes of frequency and power of EEG spectrum. At the same time, already after the first administration of Semax we registered significant increase in coherence in dominant α-rhythm in the affected hemisphere as well as between hemispheres (Plate 8). The increase in α-rhythm coherence in occipital lobes was a beneficial prognostic sign as it foreran further improvement of the basic rhythm. By day 6 of stroke in patients with moderately severe stroke treated with Semax we found complete improvement of EEG pattern versus placebo group: significant acceleration of α_1-rhythm on average by 1–2 Hz, bilateral increase in its power with regression of inter-hemispheric asymmetry (in 37.5% cases it was complete) (see Fig. 15.2). In patients (43.5%; 10/23) with no focal EEG changes on admission, no focus of δ- and θ-activity had been formed by day 6 in the projection of ischemic lesion, whereas in more than a half of patients that received placebo θ-focus was registered (4–6.9 Hz).

In 82.3% (14/17) of patients with severe stroke treated with Semax, we registered early focal increase in δ- and θ-activity power in the projection of ischemic lesion already within the first study. In 23.5% (4/17) patients, unstable frequency EEG spectrum with appearance of bilateral synchronic activity in α- and β-ranges was determined, this suggesting irritation of deep mesodiencephalic structures. After ending of treatment with Semax, in all patients with severe stroke stable frequency EEG spectrum as well as the reduction of total energy level of spectrum were registered. Along with this, in 41.2% (7/17) patients with severe stroke focal changes persisted during all the acute period of stroke, this correlating with the formation of a permanent focal deficit in the neurological status.

Multi-let SSEP of different modalities revealed the predominant influence of Semax on the state of nonspecific brain structures. Its influence prevailed in patients with moderately severe stroke and prognostically beneficial EEG pattern. The first administration of Semax caused a significant increase in amplitudes and decrease in latent periods of middle- and long-latent SSEP components in both the ischemized and "intact" hemispheres. By day 6 after the stroke onset these tendencies remained and elevated and significant improvement of zonal distribution of long-latency peaks was observed (Plate 9).

Thus, our neurophysiological investigation demonstrated faster and more complete improvement of functional state of brain in patients with acute carotid ischemic stroke treated with Semax. The analysis of the dynamics of bioelectrical brain activity revealed significant positive changes in frequency and power characteristics, as well as in zonal distribution of basic α-rhythm by day 6 in patients from the Semax group versus the placebo group (Plate 10). At the same time, the reduction of local δ- and θ-activity power in the projection of the ischemic lesion was found under the Semax treatment. Moreover, in 43.5% of severely affected patients, the full prevention of the formation of slow activity focus was demonstrated, which is the basic marker of brain infarction. Significant improvement of middle- and long-latent SSEP components after treatment with Semax reflected the beneficial effect of the drug on the nonspecific brain pathways that was clinically manifested by powerful nootropic effects.

So, the normalization of neurophysiological findings, which have favorable prognostic significance [89, 90], confirms the positive influence of Semax on the functional recovery in patients with ischemic stroke.

These first positive results of the application of Semax in acute ischemic stroke make necessary the arrangement of large trials of the drug.

15.6.2. Vasoactive intestinal polypeptide and its derivatives

It is interesting that neuroprotective properties, similar to those described in short fragments of ACTH, have been recently discovered in regulatory peptides of other groups. Gozes *et al.* [76, 77] and Spong *et al.* [236] showed neuroprotective effects of such a derivative of vasoactive intestinal polypeptide (VIP) as NAP neuropeptide (NAPVSIPQ), produced by glial tissue in response to the action of VIP. Combined protective action of NAP was demonstrated, which was manifested by antioxidant and anti-inflammatory effects. The neuropeptide was shown to inhibit lipid peroxidation, to depress TNF-α and IL-1β gene expression, and to cause neurotrophic effect. NAP administered 1 h after experimental ischemic stroke led to significant reduction of the infarction volume in animals. Clinical trials of the neuropeptide have been started.

15.6.3. Cerebrolysin

Cerebrolysin is a peptide preparation produced by the biotechnological standardized enzymatic breakdown of purified brain proteins extracted from young pig brain. It consists of 15% peptides with a molecular weight not exceeding 10 kDa and 85% amino acids based on the total nitrogen [1]. The solution, ready for injection or infusion, is free of proteins, lipids, and antigenic properties. One milliliter of Cerebrolysin contains 215.2 mg of a brain-derived peptide preparation (Cerebrolysin concentrate) in aqueous solution.

Brain-specific peptides are the active fraction of the drug. Their low molecular weight avoids the possibility of anaphylaxis [1] and at the same time allows easy permeation of peptides through the BBB and their active involvement in the metabolism of brain neurons [138, 263].

Basic mechanisms of the effects of active Cerebrolysin fraction are regulation of brain energy metabolism, intrinsic neurotrophic action, modulation of endogenous growth factors activity and interaction with systems of neuropeptides and neuromediators [1, 109, 135, 242, 263, 268].

Experimental studies showed that Cerebrolysin, on the one hand, decreases brain demand for oxygen, forming its increased resistance to hypoxia–ischemia, and, on the hand, increases aerobic metabolism significantly decreasing lactate level in the nervous system [126, 263, 269]. Stimulation of energy metabolism with Cerebrolysin is especially flexible [263]: it always provides optimal ATP level in neurons, prevents them from hyperproduction of macroergic charge, this allowing cells to react adequately to energy-dependent tasks such as synthesis of proteins and neurotransmitters, functioning of ion channels, etc.

Antioxidant properties of the drug were demonstrated such as inhibition of free radical oxidation and lipid peroxidation [1], as well as its positive influence on homeostasis of microelements (magnesium, potassium, selenium, manganese, vanadium) that possess antioxidant properties [82].

In the 1970s significant effect of Cerebrolysin on postnatal brain development was demonstrated. Cerebrolysin was shown to significantly increase protein and polypeptide synthesis, to activate neuronal mitochondria, and to modulate DNA/RNA ratio [230, 232]. Cerebrolysin accelerates differentiation of certain layers of brain tissue and enhances growth and formation of the capillary net [126]. The study of mechanisms of Cerebrolysin effects showed that each of the drug peptides fulfilled specific trophic and modulatory function [1]: some peptides possessed intrinsic neurotrophic properties, and others induced synthesis of endogenous nerve growth factor in astrocytes. Along with this, Cerebrolysin peptides actively interacted with neurotransmitter systems, increasing functional interrelation between neurons and glial cells, improving synaptic transduction and cerebral tissue plasticity. Neurotrophic effects of the drug prolonged survival of brain cells under conditions of deficiency of neurotrophic support coming from target cells and oxygen/glucose deficiency [269].

Studies conducted in recent years have demonstrated the property of Cerebrolysin to increase expression of the glucose transporter gene (*GLUT-1*) responsible for glucose transport via the BBB, thus increasing its delivery to brain cells under conditions of experimental ischemia [22–24, 26]. The stabilization of GLUT-1 mRNA, as well as stabilization of p88 TAF, cellular trans-acting factor, which participates in the post-transcriptional regulation of the *GLUT-1* gene were revealed under the influence of Cerebrolysin [24, 25].

It was also shown that neurotrophic properties of Cerebrolysin are connected with protection of neuronal cytoskeleton due to inhibition of calcium dependent proteases including calpain and to increase in microtubular acidic protein 2 (MAP2) expression [109, 110, 216]. Along with this, Cerebrolysin increases affinity of BDNF binding to its receptors [99]. The drug's effect on trk-B receptors to neurotrophins suggest its involvement in regulation of natural growth factors. The capacity of Cerebrolysin to avert hyperactivation of microglia and to decrease production of IL-1β and other pro-inflammatory cytokines was shown in experiments [160], this reflecting the influence of the drug on the extent of local inflammatory response and oxidative stress in the ischemized area.

It was demonstrated that Cerebrolysin effects are more prominent for cholinergic neurons [126]. An increase in acetylcholinesterase level was registered under the influence of Cerebrolysin [126]. Cholinergic influence is a possible mechanism of the nootropic effect of the drug. It was also confirmed that Cerebrolysin stimulates complex formation of hormones in adenohypophysis [231], thus modulating molecular mechanisms of stress-mediating system triggering.

The study of Cerebrolysin safety profile in healthy volunteers and limited groups of patients showed the absence of toxicity and rather rare incidence of the side effects of the drug, which mainly occur when the drug was administered in bolus injections. After infusion of high doses of Cerebrolysin (30–50 ml), the most common adverse events included vertigo, agitation, and feeling hot. All adverse events were mild and only transiently present. There were no changes in the vital signs of the patients or in any of the lab parameters [206, 211].

The multifaceted metabolic effects of Cerebrolysin, its neurotrophic, neuromodulatory, anti-hypoxic, and antioxidant properties, as well as proven safety and good tolerability of the drug, called for its clinical application [1, 263]. First of all, the nootropic efficacy of the drug was determined, which was successfully used in treatment of patients with chronic brain ischemia [20, 99, 139, 188, 192, 233]. In patients with atherosclerotic and/or hypertonic encephalopathy, Cerebrolysin increased the efficacy of associative processes, improved mental abilities, memory, attention, and stabilized emotional background [75, 85, 100, 101].

Cerebrolysin has been used in patients with ischemic stroke since the 1970s [145, 166, 281]; however, only nootropic effects of low doses (1–2 ml) were used. As a rule, Cerebrolysin was administered at the end of the first week of stroke, so its influence on both further course of the disease and the improvement of focal neurological deficit was not found, but only decrease of apathy and depression was presumed to be a consequence of the treatment. Attempts to apply low doses of Cerebrolysin in the late recovery period of stroke also failed to show significant positive clinical dynamics [240].

In late 1980s the expediency of high dose application was first substantiated. It was believed that doses over 10 ml of Cerebrolysin would promote embodiment of neuroprotective properties of the drug. Meta-analyses of Cerebrolysin that enrolled over 2000 patients with ischemic stroke [268] showed that the drug should be administered as early as possible, as there was demonstrated an inverse correlation between its efficacy and the time span from the stroke onset till the first administration. It was shown that in the acute period of stroke the dose of the drug might be 50 ml (once a day for 2–3 weeks), in the rehabilitation period the daily dose of 30 ml can be used for 3–4 weeks. The predominant efficacy of high doses of Cerebrolysin (in comparison with doses lower than 10 ml) has been confirmed by subsequent studies [55, 136, 197, 237]. The results of multi-center Cerebrolysin trial in the acute period of carotid ischemic stroke demonstrated the advantage of daily doses of 30–50 ml versus 10–30 ml. When the drug was administered in doses of 30–50 ml, researchers registered more complete regression of neurological deficit on day 30 after the stroke onset, as well as significant improvement of functional recovery and self-service skills in the remote period [15, 143, 237].

15.6.3.1. Clinical and neurophysiological study of dose-dependent efficacy of Cerebrolysin in acute ischemic stroke

To elucidate the neuroprotective properties of Cerebrolysin and to determine optimal doses of the drug, we performed a simple pilot study of 50 patients (30 males and 20 females; average age 59.5 ± 10.6 years) with acute carotid ischemic stroke, admitted within 12 h after the event at the Neurology Clinic of the Russian State Medical University [63, 86, 92, 226].

The trial was performed after the approval of the Ethics Committee of the study center was obtained, and the patients or their legal representative gave written or witnessed informed consent to participate in the trial.

Patients with acute carotid ischemic stroke were eligible for inclusion in the trial if they (1) were within the age range 45–75 years, (2) were conscious or mildly obtunded (baseline Orgogozo score more than 15). Patients who had experienced a previous stroke with residual neurological impairment, suffered from any other disorder interfering with neurological or functional assessment, or who had a life-threatening concurrent illness were excluded from participation in the trial. Other exclusion criteria were congestive heart failure, acute myocardial infarction (within the previous 6 weeks), ECG findings of ventricular arrhythmia, and second- or third-degree atrioventricular block.

The drug was administered from the first hours of stroke via slow infusion (through a cubital catheter) in daily dose of 10, 20, or 30 ml for 5 days on the background of maximally unified therapy, including hemodilution and aspirin (in all patients), glycerol (in all cases of severe strokes), and low doses of heparin (if required). Other drugs with potential neuroprotective effects (calcium channel blockers, Glycine, piracetam, Semax, etc.) were not used.

The first infusion of Cerebrolysin was performed in two stages: 2 and 8 ml, 10 and 10 ml, 20 and 10 ml, respectively, with 1 h interval, which allowed analyzing the effects of different doses of the drug on brain functional state. Sixty patients, to whom only standardized "background" treatment (without neuroprotectors) was administered, comprised the control group. The same background therapy allowed us to compare stroke outcomes in the control and Cerebrolysin groups.

A medical history, general physical and neurological examinations, ECG, hematological and biochemical tests were included. A CT scan of the brain was performed within the first 12 h of illness. Baseline assessments involved physical and neurological examination, ECG recordings, and clinical and biochemical blood and CSF tests. Neurological status was assessed by the Original Stroke Scale [90]. The functional status was assessed by the Barthel Index (BI) [78, 161].

On admission, in 20 (40%) patients to whom Cerebrolysin was administered, we diagnosed moderately severe stroke (the total clinical score by the Original Stroke Scale ≥36), and in 30 (60%) severe stroke (the total clinical score <36).

Blood pressure, heart rate, and ECG were repeated within 6–12 h after the initiation of therapy and on days 3, 5, 7, 14, 21, and 30. Neurological assessments were made at the beginning on admission within the first 6 h, at the end of the Cerebrolysin treatment (on day 6), and at the end of the trial (on day 30). The Barthel Index was estimated on day 30.

Mortality and adverse events were followed over the entire trial period of 30 days. Safety was followed by the Safety Monitoring Committee, which reviewed all reports of death and adverse events.

Protocol-specified study end points were the safety, neurological outcome according to the Original Stroke Scale, functional outcome according to the Barthel Index, and mortality on day 30.

To study how Cerebrolysin influences the functional state of the brain, neurophysiological monitoring of spontaneous brain bioelectrical activity using EEG and mapping EEG, as well as investigations of somatosensory evoked potentials (SSEP) with their mapping were performed on admission, after the first administration of the drug and during the following 4 h, and repetitively on days 6 (after Cerebrolysin treatment cessation) and 21 (at the end of the acute period of stroke) [63, 226]. The methods were described in Section 14.5.2.

Statistical analysis. The Wilcoxon matched-pairs signed-rank test was used to analyze the statistical significance for the changes in measured parameters and the Mann–Whitney *U* test was used for pairwise and group comparisons. Mortality rates and disability levels on the Barthel Index among the treatment groups were compared with the use of Fisher's exact test. All statistical tests were interpreted at the 5% two-tailed significance level.

There were no significant differences between the treatment groups with regard to demographic and baseline characteristics, as well as to etiology of stroke and its severity on admission. This allowed comparative study of clinical and neurophysiological dynamics.

In 90% of patients with moderately severe stroke treated with Cerebrolysin, steady neurological improvement was revealed, while in the control group regressive course of illness was registered only in 70% of patients. The significant acceleration of regression of neurological deficit (assessed by the total clinical score by the Original Stroke Scale) was found in the Cerebrolysin group on days 6 and 30 after the stroke onset ($p < 0.001$–0.005) (Table 15.10). Quantitative assessment of the dynamics of separate focal symptoms confirmed the improvement of motor ($p < 0.005$) and sensory ($p < 0.05$) disturbances on the background of Cerebrolysin administration. By the end of the observation (on day 30), in 18 out of 20 (90%) patients with moderately severe illness, good functional recovery (Barthel Index ≥ 75) was registered, whereas in the control group good functional recovery was revealed only in 63.3% (19/30) cases. Comparative analysis of Cerebrolysin efficacy in relation to daily dose in moderately severe strokes showed the tendency to predominance of the efficacy of 10 ml (Fig. 15.12). This dose made the dynamics to overrun significantly ($p < 0.01$ versus control). Along with this, higher doses used (20 and 30 ml) did not lead to improvement of recovery ($p > 0.05$ versus control). The average shifts of the total clinical score by days 6 and 30 were significantly higher ($p < 0.01$–0.05) in patients who had been treated with 10 ml of Cerebrolysin compared with those who were given doses of 20 and 30 ml.

The comparison of clinical dynamics in patients with severe strokes also demonstrated the significant acceleration of regression of neurological deficit on days 6 and 30 in the Cerebrolysin group versus the control group

Table 15.10. Change in neurological outcome as measured by the mean shift from baseline on the Original Stroke Scale

Stroke scale	Control $(n = 60)$	Cerebrolysin		
		10 ml/day $(n = 20)$	20 ml/day $(n = 15)$	30 ml/day $(n = 15)$
At day 6				
Patients with mild to moderate stroke	+2.1±0.3 $n = 30$	+5.3±0.8** $n = 7$	+2.8±1.3$^{\#}$ $n = 6$	+2.5±1.1$^{\#}$ $n = 7$
Patients with severe stroke	+1.6±1.2 $n = 30$	+6.1±1.3** $n = 13$	+6.7±1.5*** $n = 9$	+4.0±1.0*^ $n = 8$
At day 30				
Patients with mild to moderate stroke	+4.6±0.5 $n = 30$	+7.1±1.2** $n = 7$	+4.7±0.9$^{\#\#}$ $n = 6$	+3.9±1.2$^{\#\#}$ $n = 7$
Patients with severe stroke	+6.5±1.5 $n = 30$	+12.5±1.5*** $n = 13$	+13.7±1.5*** $n = 9$	+7.3±1.4$^{\#\#\wedge\wedge}$ $n = 8$

Note: Patients who died during the 30-day study period were assigned the worst score of each of the neurological scales after their death. Positive values indicate improvement.
* $p < 0.05$, ** $p < 0.01$, *** $p < 0.001$, compared with control.
$^{\#} p < 0.05$, $^{\#\#} p < 0.01$, compared with 10 ml/day Cerebrolysin.
$^{\wedge} p < 0.05$, $^{\wedge\wedge} p < 0.01$, compared with 20 ml/day Cerebrolysin.

of comparable severity ($p < 0.001$–0.05) (see Table 15.10). The acceleration of clinical improvement was especially intensive on the first days of illness while the neuroprotector was being administered. In 80% (24/30) of patients with severe strokes in the Cerebrolysin group, beneficial regressive course of stroke was registered, while in the control group it was determined in 56.7% (17/30) of patients. We also revealed the tendency to decrease of 30 day-mortality in patients with severe stroke treated with Cerebrolysin (6.7%; 2/30) versus the control group (23.3%; 7/30; $p = 0.05$). The study of dose-dependent Cerebrolysin efficacy in severe stroke showed the tendency to higher efficacy in the 10 and 20 ml/day Cerebrolysin groups (see Fig. 15.12). The administration of 30 ml led to significantly lower increase in the total clinical score ($p < 0.01$–0.05 versus 20 ml).

The study failed to show any side effects of Cerebrolysin. Neither effects on systemic blood supply parameters nor allergic reactions, transient headache, or other complaints were revealed after the administration of the drug. Cerebrolysin was well tolerated by all patients.

Visual analysis of spontaneous electrical activity of brain after the first dose administration failed to determine any significant effects produced by low (1–2 ml) doses of the drug on EEG. When the dose was increased in patients with moderately severe strokes, the enhancement of synchronizing effects was registered with additional activation in hyperventilation regimen.

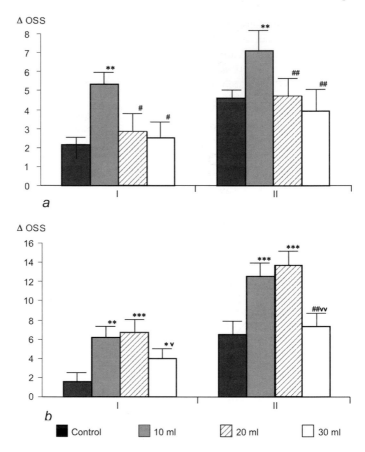

Figure 15.12. Dynamics of total clinical score (by the Original Stroke Scale – OS) by days 6 and 30 after the onset of carotid ischemic stroke during Cerebrolysin treatment: a) moderately severe stroke; b) severe stroke. I – day 6; II – day 30. Significance of the differences versus control: * $p < 0.05$, ** $p < 0.01$, *** $p < 0.001$. Significance of the differences versus 10 ml of Cerebrolysin: # $p < 0.05$, ## $p < 0.01$. Significance of the differences versus 20 ml of Cerebrolysin: v $p < 0.05$, vv $p < 0.01$.

The extent of synchronizing effect was individual; it predominated in patients with initial (on admission) hypersynchronized pattern of EEG with signs of "mesodiencephalon irritation". In cases with "flat" desynchronized low-amplitude EEG at baseline, the effect of 10 ml of Cerebrolysin was minimal according to visual evaluation and manifested only in functional exertion tests. Activation of synchronizing effects, as usual, was accompanied with smoothing of inter-hemispheric EEG asymmetry due to increase of α-index in the affected hemisphere. When higher doses of Cerebrolysin (20 and 30 ml) were applied, significant enhancement of electrical activity synchronization was observed. In 60% of cases, increase of α-index followed the administration, and in 30% of cases, short θ-wave discharges of moderate and high amplitude were observed predominantly in the frontal lobes.

When 10 ml of Cerebrolysin was administered in patients with severe strokes, it produced a short-term (only 30–40 min long) escalation of basic EEG rhythms, increased α-rhythm amplitude. At the same time, basic tendencies that had place in the background recordings remained. When the dose was elevated to 20 ml, steady increase in amplitude and frequency of α-rhythm occurred, and inter-hemispheric EEG asymmetry decreased. In 2/20 (20%) patients, repeated infusion of 10 ml (up to total dose of 20 ml) was accompanied by a significant increase of diffuse slow wave representation. When 30 ml were administered, the most characteristic responses were "slowing" of EEG potentials, an increase in δ- and θ-activity amplitudes, which was accompanied by mitigation of "imposing" of photo-stimulation rhythms on EEG. Identical EEG reactions have been described earlier when gamma-oxybutyric acid (GOBA) was used for treatment of patients with severe ischemic stroke [60].

Computer EEG analysis with mapping of bioelectrical brain activity revealed a stereotype sequence of brain reactions on an administration of Cerebrolysin and increase of its dose. The effect of the drug was prominently manifested 7–15 min after its administration by synchronizing of EEG, an increase in power of EEG frequency spectrum mainly in the initially (on admission) predominant rhythms ranges—according to baseline EEG pattern. When the dose of Cerebrolysin was "insufficient" (lower than optimal dose), these changes were found retained only for 40–60 min after the drug administration, and then EEG picture returned to the initial pattern (Plate 11). In cases of administration of optimal Cerebrolysin dose, this first stage of "quantitative" EEG changes gave way to the next stage of "qualitative" normalization of EEG pattern: in 30–40 min after the drug administration, significant acceleration of α_1-rhythm on average by 1–2 Hz with steady bilateral increase in its power and regression of inter-hemispheric asymmetry was registered, which was accompanied by significant decrease in power and limitation of zonal distribution of δ- and θ-activity in the ischemized hemisphere (see Plate 11). In these observations more complete normalization of EEG pattern was seen on day 6 and it was manifested by a significant regression of inter-hemispheric asymmetry in α-1-range and by prevention of formation of slow activity focus (or its minimal presentation) in the projection of the ischemized brain area (see

Plate 11), which clinically correlated with more complete neurological recovery. The administration of excessively high dose of Cerebrolysin led to a significant increase in power of the EEG frequency spectrum with predominant contribution of low (δ and θ) ranges, especially in the affected hemisphere, as well as to signs of mesodiencephalon irritation such as appearance of hypersynchronized bilateral θ-discharges (Plate 12). By the end of the first hour of EEG monitoring, no tendency towards EEG pattern normalization was registered. This Cerebrolysin "overdose" was especially dangerous in patients with prognostically unfavorable EEG pattern on admission [223, 224]. In such cases "fixation" of negative tendencies, those in the baseline EEG pattern, was found and retained up to day 6 of stroke. A constant slow wave focus in the projection of ischemia formed, and inter-hemispheric asymmetry in α-1-range with predominance of power of α-1-range in the "intact" occipital lobe was revealed.

The clinical and neurophysiological comparison showed that the choice of optimal dose of Cerebrolysin in acute period of carotid ischemic stroke determined efficacy of treatment and depended not only on clinical severity of stroke, but also on individual peculiarities of EEG pattern. In the majority of patients with moderately severe disease, who had average energy level of EEG spectrum of 40–100 $\mu V^2/Hz$, 10 ml of Cerebrolysin was sufficient and effective daily dose in any EEG pattern. Lower doses (2–5 ml) led to "insufficient" effect and did not alter "qualitative" structure of amplitude–frequency pattern of spontaneous bioelectrical brain activity. Higher doses (20 and 30 ml) did not show significant advantages over control group and yielded to 10 ml efficacy. In patients with high average energy level of EEG spectrum (over 100 $\mu V^2/Hz$), labile type of spectrogram, and presence of bilaterally synchronized α-θ-activity, the administration of 20 or 30 ml of Cerebrolysin led to undesirable general increase in power of slow frequency ranges with tendency towards their "slowing" in the projection of the ischemic area. These changes remained up to day 6 after the stroke onset and indicated formation of permanent focal morphological defect. Only in patients with low average energy level of EEG spectrum (<30 $\mu V^2/Hz$) and steady pattern of spectrogram with absence of any signs of mesodiencephalon irritation, the dose of 20 ml caused more complete normalization of EEG pattern than 10 ml and was preferable.

In the majority of patients with severe strokes, who had average energy level of EEG spectrum of 20–100 $\mu V^2/Hz$, 20 ml of Cerebrolysin appeared to be the optimal daily dose; the administration of 10 ml caused "insufficient" effects on EEG; the increase of the dose up to 30 ml led to general (non-focal) elevation of slow activity power. After single bolus injection of 30 ml of Cerebrolysin, changes were seen already 5 min later manifested as reconstitution of EEG rhythmics with prominent slowing of dominant part of EEG spectrum (GOBA-like effect); 60–90 min later, general increase of δ- and θ-activity power was replaced by its vivid shifting onto the ischemized hemisphere. On day 6 of stroke in these cases powerful focus of slow activity was registered in the projection of ischemic lesion, which was accompanied by significant depression of normal rhythms in the ischemized hemisphere. The clinical analog of such EEG reconstitution was the formation of a permanent focal morphological defect, resistant to therapy. Only in severe stroke patients with desynchronized low-amplitude EEG at baseline with low average energy level of EEG spectrum, the administration of 30 ml of Cerebrolysin was optimal as it produced positive dynamic changes of bioelectrical brain activity, compared with 20 ml. Oppositely, in cases of high average energy level of EEG spectrum at baseline with signs of bilaterally synchronized activity, the most effective was the dose of 10 ml.

The comparison of averaged EEG maps recorded on days 1 and 6 in patients treated with Cerebrolysin in individually chosen optimal doses demonstrated a significant decrease in slow range power in the ischemized hemisphere ($p < 0.01$) as well as an increase in α-range power in occipital to central derivations of the affected hemisphere, which correlated with good neurological and functional recovery and acceleration of focal deficit regression.

Studying Cerebrolysin effects on SSEP we registered normalization of their parameters only in cases when evoked potentials at baseline (i.e. on admission) were relatively intact. In cases of severely disordered SSEP (such as great lowering of amplitude or absence of components and prominent changes of basic peak latencies), the administration of the drug in the doses studied did not lead to normalization of evoked activity.

Thus, the pilot clinical and neurophysiological study confirmed good safety profile of Cerebrolysin and demonstrated certain favorable effects of the drug in daily dose of 10–30 ml on neurological recovery and functional outcome in acute carotid ischemic stroke. Mortality analysis in total cohort of patients without subdivision on severity categories revealed only a tendency towards mortality decrease in patients treated with Cerebrolysin (4%, 2/50) versus the control group (15%, 9/60; $p < 0.1$). However, in patients with severe strokes, the decrease in a mortality rate was significant ($p = 0.05$). Comparative study of efficacy of different doses of the drug showed that, when there is no possibility for neurophysiological monitoring in clinic, the optimal daily dose of Cerebrolysin is 10 ml for moderately severe and 20 ml for severe stroke. However, most exact choice of optimal daily dose of Cerebrolysin may be determined with the help of individual EEG pattern.

In is worth noting that the study showed that high doses of Cerebrolysin were not always optimal and effective in acute ischemic stroke. Often such doses as 20–30 ml not only showed no advantages over lower ones, but also produced GOBA-like activity on functional state of the brain, thus retarding recovery. Such effects of the drug may be explained by an inhibition of *formatio reticularis* of the brain stem, posterior hypothalamus, nonspecific thalamic nuclei, and caudate nucleus with following predominance of ascending influences from anterior hypothalamus on cerebral cortex. Apparently, the individual susceptibility to high doses of Cerebrolysin is determined by individual peculiarities of the signal transduction, neurotrophic, and other molecular brain cell systems, which appear to be targets for neuropeptide compounds of the drug [73, 199].

Clearly, the results obtained so far are preliminary and need to be verified in further trials with a larger number of patients.

It is worthwhile to make a cautionary remark on acquisition of biological material from cattle brain for pharmaceutical aims. Patients may be infected by prions (mutant proteins), which appear to be carriers of transmissible encephalopathy and cause severe neurodegenerations in humans, such as Creutzfeldt–Jakob's disease.

15.7.   Gangliosides

Along with endogenous regulators such as neuropeptides, the signal transduction system includes receptor structures which are also able to modulate processes of neuronal damage and reparation. Receptor-mediated generation of second intracellular messengers is a central event in the signaling chain and leads to further modulation of protein activity (enzymes, ion channels, receptors). The importance of receptors in the universal response of brain tissue to ischemia gave rise to interest in study of the properties of gangliosides, which are their important structural components.

Gangliosides are related to glycosphingolipids containing a less metabolically active hydrophobic (ceramide and fatty acids) and highly metabolically active hydrophilic (oligosaccharide) portions [172, 254]. The oligosaccharide component of gangliosides is the basis of their wide variety. Over 90% of all cerebral gangliosides are represented by five types: GM1, GD1α, GD1β, GT1α, and GT1β. The highest content of gangliosides is detected in brain cortex, cerebellum, and caudate nucleus [45, 73, 146, 195, 243]; other human tissues contain minimal levels of gangliosides.

Gangliosides are marker lipids of membranes of neurons and glial cells, especially in synaptic portions, as well as one of the major components of neuronal receptors [172, 250]. They influence conformation of both lipid and protein phases of a membrane leading to increase in rigidity of its superficial layer, to decrease in the number of polar groups, and elevation of the number of hydrophobic groups, as well as to restriction of the motion of surrounding molecules [174, 255]. The rate and intensity of dynamic processes developing on the membrane surface have an influence on formation of synaptic contacts, signaling transmission, sorting, and storage of information, as well as on memory formation [254, 255].

Gangliosides have a strong interaction with calcium ions and maintain a steady content of free calcium near presynaptic membrane, promoting functional activity of a neuron even under conditions of its multiple and prolonged stimulation and participating in ion transport processes [140, 151, 170, 262]. The activating effect of gangliosides on Na^+/K^+- and Mg^{2+}-ATPase functioning has been noted, as well as their normalizing influence on activity of cytosolic and mitochondrial enzymes [160, 194].

GM4 ganglioside is a superficial marker of oligodendroglia and myelin; GD1β and GT1β are the specific antigens of neuronal plasma membranes [4, 257]. It was shown that gangliosides can bind serotonin and have an influence on dopaminergic transmission [71, 249]. Specific antigen located on cholinergic nerve terminals also has ganglioside nature. According to Avrova [11], gangliosides take part in recognition of surface of neurons, specified by the type of mediator they produce. Being part of receptor complexes, gangliosides participate in immune response as antigens of immune competent and target cells. That allows their use as specific markers of the cited cells [29, 244]. Gangliosides can implement signal transduction between various types of immune competent cells and serve as modulators of immune response [119, 222]. It was demonstrated that ganglioside GQ1β produced strong trophic effect, identical to that of NGF [116].

Receptor function of gangliosides is as well confirmed by their interaction with hormones, particularly with gonadotropin and thyroid stimulating hormone [147]. Also substantiated is a suggestion that gangliosides are in tight interaction with neuropeptides. Polak and Bloom [196] showed that gangliosides, having an influence on excitatory (substance P, vasoactive intestinal polypeptide, angiotensin), as well as on inhibitory (β-endorphin, luliberin, somatostatin) neuropeptides, can participate in the embodiment of endpoint functional effects of neuropeptides.

Summarizing the data about poly-functional properties of gangliosides, one concludes that they are recognizing membrane-associated molecules, which contribute to modification of membrane surface, and form an anion layer with prominent affinity to cations. Basic molecular functions of gangliosides are cell recognition and intercellular interactions; reception on neuronal membranes (interaction with neuromediators, hormones, antibodies, toxins, cations); regulation of ion micro-surroundings, and membrane-associated electrogenesis. Thus, gangliosides relate to all basic functions of neurons: transmission, mediation, synaptogenesis [127, 128, 209]. It is noteworthy that exogenous gangliosides, especially GM1, when introduced into the human body can incorporate into the structures of neuronal membranes and produce dynamic reconstitution of endogenous gangliosides [32].

Taking into consideration the importance of the functions of gangliosides, their participation in modulation of the most important molecular and cellular processes in nervous tissue, it was of great interest to study whether a neuroprotective effect of gangliosides exists under conditions of brain ischemia.

The first experimental studies of gangliosides in brain ischemia assessed their influence on the dynamics and the extent of reparative processes. It was demonstrated that monosialoganglioside GM1 significantly accelerated recovery of the affected activity of cholinergic and dopaminergic neurons [250, 270]; promoted fast functional recovery of experimental animals being injected 10–15 days after focal ischemic brain damage [21, 217]. However, modulating effects of gangliosides on calcium metabolism and ion channels activity, as well as their revealed property to limit "excitotoxic" effects of glutamate and aspartate without simultaneous inhibition of their physiological receptor activity [41, 163] showed that gangliosides may possess neuroprotective properties.

It was found that 30 min after the induction of acute focal brain ischemia, injection of GM1 significantly accelerated normalization of oxygen and glucose metabolism [258]. Prolonged treatment of rats with experimental ischemic stroke with GM1 derivates was accompanied by significant decrease in infarction area [219].

Seventy eight patients with acute ischemic stroke were included in the first clinical trials [17, 18] of GM1. They were administered GM1 intramuscularly in daily dose of 40 mg or placebo on the background of corticosteroids used for treatment of brain edema. The absence of side effects of the ganglioside was demonstrated, as well as its statistically significant favorable action on stroke outcome and functional brain activity assessed by EEG and visual brain evoked potentials. The results of the first studies led to a series of new randomized, placebo-controlled clinical trials (14 for the period of 1984–1994, enrolling about 1500 patients), where higher daily dose of GM1 (100 mg intramuscularly or intravenously) was used, as well treatment was started earlier (within 5–72 h after the stroke onset). The most substantiated and wide multi-center trial enrolled 502 patients with the first-time hemispheric ischemic stroke, who were randomized in 31 clinical centers within the first 12 h after the stroke onset [7]. Patients received GM1 intravenously in dose of 100 mg or placebo for

15 days. The half of patients from each group also received hemodilution. The efficacy of treatment was assessed on days 15, 21, and 120 according to such end-points as mortality rate, the extent of functional recovery (by the Rankin Scale [207]), and neurological improvement (by the Canadian Neurological Scale; CNS [42]). The extent of neurological recovery on day 15 in patients treated with GM1 was significantly higher than in the placebo group; significant influence of the ganglioside on the extent of neurological improvement was registered for the first 10 days of treatment. However, on days 21 and 120 no positive effect of GM1 treatment was registered. Clinical outcome in patients treated from the first 6 h of stroke did not differ from those receiving therapy from a later period. Initial positive effects of the drug led authors to recommend a continuation of its clinical trials. Several not so large clinical trials demonstrated variability of GM1 positive effects, however without statistical confirmation of their significance [5, 72, 117, 176, 267]. The multi-center SASS trials (Sygen Acute Stroke Study) [3] and EST (Early Stroke Trial) [150] also did not confirm high efficacy of GM1 in treatment of acute ischemic stroke. Mortality rates in groups treated with gangloisides and placebo did not show significant difference; positive effects of the drug on regression of focal neurological deficit was registered only as a tendency ($p = 0.06$). Along with this, statistically significant ($p = 0.016$) improvement of neurological and functional recovery was registered by the post-hoc analysis in those patients who were treated with the ganglioside from the first 4 h after the stroke onset compared with placebo.

The special review carried out to assess the effect of exogenous gangliosides in acute ischemic stroke and based on analysis of 12 trials involving 2265 people treated with GM1 [30], showed that at present there is not enough evidence to conclude that monosialoganglioside GM1 is beneficial in acute stroke.

Along with research of GM1 properties, attempts were made to create more effective drugs based on combination of endogenous gangliosides. One of these is Cronassial consisting of 4 groups of gangliosides: GM1 (12%), GD1α (40%), GD1β (16%), GT1β (19%), which are the most abundant in brain tissue.

15.7.1. Clinical and neurophysiological study of the effects of Cronassial in acute ischemic stroke

The study was aimed at elucidation of safety and efficacy of Cronassial in ischemic stroke. We performed a simple pilot study of 25 patients (15 males and 10 females; average age 57.5 ± 9.7 years) with acute carotid ischemic stroke (in 15 cases left-hemispherical and in 10 cases right-hemispherical) admitted to the neurological clinic of the Russian State

Medical University within the first 12 h after the event [88, 94, 88, 223, 224]. Patients were included in the study after they had given written or witnessed informed consent to participate in the trial, and the approval of the Ethics Committee of the study center had been obtained. The inclusion and exclusion criteria were identical to those used in clinical trials of other neuroprotectors (see Chapters 14 and 15).

The drug was administered from the first hours of stroke via slow infusion (through a cubital catheter) in daily dose of 100 mg on days 1, 3, and 5 of stroke. Background treatment was maximally unified and included hemodilution and aspirin (in all patients), glycerol (in all cases of severe strokes), and low doses of heparin (if required). Other drugs with potential neuroprotective effects (calcium channel blockers, Glycine, piracetam, Semax, Cerebrolysin, etc.) were not used.

Sixty patients to whom standardized "background" treatment (without neuroprotectors) was administered comprised the control group.

Baseline assessments, study design, and statistical analysis were identical to those in the Cerebrolysin trial (see Section 15.6.3.1). On admission, in 15 (60%) patients, to whom Cronassial was administered, we diagnosed moderately severe stroke (the total clinical score by the Original Stroke Scale $\geq$ 36), and in 10 (40%) severe stroke (the total clinical score <36). Protocol-specified study end points were the safety, neurological outcome according to the Original Stroke Scale, functional outcome according to the Barthel Index, and mortality on day 30.

To study how Cronassial influences the functional state of brain, neurophysiological monitoring of spontaneous brain bioelectrical activity using EEG and mapping EEG, as well as investigations of somatosensory evoked potentials (SSEP) with their mapping and estimation of the sensory central conduction time (sCCT), transcranial electrical stimulation (TES) with estimation of the motor central conduction time (mCCT), and electromyography (EMG) were performed on admission, after the first administration of the drug and during following 4 h, and repeatedly on days 6 (after the end of Cronassial treatment) and 21 (at the end of the acute period of stroke) [137]. The methods were described in Section 14.5.2.

No marked side effects of Cronassial or adverse events were registered. The Cronassial treatment had no statistically significant effects on ECG, hemorheological parameters, and clinical and biochemical blood and urine tests. No allergic reactions were observed. Cronassial was well tolerated by all patients.

The analysis of neurological dynamics assessed by the Original Stroke Scale showed that Cronassial accelerated regression of neurological deficit in patients with moderately severe stroke at baseline (Fig. 15.13). In these

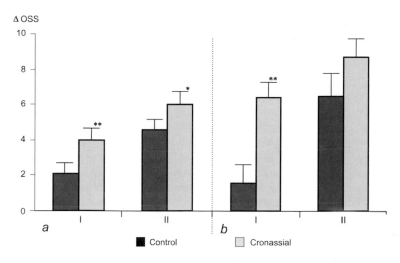

Figure 15.13. Dynamics of total clinical score (by the Original Stroke Scale – OS) by days 6 and 30 after onset of the carotid ischemic stroke during Cronassial treatment: a) moderately severe stroke; b) severe stroke. I – day 6; II – day 30. Significance of the differences versus control: $* p < 0.05$, $** p < 0.01$.

cases, significant increase in the shift of the total clinical score ($p < 0.05$ versus control) was revealed by days 6 and 30. Quantitative assessment of the dynamics of separate focal symptoms confirmed the improvement of motor ($p < 0.05$), sensory ($p < 0.01$), and speech ($p < 0.01$) disturbances on the background of Cronassial treatment. On day 30 good functional recovery (by the Barthel Index) was found in 86.7% (13/15) patients treated with Cronassial, who had moderately severe stroke at baseline, while in the control group good functional recovery was registered in 63.3% cases of comparable severity.

In severe strokes, Cronassial did not influence general severity of the patient's state, consciousness disturbances and other general cerebral symptoms, as well as mortality rate ($p > 0.05$ versus control). Only in those cases when the stroke severity was mainly indicated by the extent of focal neurological deficit (without consciousness disturbances, signs of brain edema, and other general cerebral symptoms), we could see prominent beneficial effect of Cronassial that was manifested by fast neurological recovery in direct relation to the infusion of the drug (within 30–40 min after the administration). In strokes with progressive course, connected with development of thrombosis, Cronassial could stop the increase in the extent of paresis and other focal neurological symptoms and caused functional improvement. However, statistical analysis conducted in the total group of severe strokes did not confirm the beneficial effect of the drug (see Fig. 15.13). The number of patients who participated in the study was insufficient for making a conclusion about Cronassial efficacy in patients with severe strokes.

The influence of Cronassial on spontaneous bioelectrical activity was manifested 7–10 min after its administration by a bilateral increase in power of EEG frequency spectrum, predominantly in fast frequency ranges and on the side of ischemia. Twenty to thirty minutes later, such "irritation" stage was followed by a bilateral decrease in common energy level of EEG spectrum lower than a baseline level registered on admission. This power decrease was evenly distributed in all frequency ranges. Only 60–90 min after the infusion of the drug, the energy power of the EEG spectrum elevated again and was accompanied by a tendency towards EEG pattern normalization. The basic EEG feature of Cronassial action was its directed influence on focal changes of bioelectrical brain activity, the improvement of which was evident and considerable in cases when general slow-wave (δ and θ) activity was absent in the initial EEG picture (on admission) (Plate 13). The statistical comparison of averaged EEG maps on days 1 and 6 after the stroke onset showed significant improvement of EEG pattern on day 6 in patients with moderately severe stroke treated with Cronassial. It was manifested by a decrease in δ- and θ-ranges power in the ischemized hemisphere ($p < 0.05$), and an increase in α-1-range power in occipital lobes ($p < 0.05$). At the same time, no positive EEG dynamics was registered in patients with severe strokes ($p > 0.05$).

The study of the influence of Cronassial on afferent and efferent central conduction pathways confirmed beneficial effects of the drug on the improvement of motor and sensory functions. Significant normalization of motor and sensory central conduction time (CCT) was revealed on the side of acute focal ischemia after the first infusion of the drug ($p < 0.05$), as well as on day 6 ($p < 0.05$), which had favorable prognostic significance for recovery of motor and sensory functions and preceded clinical improvement [137, 223]. Acceleration of motor and sensory central conduction was more pronounced in patients with severe strokes, who had the highest values of mCCT and sCCT at baseline.

Electroneuromyography revealed a positive influence of Cronassial on the functional state of the segmental peripheral neuromotor system. By the end of the acute period of stroke (on day 21), significant normalization of conduction velocities for motor and sensory fibers of median, ulnar, tibial, and peroneal nerves of both sides was observed ($p < 0.05$ versus controls).

Thus, the results of clinical and neurophysiological study of the safety and efficacy of combined ganglioside drug Cronassial on the whole corresponded to the results of the trials of other ganglioside drugs (GM1) in acute ischemic stroke. Good safety profile of Cronassial with absence of its side effects and adverse events was shown in all treated patients. A prevalent influence of the drug on the extent of focal neurological deficit was demonstrated. This explained why Cronassial was significantly effective only in patients with moderately severe stroke and in cases of severe strokes when the severity was mainly connected with great focal neurological deficit ($p < 0.05$). At the same time, the evaluation of Cronassial efficacy in the total stroke group or separately in all patients with severe strokes did not confirm positive clinical effects of the drug on both mortality rate and the dynamics of neurological and functional recovery. Neurophysiological study objectified the directed action of Cronassial on reparative processes in the ischemic area of the brain.

The results call for confirmation with greater patient cohort in randomized, double-blind, placebo-controlled trial. Unfortunately, the study was stopped.

15.8. Prevention of apoptosis

Studies conducted during the last decade have proved that programmed cell death takes part in the "up-formation" of brain infarction, additionally damaging the penumbral tissue. Mechanisms of apoptosis develop slower than the fast reactions of necrotic cell damage, reaching their peak on days 2–3 after the stroke onset. Close interrelation was shown between molecular triggers and signaling pathways of necrotic and apoptotic processes (see Chapter 10).

Koike and Tanaka [133] and Lee *et al.* [148] have suggested that one of the most important strategy tasks of secondary neuroprotection is to establish control over apoptosis. At present, development of anti-apoptotic defense techniques is only started and carried out in preclinical trials. Protein synthesis inhibitors were found to depress apoptosis [64]. Thus, injection of cycloheximide to experimental animals each 30 min for 24 h after the induction of acute focal ischemia decreased the volume of brain infarction by 70% by inhibiting the production of proteins [156].

It is known that native anti-apoptotic signaling pathways are activated by neurotrophic factors, certain cytokines, and stress factors [43, 44, 169]. Taking into account results of recent experimental investigations, it can be maintained that promising anti-apoptotic strategies may be connected with activation of such transcription factor as NF-kappaB, which induces expression of protective stress proteins, with enhancing HSP72 expression or using exogenous HSP72 [159, 205, 220], with induction of extracellular signal-regulated kinases cascade [113, 181, 285] and blocking PARP activity and JNK/p38 cascade [84, 162, 183, 278]; with activation of *Bcl-2* expression or using exogenous Bcl-2 [49, 157, 169], as well as with inhibition of effector caspases [35, 36, 98].

On the basis of modern achievements of molecular genetic studies, the laboratory synthesis of stable analogs of endogenous neurotrophins and molecular effectors, which "switch off" mechanisms of programmed cell death, is being conducted using recombinant DNA techniques. Genetically modified cells permanently producing neurotrophic factors are created for possible implantation in the affected brain area. In preclinical studies new methods of incorporation of genetic material into animals are developed by means of using of special viral carriers (vectors) and protein transduction domains [234, 284].

The results of such scientific research clearly present great interest and direct future therapeutic strategies. Along with this, even at present secondary neuroprotectors, such as neuropeptides with trophic, anti-inflammatory, and antioxidant properties, are being established in clinical practice. They have an influence on triggering mechanisms of delayed ischemic events including apoptotic mechanisms. Their positive clinical effects have already been demonstrated in randomized, double-blind, placebo-controlled studies. Further clinical trials are in progress.

REFERENCES

1. Akras, A., 1991, Cerebrolysin: general review. In *The 3rd Int Symp on Cerebrolysin*, Moscow, 5 (in Russian).
2. Alexidze, N. G., Balavadze, M. V., Ponomaryova-Stepnaya, M. A., *et al.*, 1983, *Bull Exp Biol Med.* **96** (7): 24 (in Russian).
3. Alter, M. S., for the SASS investigators, 1994, *Stroke.* **25**: 1141–1148.
4. Ando, S., Chang, N. C., Yu, R. K., 1978, *Analyt Biochem.* **89**: 437–441.
5. Angeleri, F., Scarpino, O., Martinazzo, C., *et al.*, 1992, *Cerebrovsc Dis.* **2**: 163–168.
6. Arefieva, I. A., 1992, *The Study of ACTH Analog Effect and Tuftsine on Mammalian Nervous System Cells*. Candidate's dissertation, Moscow (in Russian).
7. Argentino, C., Sacchetti, M. L., Toni, D., *et al.*, 1989, *Stroke.* **20**: 1143–1149.
8. Ashmarin, I. P., Kamensky, A. A., Myasoedov, N. F., Skvortsova, V. I., 2000, *Regulatory Peptides.* **89**: 51.
9. Ashmarin, I. P., Nezavibatko, V. N., Myasoyedov, N. F., *et al.*, 1997, *J Higher Nerv Activ.* **47** (2): 420–430 (in Russian).
10. Ashmarin, I. P., Stukalov, P. V., 1996, *Neurochemistry*. Institute of Biomedical Chemistry Publishing House, RAMS, Moscow (in Russian).
11. Avrova, N. F., 1984, *Ukr Biochem J.* **56** (3): 245–253 (in Russian).
12. Ay, H., Ay, I., Koroshetz, W. J., Finklestein, S. P., 1999, *Cerebrovasc Dis.* **9** (3): 131–135.
13. Ay, I., Sigimori, H., Finklestein, S. P., 2001, *Brain Res Mol Brain Res.* **87** (1): 71–80.
14. Banati, R. B., Graeber, M. B., 1994, *Dev Neurosci.* **16**: 114–127.

15. Barolin, G. S., 1996, *EuroRehab.* **3**: 135–143.
16. Bashkatova, V. G., Koshelev, V. B., Fadyukova, O. E., *et al.*, 2001, *Brain Res.* **894** (1): 145–149.
17. Bassi, S., Albizzati, M. G., Sbacchi, M., *et al.*, 1984, *J Neurosci Res.* **12**: 493–498.
18. Battistin, L., Cesari, A., Galligioni, F., *et al.*, 1985, *Eur Neurol.* **24**: 343–351.
19. Betz, A. L., Yang, G. Y., Davidson, B. L., 1995, *J Cerebr Blood Flow Metab.* **15**: 547–551.
20. Birkmayer, W., 1953, *Neural Med.* **2**: 100.
21. Bjorklund, A., Stenevi, U., 1979, *Physiol Rev.* **59**: 62–100.
22. Boado, R. J., 1998, *Neurosci Lett.* **255** (3): 147–150.
23. Boado, R. J., 1999, *Brain Res Mol Brain Res.* **63** (2): 371–374.
24. Boado, R. J., 2000, *J Neural Transm Suppl.* **59**: 255–262.
25. Boado, R. J., 2001, *Neurosci Res.* **40** (4): 337–342.
26. Boado, R. J., Wu, D., Windisch, M., 1999, *Neurosci Res.* **34** (4): 217–224.
27. Bowes, M. P., Zivin, J. A., Rothlein, R., 1993, *Exp Neurol.* **119**: 215–219.
28. Brenneman, D. E., Schultzberg, M., Bartfai, T., Oozes, I., 1992, *J Neurochem.* **58**: 454–460.
29. Cahan, L. D., 1982, *Proc Natl Acad Sci USA.* **79** (24): 7629–7633.
30. Candelise, L., Ciccone, A., 2001, *Cochrane Database Syst Rev.* **4**: CD000094.
31. Cao, X., Phillis, J. W., 1994, *Brain Res.* **644**: 267–272.
32. Carolei, A., Fieschi, C., Bruno, R., Toffano, G., 1991, *Cerebrovasc Brain Metab Rev.* **3**: 134–157.
33. Chen, C. C., Manning, A. M., 1996, *Cytokine.* **8** (1): 58–65.
34. Chen, H., Chopp, M., Zhang, R. L., *et al.*, 1994, *Ann Neur.* **35**: 458–463.
35. Chen, J., Li, Y., Wang, L., *et al.*, 2002, *J Neurol Sci.* **199** (1-2): 17–24.
36. Chen, J., Nagayama, T., Jin, K., *et al.*, 1998, *J. Neurosci.* **18** (13): 4914–4928.
37. Chen, L., Haught, W. H., Yang, B., *et al.*, 1997, *J Am Coll Cardiol.* **30**: 569–575.
38. Chopp, M., Zhang, R. L., Chen, H., *et al.*, 1994, *Stroke.* **25**: 869–876.
39. Cockcroft, K. M., Meistrell, M., Zimmerman, G. A., *et al.*, 1996, *Stroke.* **27**: 1393–1398.
40. Collart, M. A., Baeuerle, P., Vassali, P., 1990, *Mol Cell Biol.* **10**: 1498–1506.
41. Collingridge, G. L., Bliss, T. V., 1987, *Trends Neurosci.* **10**: 288–293.
42. Cote, R., Hachinski, V. C., Shurvell, B. L., *et al.*, 1986, *Stroke.* **17**: 731–737.
43. Culmsee, C., Vedder, H., Ravati, A., *et al.*, 1999, *J Cerebr Blood Flow Metab.* **19** (11): 1263–1269.
44. Culmsee, C., Zhu, Y., Krieglstein, J., Mattson, M. P., 2001, *J Cerebr Blood Flow Metab.* **21** (4): 334–343.
45. Dal Toso, R., Skaper, S. D., Ferrari, G., *et al.*, 1988, Ganglioside involvement in membrane-mediated transfer of trophic information: relationship to GM1 effects following CNS injury. In *Pharmacological Approaches to the Treatment of Brain and Spinal Cord Injury* (Stein, D. G., Sabel, B. A., eds.) Plenum Press, New York, pp. 143–165.
46. Dag, K. J. E., 1992, *Eur J Pharmacol.* **219**: 153–158.
47. Dawson, D. A., Martin, D., Hallenbeck, J. M., 1996, *Neurosci Lett.* **218**: 41–44.
48. Dawson, D. A., Masayasu, H., Graham, D. I., *et al.*, 1995, *Neurosci Lett.* **185**: 65–69.
49. De Bilbao, F., Guarin, E., Nef, P., *et al.*, 2000, *Eur J Neurosci.* **12** (3): 921–934.
50. Devuyst, G., Bogousslavsky, J., 1999, *J Neurol Neurosurg Psychiatr.* **67**: 419–427.
51. De Wied, D., 1987, *Progr Brain Res.* **72**: 93–108.
52. Di Napoli, M., Papa, F., 2001, *Stroke.* **32** (10): 2446–2447.
53. Ding, A., Nathan, C. F., Graycar, J., *et al.*, 1990, *J Immunol.* **145**: 940–944.

54. Dolotov, O. V., Grivennikov, I. A., 1997, Application of primary cultures of rat fetal neurons to the study of neurotrophic action of peptides. In *15th ESACT Meet,* Tours.

55. Dostoynova, T. V., 1997, Cerebrolysin efficacy in ischemic stroke. In *Actual Questions of Clinical Angioneurology*, St. Petersburg, pp. 21–22 (in Russian).

56. Duval, D., 2000, *Stroke.* **31** (4): 989–990.

57. Emmanuel, N. M., Denisov, E. T., Mayzus, E. K., 1965, *Chain Reactions of Hydrocarbons Oxidation in Liquid Phase*. Moscow (in Russian).

58. Endres, M., Laufs, U., Huang, Z., *et al.*, 1998, *Proc Natl Acad Sci USA.* **95**: 8880–8885.

59. Enlimomab Acute Stroke Trial Study Group, 1997, *Neurology.* **48**: A270.

60. Erokhin, O. Y., 1984, *Monitoring of Brain Bioelectrical Activity and Energy Brain Metabolism in Acute Ischemic Stroke.* Candidate's dissertation, Moscow, 231 (in Russian).

61. Escott, K. J., Beech, J. S., Haga, K. K., *et al.*, 1998, *J Cerebr Blood Flow Metab.* **18**: 281–287.

62. FIBLAST Safety Study Group, 1998, *Stroke.* **29**: 287.

63. Fidler, S. M., 1993, *Clinical and Electrophysiological Study of Cerebrolysin Effects on Functional State of Brain in Acute Period of Ischemic Stroke.* Candidate's dissertation, Moscow (in Russian).

64. Fisher, M., 1995, *Eur Neurol.* **35**: 3–7.

65. Fisher, M., Bogousslavsky, J., 1998, *JAMA.* **279**: 1298–1303.

66. Fisher, M., Shebitz, V., 2000, Review on therapeutic approaches to ischemic stroke: past, present, future. *Korsakoff J Neurol Psychiatr (Stroke).* **1**: 21–33 (in Russian).

67. Floyd, R. A., Hensley, K., 2000, *Ann NY Acad Sci.* **899**: 222–237.

68. Fujiwara, K., Date, I., Shingo, T., *et al.*, 2001, *Cell Transplant.* **10** (4–5): 419–422.

69. Garcia, J. H., Liu, K.-F., Bree, M. P., 1996, *Am J Pathol.* **148**: 241–248.

70. Garcia, J. H., Liu, K. F., Yoshida, Y., *et al.*, 1994, *Am J Pathol.* **195**: 721–740.

71. Gielen, W., 1968, *Z Nature.* **23**: 1007–1008.

72. Giraldi, C., Masi, M. C., Manetti, M., *et al.*, 1990, *Acta Neurol Nap.* **12**: 214–221.

73. Glebov, R. N., Kryzhanovsky, G. N., 1978, *Functional Biochemistry of Synapses.* Meditsina, Moscow (in Russian).

74. Goldberg, M. P., 1997, *Stroke Trials Database, Internet Stroke Center at Washington University* (cited 5 April 1999) (http://www.neuro.wustl.edu/stroke).

75. Gothe, M., 1974, *ZFA.* **50**: 588.

76. Gozes, I., *et al.*, 1999, *J Neurochem.* **72**: 1283.

77. Gozes, I., Giladi, E., Pinhasov, A., *et al.*, 2000, *Regulatory Peptides.* **89** (1–3): 62.

78. Grander, C. V., Hamilton, B. R., Gresham, G. E., *et al.*, 1989, *Arch Phys Med Rehabil.* **70**: 100–103.

79. Grilli, M., Barbieri, I., Basudev, H., *et al.*, 2000, *Eur J Neurosci.* **12** (7): 2265–2272.

80. Grivennikov, I. A., Dolotov, O. V., Goldina, Ju. I., 1999, *Mol Biol.* **33** (1): 120–126 (in Russian).

81. Grivennikov, I. A., Dolotov, O. V., Myasoedov, N. F., *et al.*, 1997, Effects of a new behaviorally active ACTH analog, Semax, on cholinergic basal forebrain neurons. *Soc Neurosci. 27th Ann Meet New Orlean.* **23** (1): 891.

82. Gromova, O. A., Avdeyenko, T. V., Burtsev, Ye. M., 1998, *Korsakoff J Neurol Psychiatr.* **1**: 36–37 (in Russian).

83. Gross, C. E., Bednar, M. M., Howard, D. B., Sporn, M. B., 1993, *Stroke.* **24**: 558–562.

84. Gu, Z., Jiang, Q., Zhang, G., 2001, *Neuroreport.* **12** (16): 3487–3491.

85. Gurlenya, A. M., Talapin, V. I., 1978, *Pharmacotherapy of Nervous Diseases.* Vysshaya Shkola, Minsk, 288 (in Russian).

86. Gusev, E. I., 1992, *Ischemic brain disease*. Assembly speech, Moscow (in Russian).
87. Gusev, E. I., Burd, G. S., Skvortsova, V. I., *et al.*, 1994, *Korsakoff J Neurol Psychiatr.* **1**: 9–13 (in Russian).
88. Gusev, E. I., Burd, G. S., Skvortsova, V. I., 1995, *Herald RSMU.* **1** (1): 21–28 (in Russian).
89. Gusev, E. I., Burd, G. S., Skvortsova, V. I., Gekht, A. B., 1996, The system of neuroprotective therapy of ischemic stroke. In *Modern Techniques of Diagnosis and Treatment of Nervous Diseases*, Ufa, pp. 107–118 (in Russian).
90. Gusev, E. I., Skvortsova, V. I., 1990, *Case Record Form for the Examination and Treatment of Patients with Ischemic Stroke*. Moscow, pp. 1–44 (in Russian).
91. Gusev, E. I., Skvortsova, V. I., Myasoedov, N. F., *et al.*, 1999, *Cerebrovasc Dis.* **9**: 126.
92. Gusev, E. I., Skvortsova, V. I., Rayevsky, K. S., *et al.*, 1995, Detection of neurotransmitter amino acid content in CSF of patients with ischemic stroke. In *Functional Studies as a Basis for Drug Creation*, Moscow, pp. 133–134 (in Russian).
93. Gusev, E. I., Skvortsova, V. U., Raevsky, K. S., *et al.*, 1998, *Aktuelle Neurologie.* **25** (3): 132.
94. Gusev, E. I., Skvortsova, V. I., Zhuravleva, E. Yu., *et al.*, 1997, Neuropeptide Semax (ACTH 4–10) in therapy of acute ischemic stroke. In *2nd Int Santa Margherita Ligure Symp on New Therapeutic Strategies in Ischemic Stroke*, Genoa. **11**: 55–56.
95. Gusev, E. I., Skvortsova, V. I., Zhuravlyova, E. Yu., Yakovleva, E. A., 1999, *Int Med J.* **5** (1): 45–51 (in Russian).
96. Gwag, B. J., Canzoniero, L. M., Sensi, S. L., 1999, *Neuroscience.* **90** (4): 1339–1348.
97. Haley, E. C., On behalf of the RANTTAS II investigators, 1998, *Stroke.* **29**: 1256–1257.
98. Hara, H., Fink, K., Endres, M., *et al.*, 1997, *J Cerebr Blood Flow Metab.* **17**: 370–375.
99. Harrer, G., 1954, *Dtsch Med Wochenschr.* **79**: 983.
100. Harrer, G., 1972, *Arztliche Praxis.* **24**: 3595.
101. Hebenstait, G. F., 1991, Efficacy of amino acid peptide compound in cerebral dysfunction with especial reference to gerontopsychiatry. In *Abst Int Symp on Cerebrolysin Application*, Moscow, pp. 16–20 (in Russian).
102. Henric-Noack, P., Prehn, J. H. M., Kreiglestein, J., 1996, *Stroke.* **27**: 1609–1615.
103. Herin, G. A., Du, S., Aizenman, E., 2001, *J Neurochem.* **78** (6): 1307–1314.
104. Hernandez, L. A., Grisham, M. B., Twohic, B., *et al.*, 1987, *Am J Physiol.* **253**: H699–H703.
105. Hershkovitz, M., 1983, *Biochim Acta.* **692**: 495–497.
106. Hershkovitz, M., Zwiers, H., Gispen, W. H., 1982, *Biochim Acta.* **692**: 495–497.
107. Hockenbery, D. M., Oltvai, Z. N., Xiao-Ming, Y., *et al.*, 1993, *Cell.* **75**: 241–251.
108. Hussein, O., Schlezinger, S., Rosenblat, M., *et al.*, 1997, *Atherosclerosis.* **128**: 11–18.
109. Hutter-Paier, L., Grygar, E., Windisch, M., 1996, *J Neural Transm Suppl.* **47**: 267–273.
110. Iadecola, C., Zhang, F., Xu, X., 1995, *Am J Physiol.* **268**: 286–292.
111. Ichiama, T., Campbell, I. L., Furukawa, S., *et al.*, 1999, *J Neurosci Res.* **58**: 684–689.
112. Imai, H., Masayasu, H., Dewar, D., *et al.*, 2001, *Stroke.* **32** (9): 2149–2154.
113. Irving, E. A., Barone, F. C., Reith, A. D., *et al*, 2000, *Brain Res Mol Brain Res.* **77** (1): 65–75.
114. Irving, E. A., Hadingham, S. J., Roberts, J., *et al.*, 2000, *Neurosci Lett.* **288** (1): 45–48.
115. Issawdeh, S., Lorentien, J. C., Mustafa, M. I., *et al.*, 1996, *J Neuroimmunol.* **69** (1–2): 103–115.
116. Isuji, S., Arita, M., Nagai, G., 1983, *J Biochem.* **94** (1): 303–306.

117. Jamieson, D. G., Reivich, M., Alves, W., *et al.*, 1989, *J Cerebr Blood Flow Metab.* **9** (1): S602.
118. Jean, W. C., Spellman, S. R., Nussbaum, E. S., Low, W. C., 1998, *Neurosurgery.* **43**: 1382–1396.
119. Jensen, M. L., Henriksen, U., Dahl, B. J., *et al.*, 1986, *Allergy.* **41** (2): 151–156.
120. Kadoya, C., Domino, E. F., Yang, G. Y., *et al.*, 1995, *Stroke.* **26**: 1035–1038.
121. Kamkin, A. G., Kiseleva, I. S., Kositskii, I. G., *et al.*, 1985, *USSR Acad Sci Reports.* **284**: 245–248.
122. Kane, D. J., Ord, T., Anton, R., *et al.*, 1995, *J Neurosci.* **40**: 269–275.
123. Kaplan, A. Ya., Kamensky, A. A., Ashmarin, I. P., *et al.*, 1991, Anti-ischemic properties of neuropeptide Semax: EEG analysis. In *Application of Minor Regulatory Peptides in Anesthesiology and Intensive Therapy,* Moscow, pp. 78–87 (in Russian).
124. Kaplan, A. Ya., Kochetova, A. G., Nezavibatko, V. N., *et al.*, 1996, *Neurosci Res Commun.* **19** (2): 115–123.
125. Kaplan, A. Ya., Koshelev, V. B., Nezavibatko, V. N., Ashmarin, I. P., 1992, *Human Physiol.* **18** (5): 104–107 (in Russian).
126. Karasek, F., 1975. In *2nd Sowjet-Osterrich Symp,* 34.
127. Karpiak, S. E., Hanadik, S. P., Wakade, C. G., *et al.*, 1990, *Crit Rev Neurobiol.* **5** (3): 221–237.
128. Karpiak, S. E., Li, Y. S., Hanadik, S. P., 1986, *Clin Neuropharmacol (CNK).* **9**: 338–348.
129. Kassel, N. F., Haley, E. C., Appersonhausen, C., *et al.*, 1996, *J Neurosurg.* **84** (2): 221–228.
130. Kawamata, T., Dietrich, D. W., Schallert, E., *et al.*, 1997, *Proc Natl Acad Sci USA.* **94**: 8179–8184.
131. Kawamata, T., Ren, J., Chan, T. C., *et al.*, 1998, *Neuroreport.* **9**: 1441–1445.
132. Kinouchi, H., Mizui, T., Carlson, E., *et al.*, 1991, *J Cerebr Blood Flow Metab.* **21**: 423.
133. Koike, T., Tanaka, S., 1991, *Proc Natl Acad Sci USA.* **88**: 3892–3896.
134. Kolesnick, R., Golde, D. W., 1994, *Cell.* **77**: 325–328.
135. Koper, D., 1997, Cerebrolysin: unique neurotrophic effect and clinical use. In *1st Int Mondsee Med Meet.* Unterach, Austria, 6.
136. Koppi, S., *et al.*, 1996, *Wien Med Wschr.* **145**: 555–567.
137. Kovalenko, A. V., 1992, *Functional State of Efferent and Afferent Pathways of CNS in Acute Ischemic Stroke on the Background of Neuroprotective Therapy.* Candidate's dissertation, Moscow (in Russian).
138. Kovalev, G. V., 1985, Pharmacology and clinical application of neuroreactive amino acids and their analogs. In *Works Volgograd Med Inst*, Volgograd. **37** (5): 295 (in Russian).
139. Krammer, F., 1959, *Prakt Arzt.* **13**: 649.
140. Kreps, E. M., 1981, *Lipids of Cellular Membranes.* Nauka, Leningrad, 339 (in Russian).
141. Kryzhanovsky, G. N., Lutsenko, V. K., 1995, *Adv Contemp Biol.* **115** (1): 31–49 (in Russian).
142. Kunsch, C., Lang, R. K., Rosen, C. A., Shannon, M. F., 1994, *Immunol.* **153**: 153–164.
143. Ladurner, G., 2001, *Stroke.* **32**: 323.
144. Laufs, U., Gertz, K., Dirnagl, U., *et al.*, 2002, *Brain Res.* **942** (1–2): 23–30.
145. Lebedeva, N. V., 1975, The experience of Cerebrolysin application in stroke patients. In *Works 2nd Moscow Med Inst*, Moscow. **48** (2): 80–85 (in Russian).
146. Ledeen, R. W., 1983, Gangliosides. In *Handbook of Neurochemistry,* Plenum Press, New York. **3**: 41–90.

147. Lee, A.-S., 1976, *Prakt Arzt.* **24**: 907.
148. Lee, J. M., Zipfel, G. J., Choi, D. W., 1999, *Nature.* **399** (24): 7–14.
149. Lees, K. R., 1998, *Lancet.* **351**: 1447–1448.
150. Lenzi, G. L., Grigoletto, F., Gent, M., *et al.*, 1994, *Stroke.* **25**: 1552–1558.
151. Leskawa, K. C., Rosenberg, A., 1980, *Adv Exp Med Biol.* **125**: 125–135.
152. Liao, J. K., 2002, *Atheroscler Suppl.* **3** (1): 21–25.
153. Liao, S., Chen, W., Kuo, J., Chen, C., 2001, *Neurosci Lett.* **297** (3): 159–162.
154. Libermann, T. A., Baltimore, D., 1990, *Mol Cell Biol.* 2327–2334.
155. Lin, S. Z., Hoffer, B. J., Kaplan, P., Wang, Y., 1999, *Stroke.* **30**: 126–133.
156. Linnik, M. D., Zobrist, R. H., Hatfield, M. D., 1993, *Stroke.* **24**: 2002–2008.
157. Lipton, P., 1999, *Physiol Rev.* **79** (4): 1431–1568.
158. Loddick, S. A., *et al.*, 1998, *Proc Natl Acad Sci USA.* **95**: 1894–1898.
159. Lu, A., Ran, A., Parmentier-Batteur, S., *et al.*, 2002, *J Neurochem.* **81** (2): 355–364.
160. Mahadic, S. P., Hawver, D. B., Hangung, B. K., *et al.*, 1989, *J Neurosci Res.* **24** (3): 402–412.
161. Mahoney, F. I., Barthel, D. W., 1965, *Md State Med J.* **14**: 61–65.
162. Mandir, A. S., Poitras, M. F., Berliner, A. R., *et al.*, 2000, *J Neurosci* **20** (21): 8005–8011.
163. Manev, H., Costa, E., Wroblewski, J. T., Guidotti, A., 1990, *FASEB J.* **4**: 2789–2797.
164. Manna, S. K., Aggarwal, B. B., 1998, *J Immunol.* **161**: 2873–2880.
165. Marino, M. W., Dunn, A., Crail, D., *et al.*, 1997, *Proc Natl Acad Sci USA.* **94**: 8093–8098.
166. Markova, E. D., Ivanova-Smolenskaya, I. A., *et al.*, 1975, In *Theoretical and Practical Basis of Treatment with Cerebrolysin. Abst 2nd Symp*, Moscow (in Russian).
167. Martin-Villalba, A., Hahne, M., Kleber, S., *et al.*, 2001, *Cell Death Differ.* **8** (7): 659–661.
168. Mattioli, R., Huston, J. P., Spieler, R. E., 2000, *Neural Plast.* **7** (4): 291–301.
169. Mattson, M. P, Culmsee, C., Yu, Z. F., 2000, *Cell Tissue Res.* **301** (1): 173–187.
170. McDaniel, R., 1985, *Biochim Biophys Acta.* **819** (2): 153–160.
171. McGrow, C. P., 1977, *Arch Neurol.* **34**: 334–336.
172. Miceli, G., 1977, *Acta Psychiatr Scand.* **55**: 103–110.
173. Mihajlovic, R., Jovanovic, M., Djordjevic, D., Jovicic, A., 1996, *Cerebrovasc Dis.* **6**: 13.
174. Mirzoyan, S. A., Sekoyan, E. S., Sotsky, O. P., 1984, *Bull Exp Biol Med.* **97** (6): 681–683 (in Russian).
175. Mizui, T., Kinouchi, H., Chan, P. H., 1992, *Am J Physiol.* **262** (2): 313–317.
176. Monaco, P., Pastore, L., Cottone, S., *et al.*, 1991, Early treatment of patients with ischaemic stroke: a double-blind study with monosialotetraesosigangliolode (GM1). *First Int Conf on Stroke,* 7.
177. Myasoyedov, N. F., Skvortsova, V. I., Nasonov, E. L., *et al.*, 1999, *Korsakoff J Neurol Psychiatr.* **5**: 15–19 (in Russian).
178. Nakajima, Y., Mori, A., Maeda, T., Fujimiya, M., 1997, *Brain Res.* **765**: 113–121.
179. Namura, S., Nagata, I., Takami, S., 2001, *Stroke.* **32** (8): 1906–1911.
180. Nanri, K., Montecot, C., Springhetti, V., *et al.*, 1998, *Stroke.* **29**: 1248–1253.
181. Nozaki, K., Nishimura, M., Hashimoto, N., 2001, *Mol Neurobiol* **23** (1): 1–19.
182. Ogawa, A., Yoshimoto, T., Kikuchi, H., *et al.*, 1999, *Cerebrovasc Dis.* **9**: 112–118.
183. Okamoto, S., Li, Z., Ju, C., *et al.*, 2002, *Proc Natl Acad Sci USA.* **99** (6): 3974–3979.
184. Oktem, I. S., Menku, A., Akdemir, H., *et al.*, 2000, *Res Exp Med.* **199** (4): 231–242.
185. Onal, M. Z., Fisher, M., 1996, *Drugs Today.* **32** (7): 573–592.

186. Onal, M. Z., Fisher, M., 1997, *Eur Neurol.* **38**: 141–154.
187. Orgogozo, J. M., Dartigues, J. F., 1986. In *Acute Brain Ischemia/Medical and Surgical Therapy* (Battistini, N., ed.) Raven Press, New York, pp. 282–289.
188. Ortner, S., 1956, *Prakt Atzt.* **10**: 360.
189. Pahan, K., Sheikh, F. G., Namboodiri, A. M., Singh, I., 1997, *J Clin Invest.* **100**: 2671–2679.
190. Panetta, T., Marcheselli, V. L., Braquet, P., Bazan, N. G., 1989, *Ann NY Acad Sci.* **559**: 340–351.
191. Parnham, M., Sies, H., 2000, *Expert Opin Investig Drugs.* **9** (3): 607–619.
192. Panse, F., 1955, *Med Klin.* **50**: 942.
193. Parnham, M., Sies, H., 2000, *Expert Opin Investig Drugs.* **9** (3): 607–619.
194. Partington, C. R., 1979, *Mol Pharmacol.* **15** (3): 424–491.
195. Peters, G. R., Hwang, L.-J., Musch, B., *et al.*, 1996, *Stroke.* **27**: 195.
196. Polak, J. M., Bloom, S. R., 1979, *J Histochem Cytochem.* **27** (10): 1398–1400.
197. Polischuk, N., 1997, Cerebrolysin in acute head injury and acute stroke. In *1st Int Mondsee Med Meet*, Unterach, Austria, 7.
198. Ponomaryova-Stepnaya, M. A., Alfeyeva, L. Yu., Maximova, L. A., *et al.*, 1981, *Chem-Pharm J.* **10**: 37–42 (in Russian).
199. Ponomaryova-Stepnaya, M. A., Porunkevich, E. A., Skuinsh, A. A., *et al.*, 1984, *Bull Exp Biol Med.* **51**: 82 (in Russian).
200. Porciuncula, L. O., Rocha, J. B., Boeck, C. R., *et al.*, 2001, *Neurosci Lett.* **299** (3): 217–220.
201. Potaman, V. N., Alfeeva, L. Y., Kamensky, A. A., *et al.*, 1991, *Biochem Biophys Res Commun.* **176** (2): 741–746.
202. Potaman, V. N., Antonova, L. V., Dubynin, V. A., *et al.*, 1991, *Neurosci Lett.* **127**: 133–136.
203. Prehn, J. H., Backhauss, C., Krieglstein, J., 1993, *J Cerebr Blood Flow Metab.* **13**: 521–525.
204. Prehn, J. H. M., Bindokas, V. P., Marcuccilli, C. J., *et al.*, 1994, *Proc Natl Acad Sci USA.* **91**: 12599–12603.
205. Rajdev, S., Hara, K., Kokubo, Y., *et al.*, 2000, *Ann Neurol.* **47** (6): 782–791.
206. Rainer, M., 2000, Open-label study with Cerebrolysin – a neurotrophic agent for the treatment of dementia. In *Proc 6th Int Stochholm Springfield Symp on Advances in Alzheimer's Disease,* **4**: 225.
207. Rankin, J., 1957, *Prognosis Scot Med J.* **2**: 200–215.
208. RANTTAS investigators, 1996, *Stroke.* **27**: 195.
209. Rotondo, G., Meniero, G., Toffano, G., 1990, *Aviat Space Enviton Med.* **61** (2): 162–164.
210. Rudzinski, W., Krejza, J., 2002, *Neurol Neurochir Pol.* **36** (1): 143–156.
211. Ruether, E., Husmann, R., Kinzler, E., *et al.*, 2001, *Int Clin Psychopharmacol.* **16** (5): 253–263.
212. Saleh, T. M., Cribb, A. E., Connell, B. J., 2001, *Am J Physiol Regul Integr Comp Physiol.* **281** (5): 1531–1539.
213. Scandinavian Stroke Study Group, 1985, *Stroke.* **16**: 885–890.
214. Schabitz, W. R., Schwab, S., Spranger, M., Hacke, W., 1997, *J Cerebr Blood Flow Metab.* **17** (5): 500–506.
215. Schmid-Elsaesser, R., Zausinger, S., Hungerhuber, E., *et al.*, 1999, *Neurosurgery.* **44** (1): 163–171.

216. Schwab, M., Antonow-Schlorke, I., Zwiener, U., Bauer, R., 1998, *J Neural Transm Suppl.* **53**: 299–311.
217. Seifert, W., 1981, Gangliosides in nerve cell cultures. In *Gangliosides in Neurological and Neuromuscular Function, Development and Repair* (Rapport, M. M., Gorio, A., eds.) Raven Press, New York, pp. 99–117.
218. Sen, S., Phyllis, J. W., 1993, *Free Rad Res Commun.* **19**: 255–265.
219. Seren, M. C., Rubini, R., Lassaro, A., *et al.*, 1990, *Stroke.* **21**: 1607–1612.
220. Sharp, F. R., Massa, S. M., Swanson, R. A., 1999, *Trends Neurosci.* **22** (3): 97–99.
221. Silver, B., Weber, J., Fisher, M., 1995, *Clin Neuropharmacol.* **19**: 101–128.
222. Sinitsyna, E. V., Limenovskaya, A. F., Kluchareva, T. E., *et al.*, 1985, The influence of gangliosides on natural killers activity in Syrian hamsters. In *Abst 5th All-Union Symp of Biochemists*, Nauka, Moscow. **3**: 132–133 (in Russian).
223. Skvortsova, V. I., 1993, *Clinical and Neurophysiological Monitoring and Neuroprotective Therapy in Acute Ischemic Stroke*. Doctoral dissertation, Moscow (in Russian).
224. Skvortsova, V. I., 1994, *News Med Pharm.* **4**: 26–31 (in Russian).
225. Skvortsova, V. I., 2000, *Int J Med Practice.* **4**: 33–35 (in Russian).
226. Skvortsova, V. I., Fidler, S. M., 1983, Clinical and electrophysiological study of Cerebrolysin effects on functional state of brain in acute ischemic stroke. In *Vascular Diseases of Nervous System*, Moscow, pp. 61–68 (in Russian).
227. Skvortsova, V. I., Gusev, E. I., Ashmarin, I. P., *et al.*, 2000, *Regulatory Peptides.* **89**: 81.
228. Skvortsova, V. I., Nasonov, E. L., Zhuravlyova, E. Yu., *et al.*, 1999, *Korsakoff J Neurol Psychiatr.* **5**: 27–32 (in Russian).
229. Smith, C. W., Rothlein, R., Highes, B. J., *et al.*, 1988, *J Clin Invest.* **82**: 1746–1756.
230. Sommer, H., 1975, In *2nd Sowjet-Osterrich Symp*, 50.
231. Sommer, H., Harrer, G., 1973, *Folia Clin Int.* **23**: 4.
232. Sommer, H., Quandt, J., 1973, *Schweiz Arch J Neurol Neurochir Psychiatr.* **112**: 373.
233. Sommer, H., Quandt, J., 1987, *Neuropsychiatry.* **1** (2): 89–93.
234. Song, B. W., Vinters, H. V., Wu, D., Pardridg, W. M., 2002, *J Pharmacol Exp Ther.* **301** (2): 605–610.
235. Spera, P. A., Ellison, J. A., Feuerstein, G. Z., Barone, F. C., 1998, *Neurosci Lett.* **251** (3): 189–192.
236. Spong, C. Y., Abebe, D., Auth, J., *et al.*, 2000, *Regulatory Peptides.* **89** (1–3): 82.
237. Stadler, C., 1997, Current drug treatment of stroke. In *1st Int Mondsee Med Meet*, Unterach, Austria, 3.
238. Staunton, D. E., Merluzzi, V. J., Rothlein, R., *et al.*, 1989, *Cell.* **56**: 849–853.
239. STIPAS investigators, 1994, *Stroke.* **25**: 418–423.
240. Stolyarova, L. G., Kadykov, A. S., Kistenyov, B. A., *et al.*, 1975, Cerebrolysin efficacy in treatment of patients with residual signs of stroke. In *Theoretical and Practical Basis of Treatment with Cerebrolysin. Abst 2nd Symp*, Moscow, pp. 22-26 (in Russian).
241. Sugimori, H., Speller, H., Finklestein, S. P., 2001, *Neurosci Lett.* **300** (1): 13–16.
242. Sugra, Y., *et al.*, 1993, *J Brain Nerve.* **45** (4): 325–331.
243. Svennerholm, L., 1984, Biological significance of gangliosides. In *Cellular and Pathological Aspects of Glycoconjugate Metabolism* (Dreyfus, H., *et al.*, eds.) INSERM, Paris, pp. 21–44.
244. Tai, T., 1983, *Proc Natl Acad Sci USA.* **80** (17): 5392–5396.
245. Takasago, T., Peters, E. E., Graham, D. I., *et al.*, 1997, *Br J Pharmacol.* **122**: 1251–1256.

246. Takeda, A., Onodera, H., Yamasaki, Y., *et al.*, 1992, *Brain Res.* **569**: 177–180.
247. The Enlimomab Acute Stroke Trial Investigators, 2001, *Neurology.* **57** (8): 1428–1434.
248. The Tirilazad International Steering Committee Stroke, 2000. **31** (9): 2257–2265.
249. Toffano, G., 1984, *Brain Res.* **256** (2): 233–239.
250. Toffano, G., Savoini, G. E., Moroni, F., *et al.*, 1983, *Brain Res.* **261**: 163–166.
251. Toung, T. K., Hurn, P. D., Traystman, R. J., Sieber, F. E., 2000, *Stroke.* **31** (11): 2701–2706.
252. Toung, T. K., Traystman, R. J., Hurn, P. D., 1998, *Stroke.* **29**: 1666–1670.
253. Tsunawasi, S., Sporn, M., Ding, A., Nathan, C., 1988, *Nature.* **334**: 260–262.
254. Tumanova, S. Yu., 1985, Poly-functional role of gangliosides on cellular surface. In *Abst 5th All-Union Symp of Biochemists*, Nauka, Moscow. **1**: 225 (in Russian).
255. Tumanova, S. Yu., Prokhorova, M. I., 1982, *Neurochemistry.* **1** (2): 184–199 (in Russian).
256. Twichell, T. E., 1951, *Brain.* **74**: 443–480.
257. Urban, P. F., Harth, S., Frey, L., *et al.*, 1980, *Adv Exp Med Biol.* **125**: 149–157.
258. Urbanics, R., Greenberg, J. H., Toffano, G., *et al.*, 1989, *Stroke.* **20**: 795–802.
259. Van Der Worp, H. B., Kappelle, L. J., Algra, A., *et al.*, 2002, *Neurology.* **58** (1): 133–135.
260. Vaughan, C. J., Delanty, N., 1999, *Stroke.* **30** (9): 1969–1973.
261. Vaughan, C. J., Delanty, N., Basson, C. T., 2001, *CNS Drugs.* **15** (8): 589–596.
262. Veh, R., Avrova, F., 1983, *J Evolut Biochem Physiol.* **19** (5): 507–510 (in Russian).
263. Vindish, M., 1991, Cerebrolysin – the last results in evaluation of multi-modal effect of the drug. In *The 3rd Int Symp on Cerebrolysin*, Moscow, pp. 81–86 (in Russian).
264. Volkov, A. V., Zarzhetsky, Yu. V., Postnov, A. Yu., *et al.*, 1992, The results of regulatory peptides application in resuscitation after experimental cardiac arrest. In *Terminal Stages and Post-resuscitation Pathology of the Organism: Pathophysiology, Clinical Manifestations, Prevention and Treatment*, Institute of General Resuscitation, Russian Academy of Medical Sciences, Moscow, pp. 69–76 (in Russian).
265. Wahlgren, N. G., 1997, A review of earlier clinical studies on neuroprotective agents and current approaches. In *Neuroprotective Agents and Cerebral Ischemia* (Green, R., Cross, A. J., eds.) Academic Press, San Diego-London-Boston-New York-Sydney-Tokyo-Toronto, pp. 337–363.
266. Waters, C., 1997, *RBI Neurotransmissions, Newsletter for Neuroscientist.* **XIII** (2): 2–7.
267. Wender, M., Mularek, J., Godlewski, A., *et al.*, 1993, *Neurol Neurochir Pol.* **27**: 31–38.
268. Windisch, M., 1994, Cerebrolysin. In *EBEVE Pharmaceuticals,* Austria, pp. 14–22.
269. Windisch, M., Piswanger, A., 1987, *Neuropsychiatry.* **1** (2): 83–88.
270. Wojcik, M., Ulas, J., Oderfield-Nowak, B., 1982, *Neuroscience.* **7**: 495–499.
271. Wong, M.-L., Bongiorno, P. B., Rettori, V., 1997, *Exp Med J.* **163**: 740–745.
272. Xue, D., Slivka, A., Buchan, A. M., 1992, *Stroke.* **23**: 894–899.
273. Yakovleva, E. V., Kuzenkov, V. S., Fedorov, V. N., *et al.*, 1999, *Bull Exp Biol Med.* **128** (8): 172–174 (in Russian).
274. Yamaguchi, T., Sano, K., Takakura, K., *et al.*, 1998, *Stroke.* **29**: 12–17.
275. Yamasaki, Y., Matsuo, Y., Matsuura, N., *et al.*, 1995, *Stroke.* **26**: 318–323.
276. Yamashita, K., Wiessner, C., Lindholm, D., *et al.*, 1997, *Metab Brain Dis.* **12** (4). 271–280.
277. Yanamoto, H., Nagata, I., Sakata, M., *et al.*, 2000, *Brain Res.* **859** (2): 240–248.
278. Yu, S. W., Wang, H., Poitras, M. F., *et al.*, 2002, *Science.* **297** (5579): 259–263.
279. Yue, T. L., Gu, J. L., Lysko, P. G., *et al.*, 1992, *Brain Res.* **574**: 193–197.

280. Yun, H. Y., Dawson, V. L., Dawson, T. M., 1999, *Diabetes Res Clin Pract.* **45** (2–3): 113–115.
281. Zavalishin, I. A., Niyazbekov, A. S., 1975, In *Theoretical and Practical Basis of Treatment with Cerebrolysin. Abst 2nd Symp*, Moscow (in Russian).
282. Zhang, R. L., Chopp, M., Jiang, N., *et al.*, 1995, *Stroke.* **26**: 1438–1442.
283. Zhang, R. L., Chopp, M., Zaloga, C., *et al.*, 1994, *Neurology.* **44**: 1747–1751.
284. Zhang, Y., Pardridge, W. M., 2001, *Brain Res.* **889**: 49–56.
285. Zhu, Y., Yang, G. Y., Ahlemeyer, B., *et al.*, 2002, *J Neurosci.* **22** (10): 3898–3909.
286. Zhuravlyova, E. Yu., 1998, *The Role of Peptide-ergic Neurotransmitter Systems in the Pathogenesis and Neuroprotective Therapy of Acute Ischemic Stroke (Clinical and Biochemical Study)*. Candidate's dissertation, Moscow, 191 (in Russian).

Chapter 16

Reparative Therapy

Infarction begins to form from the first minutes of acute focal ischemia and 3–6 h later there already exists irreversible morphological damage in brain tissue. Thus, from the first hours after stroke onset it is necessary to proceed with treatment aimed at the improvement of reparative and regenerative processes, plasticity of brain tissue, and the formation of new associative pathways, along with neuroprotection aimed at prevention of further brain damage and resulting enlargement of infarction volume.

It is difficult to define the borders between neuroprotection and reparative therapy. The majority of secondary neuroprotectors possess reparative properties. All neurotrophic factors, modulators of functional state of membranes and receptors (gangliosides), and endogenous neuropeptide regulators have prominent influence on the extent of regenerative and reparative processes. On the other hand, "predominantly reparative" drugs may have some neuroprotective effects. Thus, the subdivision of "reparative therapy" is highly conditional and only emphasizes the theoretical importance of this therapeutic strategy.

Derivates of GABA and choline and growth phosphoproteins are related to the drugs with regenerative and reparative effects in addition to the drugs described in Chapter 15. All of them have a complex effect on brain metabolism and cerebral blood supply.

16.1. Piracetam

Piracetam (2-oxo-1-pyrrolidine acetamide; Nootropil, Normabrain, Pyramem) is the first representative of nootropic drugs, which gave the name to the whole drug group. The structural basis of piracetam is a cyclic form of GABA (2-pyrrolidone), but this drug is not a source of a metabolically active

GABA. Piracetam permeates through the blood brain barrier (BBB) reaching peak content in CSF after 3 h. Its half-life period is 7.5 h [52].

Metabolic effects of the drug are various and can be subdivided into two types: fast influence on oxidation–reduction and energy processes, and delayed reparative and regenerative action [16, 62].

It has been demonstrated that piracetam improves oxidative metabolism under conditions of brain ischemia, preventing an abrupt decrease in ATP level, increasing the rate of ATP turnover, and activating the synthesis of creatine phosphate. The drug stimulates the activity of adenylate kinase, facilitating anaerobic glucose metabolism without generation of lactate. It promotes the maintenance of protein synthesis by preserving intracellular ribosomal system and activated synthesis of DNA, RNA, and phospholipids [21, 28, 35, 41–43]. Piracetam was shown to have GABA-mimetic post-synaptic effects as well as to enhance compensatory GABA-shunting, which promotes an alternative mechanism of α-ketoglutarate to succinate transformation [47]. Under the metabolic role of GABA-shunting lie additional delivery of rapidly oxidized succinate or production of succinic semialdehyde, which can reduce with formation of GOBA and take a certain number of protons and electrons under conditions of oxygen deficiency [49]. PET studies confirmed that piracetam improved glucose uptake in the ischemized brain area in patients with ischemic stroke [26, 27]. Antioxidant properties of the drug, connected with a decrease in free radical oxidation intensity, were revealed.

Another important aspect of the metabolic action of piracetam is its influence on neurotransmission and brain tissue plasticity. These effects are manifested in the remote period, several days after the beginning of treatment. Piracetam does not bind to receptors, but causes membrane-stabilizing effects, improves liquid properties of neuronal membranes [19, 22, 35, 60], increases density of cholinergic and glutamate receptors, modulates activity of neurotransmitter processes, and normalizes trans-callosal (inter-hemispheric) information exchange [14, 39, 45].

Along with metabolic effects, piracetam has influences on microcirculation and cerebral blood supply. The drug was shown to decrease adhesion and aggregation of platelets, increase deformation capacity of erythrocyte membranes, decrease adhesion of erythrocytes to endothelium and to each other, and normalize blood viscosity [55]. Being injected to experimental animals with acute focal brain ischemia, piracetam moderately decreased systemic blood pressure and general peripheral resistance of cerebral vessels, as well as increased local blood supply [1]. The drug was shown to prevent spasm of brain arteries [34]. Preliminarily administered piracetam (before the induction of ischemia) inhibited and prevented the development of no-reflow phenomenon [29, 33].

Experimental studies, made in rats prior to, and from 1 to 12 h after, induction of a focal brain ischemia with lesion in the primary somatosensory cortex, showed that piracetam exerts a protective function on the physiological response properties of cortical neurons after ischemic injury [12].

In Russia, piracetam was one of the first drugs influencing brain cell metabolism that received access to clinical application in acute ischemic stroke. In the 70s the drug was used in daily doses of 1.4 to 4 g. Application of piracetam in such doses during the first days after the stroke onset neither led to significant changes in the patient's status nor influenced on outcome of stroke. Piracetam did not cause an "awakening" effect in patients. A correlation was revealed between the efficacy of the drug and the type of clinical syndrome in patients: better results were observed in cortical location of ischemic lesion, when cognitive disturbances and aphasic syndrome were the most important clinical manifestations. There was a less prominent effect of piracetam on the improvement of motor and sensory functions [37]. Its predominant effect on the extent of neuropsychological symptoms was confirmed in studies by Buklina [10], Poeck *et al.* [46], and Enberby *et al.* [17]. Especially beneficial effects of piracetam were demonstrated in patients with post-stroke aphasia [23, 31, 56]. Piracetam treatment was shown to result in significant improvement of speech in its different domains, as well as caused marked shift in the α-rhythm from frontal to occipital regions. The authors explained such EEG effects of the drug by a restitution of corticothalamic circuits involved in the generation of α-activity. Parallel investigation of neuropsychological tests and activation PET measurement of cerebral blood flow found that in patients treated with piracetam, recovery of various language functions was accompanied by a significant increase in task-related flow activation in special areas of the left hemisphere, such as the left transverse temporal gyrus, left triangular part of inferior frontal gyrus, and left posterior superior temporal gyrus [31]. Piracetam has now reached general acceptance as a cognition enhancer, the drug able to facilitate attentional abilities and acquisition, storage, and retrieval of information, and to attenuate the impairment of cognitive and speech functions associated with stroke [23].

Pilot clinical trials of piracetam in acute carotid ischemic stroke showed that the drug in daily dose exceeding 6 g [28, 30, 45] confirmed its beneficial effects on stroke course and substantiated the need for randomized multi-center trials. However, such a study, enrolling 927 patients, who received piracetam from the first 12 h of stroke for the following 12 weeks (on admission piracetam was administered in bolus injection in the dose of 12 g, then for 4 weeks piracetam was intravenously infused in daily dose of 12 g, then for 8 weeks 4.8 g/day were given orally), did not reveal improvement of stroke outcome (by the Orgogozo Stroke Scale) and decrease of the mortality rate. Only post hoc analysis found a tendency towards the improvement of neurological deficit in patients to whom piracetam was administered not later than 7 h after stroke onset ($p = 0.07$) [13, 15]. A special randomized, placebo-controlled trial of piracetam (PASS II) assessing treatment within first 7 h of stroke onset was stopped prematurely on the grounds of futility [8]. Recent systematic review of all randomized controlled trials of piracetam in acute stroke [48] concluded that there is no evidence in favor of routine administration of the drug in patients with acute brain ischemia. However, since a possible beneficial effect could not be completely ruled out, the conclusion was made that further controlled trials would be warranted.

16.1.1. Clinical and neurophysiological monitoring study of dose-dependent efficacy of piracetam (Nootropil) in acute carotid ischemic stroke

To study safety and dose-dependent efficacy of piracetam (Nootropil), we performed a simple pilot study of 115 patients (60 males and 55 females; average age 63.7 ± 9.7 years) in acute carotid ischemic stroke, admitted at the Neurology Clinic of the Russian State Medical University within 12 h after the event [24, 53]. The trial was performed after the approval of the Ethics Committee of the study center was obtained, and the patients or their legal representative gave written or witnessed informed consent to participate in the trial. The inclusion and exclusion criteria were identical to those used in clinical trials of neuroprotectors (see Chapters 14 and 15).

Piracetam was administered on the background treatment, which was maximally unified (see Chapters 14 and 15). Neither drugs with potential reparative and regenerative effects, nor any neuroprotectors were used. Sixty patients to whom standardized "background" treatment (without reparative and neuroprotective drugs) comprised the control group.

Patients of the piracetam and control groups were comparable by sex and age and location and severity of vascular damage. Males predominated in both groups: 53.7 and 51.7% in the control and piracetam groups. The majority of patients were older than 55 years (78.6 and 77.7%, respectively). In 55% (33/60) patients from the control group and in 52.7% (29/55) in the piracetam group stroke developed in the left hemisphere.

Baseline assessments, study design, and statistical analysis were identical to those in the Cerebrolysin and Cronassial trials. On admission, in 38/60 patients of the control group (63.3%) and in 35/55 (63.6%) patients of the piracetam group we diagnosed moderately severe stroke (total clinical score by the Original Stroke Scale ≥ 36); severe stroke (total clinical score < 36) was found in 36.7 and 36.4% of patients, respectively. Protocol-specified study end points were the safety, neurological outcome according to the Original Stroke Scale, functional outcome according to the Barthel Index, and mortality on day 30.

The influence of piracetam on the functional state of brain was studied by neurophysiological monitoring of spontaneous electrical activity using EEG with amplitude–frequency analysis and mapping EEG, and by investigations of somatosensory (SSEP) and short-latency acoustic brain stem evoked potentials (AEP). The complex of examinations was performed on admission, after the first administration of the drug and during following 4 h, and repeatedly on days 2, 6, 14, 21, and 30. The methods were described in Section 14.5.2.

The study was performed in 2 stages. The aim of *Stage I* was to conduct comparative analysis of piracetam efficacy in patients with acute ischemic stroke in relation to its dose and the rate of infusion. The drug was administered in daily dose 0.05–2 g/kg (4–12 g daily) via intravenous infusion or bolus injection for the first 5 days. In *Stage II* we conducted continuous (for all 30 days of the study) treatment with piracetam in optimal daily doses established in Stage I.

The results of *Stage I* indicated steady improvement of general cerebral and focal neurological symptoms in the majority of patients treated with piracetam (75%) and in the control group (73.3%). Progressive course of the disease was observed in 7.5 and 8.3% patients, respectively. Piracetam was shown to be the most effective in patients with moderately severe stroke and especially in cases when cortical focal symptoms prevailed in the clinical picture. In 19.2% patients with moderately severe stroke, the first infusion of piracetam caused subjective improvement of patient's condition (assessed by the patient), a decrease in aphasic disorders (assessed by the Original Scale), and some increase in general motor activity. However, statistical comparison of shifts of the total clinical score by days 6 and 30 did not reveal any differences between the groups. Only in patients with moderately severe strokes of left-hemisphere location and with clinical predominance of aphasic symptoms, the piracetam treatment caused significant acceleration of neurological improvement by day 6 (at the end of the drug treatment): the total clinical score markedly exceeded the control values (by 88.7%, $p < 0.002$). Quantitative assessment of the dynamics of separate focal neurological symptoms confirmed significant improvement of cognitive functions, especially speech in its different domains in patients treated with piracetam, compared with the control group ($p < 0.05$). There were no significant differences in the dynamics of other focal (motor, sensory, visual, etc.) symptoms ($p > 0.05$). In patients with moderately severe stroke

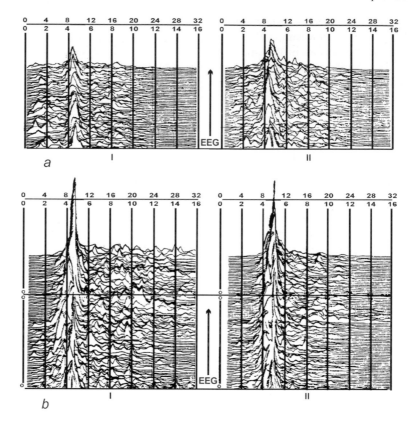

Figure 16.1. Compressed spectral EEG analysis of patient K., 55-years-old, with ischemic stroke in cortical branches of the right middle cerebral artery, treated with piracetam: a) baseline recording (the first 10 h after the onset of stroke); b) after the first intravenous bolus injection of piracetam in a dose of 3 mg (0.05 g/kg). I – right hemisphere; II – left hemisphere. Administration of piracetam causes: increase in the total power of EEG spectrum predominantly due to elevation of power of basic α-1-range; regress of inter-hemispheric asymmetry (IHA).

of the right-hemisphere location, only tendencies towards acceleration of neurological improvement were revealed after the end of treatment with piracetam ($p = 0.09$).

It is noteworthy that in short (5 day-long) treatment with piracetam, the revealed positive effects of the drug on neurological recovery completely vanished by day 30. No differences in the increase in the total clinical score were revealed between the groups by the end of the acute period of stroke: in the left hemispheric strokes the accretion comprised 4.50 ± 0.50 and 4.59 ± 0.48 in the piracetam and control groups, respectively, in the right hemispheric strokes it was 3.60 ± 1.56 and 4.76 ± 0.66, respectively ($p >$ 0.05). This suggested that the 5-day course of the piracetam treatment is not sufficient for development of long-term effects of the drug.

The presented positive clinical dynamics in patients with moderately severe stroke treated with piracetam was accompanied by normalization of spontaneous bioelectrical brain activity such as significant increase in the total power of EEG frequency spectrum, predominantly in α-1-range in the ischemized hemisphere (Fig. 16.1). This response of EEG was registered already 60 min after the first administration of the drug and persisted on days 2 and 3 (Fig. 16.2). At the same time, according to short-latency AEP, a decrease in extent of mesodiencephalic dysfunction was revealed in patients from the piracetam group, this manifested by a tendency towards normalization of latencies of N5 and N3 AEP components, which had been elongated on admission (before the beginning of treatment), and towards shortening of inter-peak N3–N5 interval.

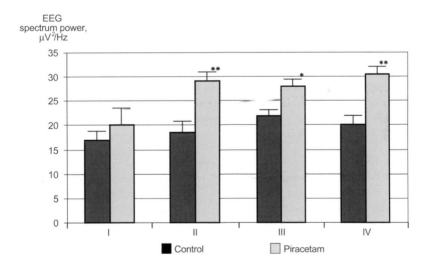

Figure 16.2. Piracetam effects on the total power of EEG spectrum on the first days after ischemic stroke onset. I – before treatment was started; II – 1 h after piracetam injection; III – on day 2; IV – on day 3. Significance of the differences versus controls: * $p < 0.01$, ** $p < 0.001$.

In severe strokes with clinical signs of brain edema, consciousness disturbances, and autonomic trophic dysfunction, piracetam was not effective. After the first administration of the drug, we did not register an "awakening" effect or other positive clinical dynamics in any patients. In 3 patients with atherothrombotic lesion of the middle cerebral artery and subtotal hemispheric damage, clinically accompanied by stupor, autonomic disturbances and severe focal deficit, intravenous bolus injection of piracetam in the dose of 0.1 g/kg provoked psychomotor anxiety and paroxysmal fluctuations of muscle tone. The dynamics of the total clinical score by day 6 in patients with severe strokes, treated with piracetam, corresponded to the control value ($p > 0.05$). Quantitative assessment of the dynamics of separate focal neurological symptoms revealed a tendency towards retardation in the regression of motor and speech disturbances in the piracetam group versus control. No differences in mortality rates between the groups were found ($p > 0.05$).

Presumably, some negative effects of piracetam on the course of severe stroke might be connected with inadequate hyperactivation of brain energy metabolism. This was confirmed by provocation of instability of EEG spectrum with appearance of discharge activity in different frequency ranges (Fig. 16.3), which developed 10–30 min after piracetam injection in 33.3% patients with severe strokes and was accompanied by psychomotor anxiety in the clinical picture. It should be noted that in such cases focal EEG changes were preserved during the full period of observation (until day 30) without any trends towards EEG pattern normalization. Steady θ-focus in the projection of ischemic lesion and inter-hemispheric asymmetry in power of α-activity with its predominance in the "intact" occipital lobe were typical. According to short-latency AEP, a tendency towards worsening in the functional state of mesodiencephalic structures was also registered.

Comparative study of piracetam efficacy in relation to mode of its administration (intravenous infusion or bolus) and frequency of its intake (2 or 4 times a day) failed to reveal any significant differences.

Clinical and neurophysiological comparison of piracetam efficacy in relation to its daily dose showed that low doses (<0.1 g/kg) did not improve clinical symptoms and caused no effect on EEG. Only a short-term (for 40–90 min after the drug administration) increase in the total power of EEG frequency spectrum was found in these cases. Doses of 0.1–0.2 g/kg (6–12 g) administered to patients with moderately severe strokes caused significant acceleration of neurological recovery compared with the control group ($p < 0.05$) and the low doses piracetam group ($p < 0.05$), as well as improvement of functional activity of the brain according to EEG, mapping EEG, and AEP.

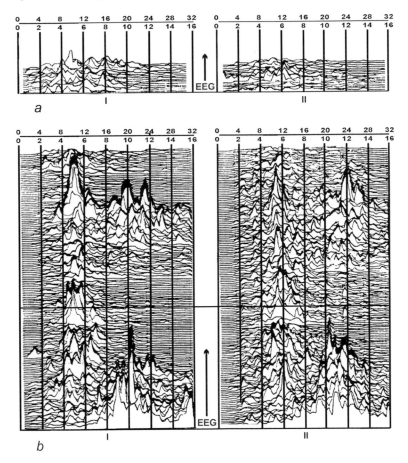

Figure 16.3. Compressed spectral EEG analysis of patient R., 61-year-old, with severe ischemic stroke in the territory of the right internal carotid artery, treated with piracetam: a) baseline recording (first 6 h after the onset of stroke); b) 20 min after the first intravenous bolus injection of piracetam at dose 3 mg (0.05 g/kg) – provocation of unsteady EEG spectrum. I – right hemisphere; II – left hemisphere.

Taking into consideration the results acquired in Stage I, in *Stage II* piracetam was administered only to patients with moderately severe stroke at baseline, and the drug treatment was prolonged until day 30 after the stroke onset. On admission piracetam was intravenously injected in dose of 12 g, then for 15 days the drug was administered via intravenous infusions in daily dose of 12 g, after that, up to day 30 orally in daily dose of 4.8 g.

In all cases piracetam was well tolerated and did not cause any significant side effects.

Statistical analysis performed in the piracetam and control groups revealed in patients treated with piracetam significant improvement of clinical dynamics with more pronounced increase in the total clinical score assessed by the Original Stroke Scale by days 6 and 30 ($p < 0.05$ in the left hemispheric and $p = 0.05$ in the right hemispheric strokes versus control). By day 30 in the piracetam group, statistically significant increase in the quota of patients with good functional recovery (Barthel Index ≥ 75) was shown ($p < 0.05$ versus control).

The study of the dynamics of cognitive functions by neuropsychological tests demonstrated that more prominent recovery was that topically connected with a focal lesion, independently of its location. The cognitive improvement was especially pronounced in patients with left hemispheric damage. The effect of piracetam on the "intact" hemisphere occurred to be less marked.

EEG and mapping EEG confirmed the positive influence of the drug on spontaneous bioelectrical brain activity, which was vividly manifested already 20–30 min after the first piracetam administration by the increase in the total power of EEG frequency spectrum, mainly due to elevation of power of basic α-1-rhythm, as well as by the increase in frequency of basic α-1-rhythm by 1–2 Hz and a tendency towards normalization of its zonal distribution. At the same time, the power of slow δ- and θ-ranges also slightly increased within the first hour of monitoring, but then returned to the baseline level. On day 6 we noted partial normalization of EEG pattern without significant changes of its frequency/amplitude ratio parameters in patients treated with piracetam ($p > 0.05$ versus control). However, at day 30 significant increase in α-range power ($p < 0.01$, compared with the control group) was found in both "intact" and affected hemispheres with marked smoothing of inter-hemispherical asymmetry of α-activity in occipital and central regions (Plate 14).

In 80% of patients treated with piracetam, the positive dynamics of SSEP was registered. The most significant changes were found in pre-Rolandic component N30, the generators of which are projected onto the 4th and the 6th Brodmann's fields (Plate 15). On day 6 significant increase in N-30 amplitude and a tendency towards normalization of its latency and zonal distribution (bilaterally in frontal regions with predominance in the hemisphere contralateral to stimulation) were revealed. In several cases N30 normalization foreran the normalization of EEG pattern. Having in mind modern concepts about N30 as of the first trans-callosal component of SSEP, we suppose that the improvement of associative inter-hemispheric contacts, which manifested several days after piracetam treatment was started, created a basis for further regression of focal deficit and for its more complete functional compensation. Possibly, this accelerated functional normalization of forebrain structures under piracetam treatment partially explains the predominant effects of the drug on the improvement of speech and motor functions.

Along with positive dynamics of N30, the normalization of long-latency SSEP components (N150, P240) was shown, this reflecting the beneficial effect of the drug on nonspecific brain pathways. The positive dynamics of evoked brain activity was mostly pronounced between days 6 and 21 after the stroke onset. The most significant dynamics was noted for the majority of SSEP components in temporal and temporo–parietal regions, that, according to experimental data, correlates with speech and memory improvement [5].

The study of short-latency AEP confirmed the improvement in functional state of the mesodiencephalic brainstem structures after piracetam administration, which was manifested by bilateral significant decrease in latent periods of N3 ($p < 0.05$) and N5 ($p < 0.01$) components, as well as in inter-peak interval N3–N5 ($p < 0.01$), compared with initial (on admission) values.

Thus, the complex clinical and neurophysiological study confirmed safety and good tolerability of piracetam (Nootropil) in acute carotid ischemic stroke of moderately severity. In such cases administration of the drug for the first 15 days intravenously in daily dose of 0.1–0.2 g/kg (6–12 g) and for the following 15 days orally in daily dose of 4.8 g improved neurological and functional recovery in patients by day 30 after the onset of stroke. We confirmed the predominant effect of the drug on normalization of speech functions. The expediency of prolonged administration of the drug is apparently connected with its remote metabolic effects, such as improvement of brain tissue plasticity and inter-hemispheric contacts and enhancement of reparative and regenerative processes.

The study demonstrated inexpediency of piracetam administration in severe strokes, when consciousness disturbances, autonomic and trophic disturbances, and signs of brain edema are present. In these cases the drug did not improve recovery; moreover, in some patients it retarded regression of neurological deficit. Taking into consideration the data obtained from neurophysiological monitoring, it can be supposed that the negative effects of piracetam in patients with severe strokes might be connected with inadequate hyperactivation of energy metabolism.

16.2.   Citicoline sodium

Citicoline sodium (citidine-5'-diphosphocholine, Citicoline, CDP-choline) is a natural endogenous compound that is an intermediate link in reactions of synthesis of membrane-associated phosphatidylcholine. Citicoline sodium can stimulate brain acetylcholinesterase and Na^+/K^+-ATPase independently of acetylcholine and noradrenaline [44]. This enzymatic stimulation may be due to the transformation of CDP-choline to membrane phosphatidylcholine [44]. Citicoline sodium causes membrane stabilizing and antioxidant effects, inhibits formation of free fatty acids taking part in generation of free radicals in acute brain ischemia, and significantly induces acetylcholine synthesis in nervous tissue [16, 40].

In experimental stroke in rats, Citicoline significantly decreased the size of brain infarction and improved neurological outcome [4, 50].

Clinical studies by Spiers, *et al.* [54] and Alvarez, *et al.* [2] showed the improvement of cognitive functions and behavioral reactions in patients with memory disorders treated with Citicoline. The drug was demonstrated to be well tolerated, no side or adverse events effects being registered [58].

Trials of Citicoline sodium in acute ischemic stroke confirmed its good safety profile and found that oral administration of the drug in daily dose of 0.5 g for several weeks of the disease leads to improvement of recovery of altered neurological functions [9, 57]. In Phase III clinical trials, significant improvement of neurological status was revealed in patients with acute ischemic stroke of different severity, treated with Citicoline sodium [11]. It is noteworthy that the trial protocol permitted late start of treatment (within 24 h after the stroke onset), which appears to be a significant difference from requirements imposed on trials of neuroprotectors in acute brain ischemia. Recently an additional trial of Citicoline sodium in patients with ischemic strokes of different severity has been completed. Primary assessment of treatment outcomes failed to confirm significant efficacy of the drug [20]. Further trials are in progress.

16.3.   Gliatilin

Gliatilin (choline alfoscerate, alpha-glycerylphosphorylcholine) is a compound containing 40% of choline that is transformed into metabolically active phosphorylcholine in brain. It is able to permeate through the BBB and to activate biosynthesis of acetylcholine in pre-synaptic membranes of cholinergic neurons [51]. It was shown that Gliatilin prevents induced cholinergic deficit, averts the development of dementia, and alleviates learning and memory deficit by increasing the synthesis and release of acetylcholine in brain structures [38]. The other mechanism of Gliatilin action is an anabolic effect, manifested as an increase in membrane-associated glycerolipid synthesis due to formation of phospholipid membrane precursors from products of metabolic degradation of the drug [7]. Gliatilin activates cholinergic neurotransmission, increasing brain tissue plasticity and causing membrane stabilizing and antioxidant effect [18, 32].

Pilot clinical trial of Gliatilin in acute severe ischemic stroke (intravenous bolus injections in dose of 1 g 3–4 times a day for 5 days) revealed an "awakening" effect of the drug, as well as improvement of respiratory dysfunction and cerebral blood supply [61]. The drug caused positive dynamics of brainstem, suggesting normalization of functional activity of certain brainstem structures [61]. Favorable effect of Gliatilin on psychic activities, memory, and speech in its different domains were also shown [3, 6, 61, 63]. Overall, the drug is well tolerated. Only short-term nausea was registered as its side effect.

The first positive results suggest the need for further well-organized (randomized, placebo-controlled) studies of Gliatilin in acute stroke.

16.4.   Carnitine chloride

The Russian drug carnitine chloride (Aplegin) produces a complex effect on cerebral metabolism and blood supply. The drug restores endogenous resources of carnitine in brain structures, satisfying brain tissue demands under conditions of hypoxia/ischemia, and normalizes oxidation of fatty acids, active ionic transport, neurotransmission, and protein and lipid metabolism. Aplegin decreases the level of anaerobic glycolysis and the extent of lactic acidosis, significantly inhibits lipid peroxidation, inhibits the arachidonic acid cycle, and decreases synthesis of pro-aggregation prostaglandins. Increasing hydrocarbonic acid production, it restores basic reserves of blood and promotes the recovery of cerebral vessel auto-regulation and normalization of microcirculation in the ischemized brain area. A cholinomimetic effect of Aplegin was also registered, which reflects its reparative and regenerative properties [25, 36].

Experimental and clinical trials proved good safety profile and tolerability of the drug [26]. No side effects or allergic reactions were revealed during its administration. The study of efficacy of Aplegin in acute ischemic stroke showed that its administration in daily dose of 15 mg/kg (1 ml of the drug contains 100 mg of carnitine chloride) during the first 7–10 days of the illness improved clinical course and outcome of stroke. Positive effects of carnitine chloride were manifested even in its late administration (from days 2–3 after the stroke onset), that emphasized its not so neuroprotective, but reparative and regenerative properties. The most prominent improvement of altered neurological functions was registered in cases when the drug was administered within first 24 h after the event. Reassuring results of the pilot clinical trial of carnitine chloride call for continuation of further studies of the drug in randomized, placebo-controlled clinical trials.

At present, reparative and regenerative properties of other potentially therapeutic drugs are being intensely studied in experiments. They are amphetamines, growth phosphoproteins, which under conditions of brain ischemia significantly stimulate growth and arborization of neuronal processes, and synaptogenesis, which is manifested by behavioral and memory improvement.

Thus, the review of drugs improving reparative and regenerative properties of brain tissue speaks for expanding their range with new drugs that belong to various pharmacological groups. The effects of these drugs is basically directed to compensation of already formed morphological lesion and this does not require a strict therapeutic window for the beginning of therapy start, an essential difference from conditions under which neuroprotection should be performed. Along with this, many drugs with reparative properties also inhibit the development of ischemic processes, i.e. fulfil neuroprotective function. This explains why it is preferred to administer them within the first hours after the onset of stroke.

The introduction of the potentially reparative drugs is occurring faster than those of potential neuroprotectors; this is connected with better safety of reparative drugs and, partially, with less strict requirements for evaluation of their efficacy. At the same time, to objectively assess the expediency of their use, similar multi-center randomized, placebo-controlled clinical trials enrolling large cohorts of patients should be performed, as in studies of neuroprotectors.

REFERENCES

1. Akopyan, V. P., Badalyan, L. O., 1987, *J Pharmacol Toxicol.* **1**: 38–41 (in Russian).
2. Alvarez, X. A., *et al.*, 1997, *Meth Find Exp Clin Pharmacol.* **19**: 201–210.
3. Antonov, I. P., 1998, *Efficacy of Gliatilin Administration in Stroke Patients Being in Early Recovery Period.* Terra Medica, St. Petersburg, pp. 36–44 (in Russian).
4. Aronowski, J., Strong, R., Grotta, J. C., 1996, *Neurol Res.* **18**: 570–574.
5. Arezzo, J. C., Vaughan, H. G., Legatt, A. D., 1981, *Electroencephalogr Clin Neurophys.* **52**: 531–539.
6. Balunov, O. A., 1998, Efficacy of Gliatilin in the system of rehabilitation measures in post-stroke patients with intellectual and memory disorders. In *Cerebrovascular Pathology*, St. Petersburg, pp. 173–174 (in Russian).

7. Ban, T. A., Panzarasa, R. M., Borra, S., *et al.*, 1991, *New Trends Clin Neuropharmacol.* **5**: 1–35.
8. Bath, P., Albers, G., 2001, *Controlled Clinical Trials. Stroke.* Science Press Ltd, London, 138.
9. Bielenberg, G. W., Haun, C., Krieglstein, J., 1986, *Biochem Pharmacol.* **35** (15): 2693–2702.
10. Buklina, S. B., 1987, *Metabolic Therapy with Piracetam in Acute Period of Ischemic Stroke.* Candidate's dissertation, Moscow (in Russian).
11. Clark, W. M., Williams, B. J., Selzer, K. A., 1999, *Stroke.* **30**: 2592–2597.
12. Coq, J. O., Xerri, C., 1999, *Eur J Neurosci.* **11** (8): 2597–2608.
13. De Deyn, P. P., 1995, *Eur J Neurol.* **2**: 7.
14. De Deyn, P. P., 1996, *Piracetam Symposium.* Venice: 18.
15. De Deyn, P. P., De Reuck, J., Deberdt, W., *et al.*, 1997, *Stroke.* **28**: 2347–2352.
16. Devuyst, G., Bogousslavsky, J., 1999, *J Neurol Neurosurg Psychiatr.* **67**: 419–427.
17. Enberby, P., Broeck, J., Hospers, W., *et al.*, 1994, *Clin Neuropharmacol.* **17** (4): 320–331.
18. Fallbrook, A., Turenne, S. D., Mamalias, N., *et al.*, 1999, *Brain Res.* **10**: 207–210.
19. Fisher, H. D., Wustmann, Ch., Rudolph, E. Z., 1987, *Klin Med.* **42** (12): 1077–1080.
20. Gammans, R. E., Sherman, D. G., 2000, *Stroke.* **31**: 278.
21. Gobert, J. G., 1978, *Nootropil in Neurological and Psychiatric Practice.* Belgium, pp. 3–5.
22. Grammate, Th., Wustmann, Ch., *et al.*, 1986, *Biomed Biochem Acta.* **45** (8): 1075–1082.
23. Gualtieri, F., Manetti, D., Romanelli, M. N., Ghelardini, C., 2002, *Curr Pharm Des.* **8** (2): 125–138.
24. Gusev, E. I., Burd, G. S., Gekht, A. B., Skvortsova, V. I., *et al.*, 1997, *Korsakoff J Neurol Psychiatr.* **10**: 24–28 (in Russian).
25. Gusev, E. I., Kuzin, V. M., Kolesnikova, T. I., *et al.*, 1999, Carnitine – a leading factor in regeneration of nervous tissue. *Med Inform Report* (February), pp. 11–23 (in Russian).
26. Heiss, W. D., Hebold, I., Klinkhammer, P., 1988, *J Cerebr Blood Flow Metab.* **8**: 613–617.
27. Heiss, W. D., Ilsen, H. W., Wagner, R., *et al.*, 1983, Remote functional depression of glucose metabolism in stroke and its alteration by activating drugs. In *Positron Emission Tomography of the Brain* (Heiss, W. D., Phelps, M. E., eds.) Springer, Berlin-Heidelberg-New York, pp. 162–168.
28. Herrschaft, H., 1988, *Med Klin.* **83**: 667–577.
29. Ivashev, M. N., Petrov, V. I., Scherbackova, T. N., 1984, *J Pharmacol Toxicol.* **6**: 40–43 (in Russian).
30. Kartin, P., Povse, M., Skondia, V., 1979, *Acta Therap.* **5**: 235–243.
31. Kessler, J., Thiel, A., Karbe, H., Heiss, W. D., 2000, *Stroke.* **31** (9): 2112–2116.
32. Khaselev, N., Murphy, R. C., 2000, *J Lipid Res.* **41** (4): 564–572.
33. Kovalyov, G. V., 1985, Pharmacology and clinical application of neuroactive amino acids and their analogs. In *Works Volgograd Med Inst.* Volgograd. **37** (5): 295 (in Russian).
34. Kovalyov, G. V., Petrov, V. I., Erdnie-Goryacheva, N. M., 1982, GABA – a modulator of sympathetic nervous system. In *Abst 5th All-Union Symp on Physiology of Autonomic Nervous System*, Yerevan, 167 (in Russian).
35. Kresun, V. I., Borisyuk, B. B., Axelrod, L. B., *et al.*, 1988, Cerebral and extracerebral effects of nootropic drugs. In *Abst 6th All-Union Symp of Pharmacologists*, Tashkent, 198 (in Russian).

36. Kuzin, V. M., Koleskikova, T. M., 1996, *Med-Pharm Herald.* **1**: 17–20 (in Russian).
37. Lebedeva, N. V., Lunev, D. K., Zaretskaya, I. Kh., *et al.*, 1978, Piracetam in treatment of various forms of cerebrovascular pathology. In *Nootropil in Neurological and Psychiatric Practice*, Moscow, pp. 25–30 (in Russian).
38. Lopez, C. M., Govoni, S., Battaini, F., *et al.*, 1991, *Pharmacol Biochem Behav.* **39** (4): 835–840.
39. Muller, W. E., Hartmann, H., Koch, S., *et al.*, 1994, *Int Acad Biomed Drug Res.* **7**: 166–173.
40. Onal, M. Z., Fisher, M., 1996, *Drugs Today.* **32** (7): 573–592.
41. Ostrovskaya, R. U., 1982, *Neuropharmacology of Nootrops*. Meditsina, Leningrad, 113 (in Russian).
42. Ostrovskaya, R. U., Trofimov, S. S., Tsybina, N. M., *et al.*, 1985, *Bull Exp Biol Med.* **3**: 311–313 (in Russian).
43. Pede, I. P., Schimpfesse, L., Crokert, R., 1973, *Farm Tijidschr.* **50** (4): 298–306.
44. Plataras, C., Tsakiris, S., Angelogianni, P., 2000, *Clin Biochem.* **33** (5): 351–357.
45. Platt, D., Horn, J., Summa, J. D., *et al.*, 1992, *Die Med Welt.* **43** (2): 181–190.
46. Poeck, K., Huber, W., Horlacher, R., 1993, *Collegium Internationale Neuropsychopharmacologium*. Capri.
47. Rayevsky, K. S., Georgiev, V. P., 1986, *Mediator Amino Acids: Neuropharmacological and Neurochemical Aspects*. Meditsina, Moscow, 240 (in Russian).
48. Ricci, S., Celani, M. G., Cantisani, T. A., Righetti, E., 2000, *J Neurol.* **274** (4): 263–266.
49. Rosanov, V. A., 1989, *Anesthesiol Resuscit.* **2**: 68–78 (in Russian).
50. Schabitz, W. R., Weber, J., Takano, K., *et al.*, 1996, *J Neurol Sci.* **138**: 21–25.
51. Seleznyova, N. D., Kolykhalov, I. V., Gerasimov, N. P., *et al.*, 1998, *Soc Clin Psychiatr.* **8** (4): 93–100 (in Russian).
52. Shtock, V. N., 1984, *Drugs in Angioneurology*. Moscow, 308 (in Russian).
53. Skvortsova, V. I., 1993, *Clinical and Neurophysiological Monitoring and Neuroprotective Therapy in Acute Ischemic Stroke.* Doctoral dissertation, Moscow (in Russian).
54. Spiers, P. A., Myers, D., Hochanadel, G. S., *et al.*, 1996, *Arch Neurol.* **53**: 441–448.
55. Stockmans, F., *et al.*, 1991, *Thrombosis Haemostasis.* 1179.
56. Szelies, B., Mielke, R., Kessler, J., Heiss, W. D., 2001, *Int J Clin Pharmacol Ther.* **39** (4): 152–157.
57. Tazaki, Y., *et al.*, 1988, *Stroke.* **19**: 211–216.
58. Tazaki, Y., Omae, T., Kuromaru, S., *et al.*, 1980, *J Int Med Res.* **8** (2): 118–126.
59. Toleando, A., Bentura, M. L., 1994, *J Neural Transm Park Dis Dement Sect.* **7** (3): 195–209.
60. Voronina, T. A., 1998, Experimental approaches to study of CNS functional disorders in senility and the research for heir pharmacological correction. In *Abst 5th All-Union Symp of Gerontologists and Geriatrists*, Kiev. **4** (1): 130–131 (in Russian).
61. Voznyuk, I. A., Odinak, M. M., Kuznetsov, A. N., 1998, *Gliatilin Application in Stroke Patients*. St. Petersburg, pp. 167–172 (in Russian).
62. Wahlgren, N. G., 1997, A review of earlier clinical studies on neuroprotective agents and current approaches. In *Neuroprotective Agents and Cerebral Ischemia* (Green, R., Cross, A. J., eds.) Academic Press, San Diego-London-Boston, pp. 337–363.
63. Yanishevsky, S. N., Odinak, M. M., Voznyuk, I. A., Onischenko, L. S., 2000, Clinical and morphological assessment of metabolically active drugs in ischemic stroke. In *Modern Approaches to Treatment of Nervous and Psychiatric Disorders*, St. Petersburg, pp. 364–365 (in Russian).

Chapter 17

Strategies and Prospects for Development of Neuroprotective Therapy for Brain Ischemia

Much new information about processes of extra- and intracellular signaling, interactions between neuronal and glial cells, and mechanisms of ischemic damage to brain tissue on the genetic, molecular, biochemical, immune, and cellular levels, as well as about processes of brain cell survival and post-ischemic regeneration and reparation has become available in recent years. This information provides the basis for further development of neuroprotective strategies.

Analyzing the present state of the problem, we should note first of all the increase in interest in scientific and applied research on the protection of brain tissue. A large number of experimental and clinical studies and over 30 international multi-center trials have been devoted to testing various neuroprotectors. The abundance of findings allows summarizing the first results and drawing some conclusions.

The studies that have been performed again proved that experimental data on efficacy of neuroprotectors cannot be directly transposed onto clinical conditions. Not only cell culture studies, but also experimental models of ischemic stroke in animals fail to afford complete prediction of drug effects in humans. Thus, glutamate receptor antagonists, being the most effective in animal models of ischemia, were inapplicable in clinic due to a wide range of serious adverse effects. The extent of neuroprotective effect of many drugs also prominently varies in experiments and under natural conditions in humans. We suppose the differences to be a consequence of not only functional sophistication of the human brain compared to that in animals, but also of a variety of additional factors, such as development of stroke under natural conditions usually on the background of chronic vascular encephalopathy and chronic brain ischemia, formed beforehand; heterogeneity of etiology and pathogenetic variants of stroke; often delayed beginning of therapy (beyond the therapeutic window). Such analysis allows corrections to be made in further research activities. On one hand, it is obvious that elaboration of new experimental models of stroke is required, where during a certain period of time before the induction of focal brain ischemia in animals, scientists should create a model of arterial hypertensive or hyperlipidemic encephalopathy, as well as to use different modes of interrupting blood delivery to brain tissue. On the other hand, transposing the experience gained in experiments to clinic should take into account that the administration of neuroprotectors later than first 12–18 h after the onset of stroke is useless.

As with all promising research, elaboration of neuroprotective drugs has been accompanied by periods of disappointment connected with substantiation of "cul-de-sac" paths of problem resolution and with evidences of limited clinical capacities of the drugs earlier regarded as promising. However, the researches have already been crowned with some success, and we have preliminary but perceptible findings about the efficacy of several primary and secondary neuroprotectors related to different pharmacological groups. They are:

- **primary neuroprotectors:**
  - non-competitive antagonists of glutamate NMDA receptors (magnesium sulfate and remacemide hydrochloride);
  - activators of inhibitory neurotransmission (chlomethiazole and Glycine);

- **secondary neuroprotectors:**
  - seleno-organic compound of complex antioxidant action (ebselen);
  - basic fibroblastic growth factor (bFGF);
  - neuropeptides (Semax – ACTH 4–10 and Cerebrolysin).

Positive effects of these drugs in acute ischemic stroke vary to a considerable extent between each other, but on the whole are aimed at decrease in 30-day mortality rate, an acceleration of regression of focal neurological symptoms, and an improvement in functional recovery. Direct correlation was shown between efficacy of neuroprotectors and the time when the therapy was begun, as well as the duration of therapy. Primary neuroprotectors were most effective when administered within 2–4 h after the onset of stroke (not later than 6 h) and continued for 5 days. The effects of secondary neuroprotectors prevailed in cases when they were administered within the first 6 h and continued for not less than 7–10 days. At present, international multi-center trials are being performed to confirm preliminary findings about favorable effects of the mentioned drugs on the course and outcome of stroke.

Tight connection was found between molecular mechanisms of secondary neuroprotection and regenerative and reparative processes in brain tissue, which explains the intermingling of neuroprotective and reparative therapy. It was demonstrated that many secondary neuroprotectors (growth factors, receptor components, anti-inflammatory cytokines, neuropeptides) possess powerful reparative properties. At the same time, the drugs conventionally related to reparative therapy can have some neuroprotective effects. However, in contradistinction to real neuroprotectors, effects of the majority of reparative drugs are not so radical, most expressed in mild to moderately severe strokes with small volume of ischemic lesion and absence of general cerebral symptoms and brain edema signs in the clinical picture; they are characterized by wide range of "soft" influences on cerebral metabolism and blood supply. There is also one more essential difference: treatment with drugs with mainly reparative properties can be started beyond the time window of brain infarction formation, and it is effective even in cases when the drugs are administered 24–48 h after stroke onset.

Despite the first successes in elaboration of neuroprotective drugs for stroke treatment, the problem remains unresolved. An especially important aspect of the problem development is more differentiated application of neuroprotectors in relation to many factors, such as duration of pre-reperfusion period, degree and length of ischemia; location of ischemic lesion, etc. At present, it is known that several neuroprotectors are effective only in transient focal ischemia (tirilazad mesylate and other free radical scavengers, antibodies to inter-cellular adhesion molecules), whereas others exert their effects in stable occlusion of an artery with limited possibilities for collateral compensation as well (ebselen). It was found that neuroprotective drugs, which embody their effects via modulation of receptors and synapses (gangliosides) cannot work when white matter is affected. Because glutamate NMDA receptors are almost absent in the white matter, it is comprehensible that the application of NMDA receptor antagonists is also not substantiated in such a location of ischemic lesion. The degree and length of ischemia, which correlate with the extent of perifocal edema and microcirculation disorders, have been proved to determine severity of stroke and the choice of a neuroprotector. If in a small cortical ischemic focus a drug with predominant regenerative and reparative properties suffices, it would be necessary to use more radical neuroprotectors in more extended ischemia. These findings show the limited character of all of the clinical trials, which evaluate effects of neuroprotectors only in a total cohort of stroke patients without their differentiation. Apparently, thorough subgroup analysis is required to choose a preferable neuroprotector.

The study of the influence of the "background" therapy on the efficacy of neuroprotectors also requires additional concern. Modulatory effects of steroids, benzodiazepines, short-term action barbiturates, and certain hypotensive drugs were described in relation to mechanisms of action known for neuroprotectors.

Interesting and complex is an effect of patient body temperature on the extent and the dynamics of brain reparative processes. Thus, several authors [1, 2, 15, 33] revealed connections between the level of hyperthermia within the first hours of ischemic stroke and worsening of stroke prognosis. A neuroprotective influence of therapeutic hypothermia was shown in experimental models of stroke [9], and at present some scientists discuss the possibility of performing randomized clinical trials of hypothermia in stroke patients. At the same time, Ginsberg and Busto [9] demonstrated that it was preferable to maintain body temperature on usual levels (36.7–37.0°C) within the first days of the disease. Moreover, clinical observations by Schwab *et al.* [33] revealed that moderate hyperthermia (up to 38.0°C) within the first 48–72 h of stroke is beneficial in cases of severe disease as it reduces increased intracranial pressure and the extent of brain edema, improving stroke outcome without any serious side effects. Thus, optimal temperature regimen in patients with acute brain ischemia should still be elaborated and substantiated.

One of the most important aspects is the study of combined application of neuroprotectors with different modes of action. Having in mind the multitude and variety of reactions leading to formation of brain infarction, we suppose that simultaneous influence on different links of ischemic cascade could be used to enhance neuroprotective effect in synergy and as well to decrease doses of neuroprotectors, i.e. to decrease the risk of side effects. Experimental studies by Ginsberg *et al.* [11] and Uematsu *et al.* [39] suggested that double blocking of potential-dependent and agonist-dependent calcium channels with the use of combined application of moderate doses of such a non-competitive antagonist NMDA receptors as MK-801 and such dihydropyridine derivate as nimodipine may promote significantly higher protection of brain tissue from acute focal ischemia, than each of these drugs taken separately. This would lead to the decrease of infarction size by 80% and to complete prevention of intracellular calcium accumulation. It was also shown that combined application of low doses of citicoline sodium together with MK-801 [29] or basic fibroblastic growth factor [31] significantly decreased the infarction volume, whereas the identical doses of each of these drugs taken separately were not effective. The combination of tirilazad with magnesium sulfate was found to provide better neuroprotection from the effects of focal ischemia than did therapy with tirilazad or magnesium alone [26, 32]. Noting the differences in therapeutic targets for primary and secondary neuroprotectors and, at the same time, contiguity of their molecular effects due to common intracellular signal transduction system, it is promising to use combined treatment with representatives of both groups in acute ischemic stroke. Obviously, the creation of effective combinations of neuroprotectors will be one of the more significant therapeutic strategies in the not distant future.

Promise in further development of neuroprotection in stroke is connected not only with systematization of already obtained information and using already known drugs with maximal benefit, but, mainly, with development of basic neuroscience research and elaboration of new directions of cytoprotective and reparative therapy. Thus, experimental studies of recent years on neuronal ischemic preconditioning, a phenomenon in which brief episodes of ischemia protect against the lethal effects of subsequent period of prolonged ischemia, has deepened our knowledge of natural molecular mechanisms of tolerance to brain ischemia.

Experimental confirmation of the efficacy of zinc-mediated excitoxicity blockers, pro-inflammatory cytokines antagonists, and anti-inflammatory and neurotrophic factors in acute focal brain ischemia give hope for their clinical application in the future. Decoding of molecular triggering mechanisms of delayed neuronal and glial reactions, induced by ischemia, calls for the application of intracellular signal molecules, modulators of receptors and ion channels activity, regulatory peptides, and certain microelements in elaboration of more effective protective drugs. Special attention is paid to development of techniques that inhibit the genetic program of apoptosis. Already pre-clinical investigations have demonstrated effectiveness of some new directions of anti-apoptotic defense, such as activation of transcription factor NF-kappa B, which induces expression of protective stress proteins, enhancing HSP72 expression or using exogenous HSP72 [21, 30, 34], induction of extracellular signal-regulated kinases cascade [16, 26, 42] and blocking PARP activity and JNK/p38 cascade [12, 22, 27, 40], activation of *Bcl-2* expression or using exogenous Bcl-2 [6, 20, 23], as well as inhibition of effector caspases [4, 5, 14].

One of the most important questions is how to provide brain cells with necessity protective molecular information for only a short time period, without prolonged and perhaps harmful consequences. New methods of incorporation of genetic material into animal brain are developed by means of using of special viral carriers (vectors) and protein transduction domains [38, 41]. Thus, the results of intake of caspase gene antagonists [1] and genes encoding glucose transporters [18] lead to significant decrease in the infarction volume (on average by 50%) in experimental animals. Modern molecular genetic studies are aimed at the synthesis of DNA recombined stable analogs of endogenous neurotrophins and molecular effectors, which temporary "switch off" mechanisms of programmed cell death, as well as at the creation of genetically modified cells constantly producing neurotrophic factors for the possibility of their implantation into affected brain area.

No doubt the intervention of scientific studies into the world of the most enigmatic and complex molecular and genetic processes requires great caution and substantiation of conclusion. Thorough, long-term theoretical and experimental studies of these methods are necessary before it can be decided whether they are to be introduced into clinical practice.

Positive effects of neuroprotection only emphasize the importance and priority of reperfusion therapy of ischemic stroke, aimed at removal of the cause of ischemia. The possibilities of therapeutic reperfusion mainly determines stroke outcome. Elaboration and implication of effective thrombolysis was the most important result of the development of stroke care in recent decades. However, a 3-h therapeutic window appears to be a significant obstacle for the application of thrombolysis. Combined administration of thrombolytic drugs and neuroprotectors expands the therapeutic window, enhances thrombolyic effect, and decreases significantly the extent of additional reperfusion damage of brain tissue. The first positive results were acquired in combined administration of recombinant tissue activator of plasminogen (rtPA) with antioxidant tirilizad [24], polyamines site blocker of glutamate NMDA receptors eliprodil [19], antagonist of glutamate AMPA receptors NBQX [25], non-competitive antagonist of NMDA receptors MK-801 [43], and citicoline [17]. It was demonstrated that combined thrombolysis and neuroprotection markedly improved the degree of cytoprotective effects, significantly reducing the volume of brain damage and the neurological deficit. According to Fisher and Bogousslavsky [8], within the first hours of ischemic stroke combination of rtPA with glutamate receptor antagonists and free radical scavengers may be expedient, whereas later combination with growth factors and citicoline may give positive results in therapy of ischemic stroke. Of course, many experimental studies will be required before such a combined treatment can be introduced into clinical practice.

Experimental and clinical studies conducted in recent years [7, 13, 17, 34–37] elucidated the importance of autoimmunization processes in the development of diffuse destructive processes in brain tissue on the background of its chronic ischemia (encephalopathy), which prepares conditions for infarction formation in response to acute CBF decrease (see Chapter 9). Our clinical observations showed that neuroprotective therapy with Glycine in patients with vascular encephalopathy without strokes in anamnesis was effective for decrease of serum autoantibodies level against brain neurospecific proteins, which was accompanied by prevention of large infarction formation in cases of acute ischemic events. We suppose that neuroprotection can be applied not only in acute ischemic stroke, but long before its development, as a preventive therapy aimed at an increase in brain tissue tolerance to acute ischemia. The great medical and social importance of the problem of stroke prevention points out the need for development of methods of preventive neuroprotection that could play an important role in the general complex of preventive measures.

REFERENCES

1. Azzimondi, G., Bassein, L., Nonino, F., *et al.*, 1995, *Stroke.* **26**: 2040–2043.
2. Betz, A. L., Yang, G. Y., Davidson, B. L., 1995, *J Cerebr Blood Flow Metab.* **15**: 547–551.
3. Castillo, J., Martinez, F., Liera, R., *et al.*, 1994, *Cerebrovasc Dis.* **4**: 56–71.
4. Chen, J., Li, Y., Wang, L., *et al.*, 2002, *J Neurol Sci.* **199** (1–2): 17–24.
5. Chen, J., Nagayama, T., Jin, K., *et al.*, 1998, *J Neurosci.* **18** (13): 4914–4928.
6. De Bilbao, F., Guarin, E., Nef, P., *et al.*, 2000, *Eur J Neurosci.* **12** (3): 921–934.
7. Efremova, N. M., Skvortsova, V. I., Gruden, M. A., *et al.*, 2000, The study of S100β protein content and of primary and secondary antibodies to S100β in patients with acute focal cerebral ischemia in relation to pathogenetic variants of stroke. In *Modern Approaches to Diagnosis of Nervous and Psychiatric Disorders*, St. Petersburg, 294 (in Russian).
8. Fisher, M., Bogousslavsky, J., 1998, *JAMA.* **279**: 1298–1303.
9. Ginsberg, M. D., 1990, *Revieves.* **2**: 68–93.
10. Ginsberg, M. D., Busto, R., 1998, *Stroke.* **29**: 529–534.
11. Ginsberg, M. D., Globus, M. V. T., Busto, R., *et al.*, 1990, The potential of combination pharmacotherapy in cerebral ischemia. In *Pharmacology of Cerebral Ischemia*, pp. 499–510.
12. Gu, Z., Jiang, Q., Zhang, G., 2001, *Neuroreport.* **12** (16): 3487–3491.
13. Gusev, E. I., Skvortsova, V. I., Dambinova, S. A., *et al.*, 2000, *Cerebrovasc Dis.* **10**: 49–60.
14. Hara, H., Fink, K., Endres, M., *et al.*, 1997, *J Cerebr Blood Flow Metab.* **17**: 370–375.
15. Hindfelt, B., 1976, *Acta Neurol Scand.* **53**: 72–79.
16. Irving, E. A., Barone, F. C., Reith, A. D., *et al*, 2000, *Brain Res Mol Brain Res.* **77** (1): 65–75.
17. Khadzhiyeva, M. Kh., Skvortsova, V. I., Sherstnyov, V. V., *et al.*, 2000, The study of neurotrophic factors and their auto-antibodies in patients with chronic ischemic brain disease. In *Modern Approaches to Diagnosis of Nervous and Psychiatric Disorders*, St. Petersburg, 341 (in Russian).
18. Lawrence, M. S., Sun, G. H., Kunis, D. M., *et al.*, 1996, *J Cerebr Blood Flow Metab.* **16**: 181–185.
19. Lekieffre, D., Benavides, J., Scatton, B., Nowicki, J. P., 1997, *Brain Res.* **776** (1–2): 88–95.
20. Lipton, P., 1999, *Physiol Rev.* **79** (4): 1431–1568.
21. Lu, A., Ran, A., Parmentier-Batteur, S., *et al.*, 2002, *J Neurochem.* **81** (2): 355–364.
22. Mandir, A. S., Poitras, M. F., Berliner, A. R., *et al.*, 2000, *J Neurosci.* **20** (21): 8005–8011.
23. Mattson, M. P., Culmsee, C., Yu, Z. F., 2000, *Cell Tissue Res.* **301** (1): 173–187.

24. Meden, P., Overgaard, K., Pedersen, H., Boysen, G., 1996, *Cerebrovasc Dis.* **6**: 141–148.
25. Meden, P., Overgaard, K., Sereghy, T., Boysen, G., 1993, *J Neurol Sci.* **119**: 209–216.
26. Nozaki, K., Nishimura, M., Hashimoto, N., 2001, *Mol Neurobiol.* **23** (1): 1–19.
27. Okamoto, S., Li, Z., Ju, C., *et al.*, 2002, *Proc Natl Acad Sci USA.* **99** (6): 3974–3979.
28. Oktem, I. S., Menku, A., Akdemir, H., *et al.*, 2000, *Res Exp Med.* **199** (4): 231–242.
29. Onal, M. Z., Li, F., Tatlisumak, T., *et al.*, 1997, *Stroke.* **28**: 1060–1065.
30. Rajdev, S., Hara, K., Kokubo, Y., *et al.*, 2000, *Ann Neurol.* **47** (6): 782–791.
31. Schabitz, W. R., Li, F., Irie, K., *et al.*, 1999, *Stroke.* **30**: 427–432.
32. Schmid-Elsaesser, R., Zausinger, S., Hungerhuber, E., *et al.*, 1999, *Neurosurgery.* **44** (1): 163–171.
33. Schwab, S., Schwarz, S., Spranger, M., *et al.*, 1998, *Stroke.* **29**: 2461–2466.
34. Sharp, F. R., Massa, S. M., Swanson, R. A., 1999, *Trends Neurosci.* **22** (3): 97–99.
35. Skvortsova, V. I., Gruden, M. A., Stakhovskaya, L. V., *et al.*, 1999, Structural neurospecific proteins (S100 and basic myelin protein) and auto-antibodies to them as markers of neuroimmunopathological reactions in chronic ischemic brain disease. In *New Technologies in Neurology and Neurosurgery at the Turn of Millennium*, Stupino, pp. 180–181 (in Russian).
36. Skvortsova, V. I., Klushnik, T. P., Stakhovskaya, L. V., *et al.*, 1999, The study of auto-antibodies to NGF in patients with acute and chronic brain ischemia. In *New Technologies in Neurology and Neurosurgery at the Turn of Millennium*, Stupino, pp. 181–182 (in Russian).
37. Skvortsova, V. I., Myasoyedov, N. F., Klushnik, T. P., *et al.*, 2000, Study of NGF and its auto-antibodies content in patients with acute cerebral ischemia. In *Modern Approaches to Diagnosis of Nervous and Psychiatric Disorders*, St. Petersburg, pp. 332–333 (in Russian).
38. Song, B. W., Vinters, H. V., Wu, D., Pardridg, W. M., 2002, *J Pharmacol Exp Ther.* **301** (2): 605–610.
39. Uematsu, D., Araki, N., Greenberg, J. H., *et al.*, 1991, *Neurology.* **41** (1): 88–94.
40. Yu, S. W., Wang, H., Poitras, M. F., *et al.*, 2002, *Science.* **297** (5579): 259–263.
41. Zhang, Y., Pardridge, W. M., 2001, *Brain Res.* **889**: 49–56.
42. Zhu, Y., Yang, G. Y., Ahlemeyer, B., *et al.*, 2002, *J Neurosci.* **22** (10): 3898–3909.
43. Zivin, J. A., Mazzarella, V., 1991, *Arch Neurol.* **48**: 1235–1240.

Conclusion

So, the problem of acute and chronic brain ischemia retains its unique medical and social importance. Chronic brain ischemia, being a consequence of atherothrombogenesis and other hereditary programmed vascular pathological processes, is found in the population of middle-age groups and gains very wide distribution in the elderly. Leading to the development of encephalopathy, chronic ischemia causes a wide range of subjective and objective neurological disturbances that limit intellectual and memory-related possibilities and the physical health of a person, and decrease the patient's quality of life. Progressive ischemic brain damage is one of the most important processes underlying brain aging and changing brain tolerance to acute damage, including acute reduction of cerebral blood supply. Ischemic stroke, which in recent years has increased in frequency, remains in second or third place in the structure of population mortality and is the main reason for human disability.

Thus, the process of brain ischemia to a considerable extent determines the level of health and the duration of life in the population. That is why the importance of discoveries made in recent years relating to basic mechanisms of injuring effects of ischemia on brain is so great. Qualitatively new knowledge breaks the pessimism with regard to the problem of post-stroke treatment that dominated in the 90s. Deepening of concepts of stroke pathogenesis and determination of the chronological algorithm of ischemic cascade reactions in brain tissue created a basis for improvement of stroke management systems, revision of therapeutic strategies, and formation of a new direction in therapy, such as neuroprotection.

Along with this, the ischemic process appears to be the most universal model to resolve fundamental theoretical tasks of neurology and general medicine. Studies of brain ischemia have greatly promoted the understanding of genetic and molecular mechanisms of functioning of neurons and glial cells, as well as separate brain structures, brain as a whole, and the general neuro–immune–endocrine system, united by the signal transduction "net". It has become clearer how a common brain response to damaging stimulus is realized, and how the natural protective facilities of neurons and glia are switched on and brain tolerance to ischemia is formed.

The analysis of modern concepts of ischemic brain lesion and basic strategies of neuroprotection proves that further development of clinical neurology is possible only under condition of tight interaction with fundamental neurosciences: molecular genetics, molecular biology, neurochemistry, neurophysiology, neuropharmacology, etc. Cooperative organization of clinical and experimental studies has already led to perceptible results. Thus, introduction of certain neuroprotectors into everyday practice of neurological clinics of the Russian State Medical University has decreased 30-day mortality rate in ischemic stroke from 21–25 to 9–12% in the last 5–7 years. Increased survival of patients with severe forms of stroke was accompanied by significant improvement in neurological and functional recovery of patients.

The understanding of basic mechanisms of brain ischemic damage and post-stroke regeneration calls for the application of neuroprotective methods not only in therapeutic, but also in preventive goals—in patients with chronic ischemic brain disease—in order to improve brain tolerance to ischemic factors.

The problem of acute and chronic ischemia is so complex that, no doubt, its solution should be achieved only after many generations of research. However, the experience accumulated in the recent decades and open perspectives for the development of this problem allow us to look into the future with optimism and to expect exciting new discoveries in the third millennium.

Index

379